AF411454

Comminution in the Minerals Industry

Comminution in the Minerals Industry

Editor

Luís Marcelo Tavares

MDPI • Basel • Beijing • Wuhan • Barcelona • Belgrade • Manchester • Tokyo • Cluj • Tianjin

Editor
Luís Marcelo Tavares
Universidade Federal do Rio de
Janeiro
Brazil

Editorial Office
MDPI
St. Alban-Anlage 66
4052 Basel, Switzerland

This is a reprint of articles from the Special Issue published online in the open access journal *Journal Not Specified* (ISSN) (available at: https://www.mdpi.com/journal/minerals/special_issues/comminution).

For citation purposes, cite each article independently as indicated on the article page online and as indicated below:

LastName, A.A.; LastName, B.B.; LastName, C.C. Article Title. *Journal Name* **Year**, *Volume Number*, Page Range.

ISBN 978-3-0365-1908-1 (Hbk)
ISBN 978-3-0365-1909-8 (PDF)

Cover image courtesy of Luís Marcelo Tavares

Contents

About the Editor

Luís Marcelo Tavares is a Professor at Universidade Federal do Rio de Janeiro (UFRJ) in Brazil. He received his bachelor´s degree in Mining Engineering (honors) and his master´s degree from Universidade Federal do Rio Grande do Sul. He was awarded a Ph.D. degree in Extractive Metallurgy at the University of Utah. He has been a member of the faculty of UFRJ since 1998, where he founded and is head of the Laboratório de Tecnologia Mineral and has also served as Department Chairman. His research interests include particle breakage, advanced models of comminution and of degradation during handling, DEM, physical concentration, classification, pelletization, iron ore processing and development of pozzolanic materials. He is a founding member and currently president of the Global Comminution Collaborative (GCC) and has received numerous awards from the Brazilian Association of Metallurgy, Materials and Mining (ABM).

Editorial

Editorial for Special Issue "Comminution in the Minerals Industry"

Luís Marcelo Tavares

Department of Metallurgical and Material Engineering, Laboratory of Mineral Technology, Universidade Federal do Rio de Janeiro, Rio de Janeiro CEP 21941-598, Brazil; tavares@metalmat.ufrj.br; Tel.: +55-(21)2290-1544 (ext. 246)

Citation: Tavares, L.M. Editorial for Special Issue "Comminution in the Minerals Industry". *Minerals* **2021**, *11*, 445. https://doi.org/10.3390/min11050445

Received: 14 April 2021
Accepted: 20 April 2021
Published: 22 April 2021

Publisher's Note: MDPI stays neutral with regard to jurisdictional claims in published maps and institutional affiliations.

Size reduction processes, which encompass crushing and grinding, represent a significant part of the capital as well as the operating cost in ore processing. Advancing the understanding of and improving such processes is worthwhile, since any measurable enhancement may lead to benefits, which may materialize as reductions in energy consumption or wear or improved performance in downstream processes.

This Special Issue [1–12] contains contributions dealing with various aspects of comminution, including those intended to improve our current level of understanding and quantification of particle breakage and ore characterization techniques that are relevant to size reduction, as well as studies involving modeling and simulation techniques.

The affiliations of the authors of the articles published in this Special Issue span 14 countries around the globe, namely China, Brazil, Italy, Korea, Sweden, Colombia, Saudi Arabia, Chile, Germany, Russia, South Africa, Canada and the United States, which attests to the highly international nature of research in this field. The themes of the manuscripts also varied widely from several that are more focused on experimental studies [1–5,9,10] to those that deal, in greater detail, with the development and application of modeling and simulation techniques in comminution [6–8,11,12]. Size reduction technologies more directly addressed in the manuscripts included jaw crushing, vertical shaft impact crushing, SAG milling, stirred milling, planetary milling and vertical roller milling. Ores involved directly in the investigations included copper, lead–zinc, gold and iron ores, as well as coal, talc and quartz.

The recognition that size reduction consumes significant amounts of energy has led to a fair amount of interest in methods of ore pretreatment over the years. Odewuyi et al. [4] critically reviewed the various methods that have been studied, ranging from the fairly well-researched thermal (conduction, microwave or radiofrequency), chemical and electric shock to the less studied magnetic, ultrasonic and bio-milling methods. The work shows that the most promising technologies, considering potential of energy reduction, safety, costs, stage of application and potential downstream benefits, are the microwave and electrical pretreatment methods, with important advances in recent years.

Comminution and liberation of minerals can benefit from a detailed knowledge of the structure and texture of rocks and ores. Optical microscopy, scanning electron microscopy and X-ray computed tomography are some of the leading tools that have been used in this task. Popov et al. [2] compared the results from X-ray computed tomography (CT) to those from optical microscopy using quantitative microstructural analysis (QMA) for selected rocks and ores. While recognizing the intrinsic advantage of the direct 3D measures obtained by CT and the ease in sample preparation, the challenges associated with discriminating minerals with similar densities limit its application. On the other hand, optical microscopy of three orthogonally oriented thin sections coupled with QMA offers detailed and precise textural information, including modal composition, grain size distribution and clustering, allowing to discriminate more effectively among minerals but demanding much greater sample preparation effort.

Baldassarre et al. [10] presented a case study of comminution flowsheet development for a high-grade mixed Zn-Pb sulfide sample. Ore samples were characterized through thin

1

section observation and SEM analyses for estimating grain sizes and examining texture features, while X-ray powder diffraction analyses were performed for the definition of target mineral concentrations of comminuted product samples.

Quartz is one of the most common gangue minerals in several ores, occasionally being an ore mineral for glass production. In a study dealing with the investigation of fracture of quartz crystals, Martinelli et al. [9] found that the application of anisotropic stresses resulted in the formation of amorphous silica and, from it, cristobalite and tridymite in nanocrystalline form.

Recognizing the growing role of single-particle breakage in advanced comminution models, Bwalya and Chimwani [6] proposed an empirical breakage probability model that builds on previous works from the authors. The model, which takes into account the effects of energy input and number of impacts to compute the breakage probability, was applied to four different materials. The work also shows the existence of a size-dependent threshold energy for particle breakage.

Components in the feed of size reduction processes often present different breakage behavior; therefore, understanding the response of mixtures has attracted significant attention in the literature. With the aim of gaining insights into the performance of vertical roller (ball-and-race) mills when grinding mixtures of coals, Duan et al. [3] used an instrumented Hardgrove machine to grind anthracite and bituminous coals. The results indicated that the breakage rate and product fineness of the mixture decreased when increasing the proportion of hard anthracite in the mixture. When ground in conjunction with bituminous coal, the grindability of anthracite improved dramatically, while the opposite behavior was found for the former. Indeed, the interaction between the components resulted in a decrease in the specific energy of the mixture compared with the mass average value for separate grinding and, therefore, demonstrated a benefit of intergrinding.

Aiming to gain insights into the response of fine iron ore concentrates to size reduction under confined compression, such as that which happens in high-pressure grinding rolls, Campos et al. [5] investigated their breakage in a piston-and-die apparatus. The work showed that saturation of breakage of particles contained in the top size range occurred at a specific energy of approximately 2 kWh/t, whereas saturation in breakage of all sizes occurred at energies above approximately 6 kWh/t. An expression was proposed to characterize the propensity of a material to break under confined bed conditions. Such knowledge can provide important insights for descriptions of confined bed comminution processes in industry.

Barrios et al. [7] used a particle replacement method in the discrete element method (DEM) to simulate the breakage of a gold ore in a laboratory jaw crusher, demonstrating very good agreement between simulations and experiments. They also showed that the power demanded by the crusher and its throughput and product size varied sensibly with feed size, frequency of strokes and closed-side setting, demonstrating the potential of technology to be used as the basis of the development of improved crushing machines.

Vertical shaft impact (VSI) crushers have been used worldwide in producing crushed fines for construction and building as an alternative to natural sand, owing to the good aspect ratio and smooth surfaces of the crushed product. With the aim of assisting in assessing the suitability of the technology for any particular application, Grunditz et al. [11] proposed a modeling framework. The model, based on the theory of energy-based breakage behavior, also relies on particle collision energy data extracted from discrete element method (DEM) simulations, which are a function of rotor diameter and frequency. It was trained and validated on the basis of a dataset from 24 different sites in Sweden, demonstrating good robustness.

Application of stirred milling has been increasing steadily in the minerals industry, given its advantages in comparison to conventional tumbling mills in fine grinding. One key component of the mill that is directly responsible for media motion is the screw or stirrer. Esteves et al. [12] studied the wear of the screw liner used in a vertical stirred mill using a combination of industrial surveys and DEM simulations. The wear profile, metal loss, power consumption and particle contact information were used to gain insights into how screw wear affects grinding performance. Measurements of stirrer wear from a full-scale vertical stirred mill were then compared to predictions from simulations of a 1:10th scaled-down version of the mill, showing relatively good agreement as long as a proper scaled stirrer frequency was used in the simulations.

Ultrafine grinding also finds important applications in the size reduction of industrial minerals. Kim et al. [1] studied wet high-energy ball milling of talc, investigating the effect of ball size on its grinding response. It was shown that larger (2 mm) ball sizes resulted in faster size reduction and an increase in the specific surface area of the product, but at the expense of greater loss of crystallinity and increased agglomeration. In contrast, the use of smaller (0.1 mm) balls allowed for preserving the crystalline structure of the talc particles, with less tendency toward agglomerate formation, while still allowing for reaching particle sizes at the nanometer scale.

Geometallurgical information regarding ore breakage and grindability often populates block models. Unfortunately, such information is not always available and, even when present, may only have a limited temporal resolution to predict the performance of the comminution processes. Seeking an alternative to this approach and taking advantage of the richness of the data currently available in modern operating plants, Avalos et al. [8] proposed training long short-term memory, a deep neural network architecture, to predict operational ore relative-hardness. They then used the approach on the basis of real-time operational data from two SAG mill datasets, namely feed tonnage, spindle speed and bearing pressure, to classify a copper ore in "hardness" categories. The approach could reach over 80% accuracy in classifying the ore as either "hard" or "soft" type. The authors then proposed extending its application to other grinding and crushing machines to forecast categorical attributes that may be of relevance to downstream processes.

Although only a sample of the vast current research in comminution in the minerals industry, this ensemble of papers characterizes the lively scene of research in this area, which occupies a central role in the future and sustainability of the mining industry.

Acknowledgments: The Guest Editor would like to thank all authors, reviewers, the editor Irwin Liang and the editorial staff of *Minerals* for their timely efforts to successfully complete this Special Issue.

Conflicts of Interest: The author declares no conflict of interest.

References

1.   Kim, H.N.; Kim, J.W.; Kim, M.S.; Lee, B.H.; Kim, J.C. Effects of Ball Size on the Grinding Behavior of Talc Using a High-Energy Ball Mill. *Minerals* **2019**, *9*, 668. [CrossRef]
2.   Popov, O.; Talovina, I.; Lieberwirth, H.; Duriagina, A. Quantitative Microstructural Analysis and X-ray Computed Tomography of Ores and Rocks—Comparison of Results. *Minerals* **2020**, *10*, 129. [CrossRef]
3.   Duan, J.; Lu, Q.; Zhao, Z.; Wang, X.; Zhang, Y.; Wang, J.; Li, B.; Xie, W.; Sun, X.; Zhu, X. Grinding Behaviors of Components in Heterogeneous Breakage of Coals of Different Ash Contents in a Ball-and-Race Mill. *Minerals* **2020**, *10*, 230. [CrossRef]
4.   Adewuyi, S.O.; Ahmed, H.A.M.; Ahmed, H.M.A. Methods of Ore Pretreatment for Comminution Energy Reduction. *Minerals* **2020**, *10*, 423. [CrossRef]
5.   Campos, T.M.; Bueno, G.; Tavares, L.M. Confined Bed Breakage of Fine Iron Ore Concentrates. *Minerals* **2020**, *10*, 666. [CrossRef]
6.   Bwalya, M.M.; Chimwani, N. Development of a More Descriptive Particle Breakage Probability Model. *Minerals* **2020**, *10*, 710. [CrossRef]
7.   Barrios, G.K.; Jiménez-Herrera, N.; Fuentes-Torres, S.N.; Tavares, L.M. DEM Simulation of Laboratory-Scale Jaw Crushing of a Gold-Bearing Ore Using a Particle Replacement Model. *Minerals* **2020**, *10*, 717. [CrossRef]
8.   Avalos, S.; Kracht, W.; Ortiz, J.M. An LSTM Approach for SAG Mill Operational Relative-Hardness Prediction. *Minerals* **2020**, *10*, 734. [CrossRef]
9.   Martinelli, G.; Plescia, P.; Tempesta, E.; Paris, E.; Gallucci, F. Fracture Analysis of α-Quartz Crystals Subjected to Shear Stress. *Minerals* **2020**, *10*, 870. [CrossRef]
10.   Baldassarre, G.; Baietto, O.; Marini, P. Comminution Effects on Mineral-Grade Distribution: The Case of an MVT Lead-Zinc Ore Deposit. *Minerals* **2020**, *10*, 893. [CrossRef]

11. Grunditz, S.; Asbjörnsson, G.; Hulthén, E.; Evertsson, M. Fit-for-Purpose VSI Modelling Framework for Process Simulation. *Minerals* **2021**, *11*, 40. [CrossRef]
12. Esteves, P.M.; Mazzinghy, D.B.; Galéry, R.; Machado, L.C.R. Industrial vertical stirred mills screw liner wear profile compared to discrete element method simulations. *Minerals* **2021**, *11*, 397. [CrossRef]

Review

Methods of Ore Pretreatment for Comminution Energy Reduction

Sefiu O. Adewuyi *, Hussin A. M. Ahmed and Haitham M. A. Ahmed

Mining Engineering Department, King Abdulaziz University Jeddah, Jeddah 21589, Saudi Arabia; hussien135@gmail.com (H.A.M.A.); hmahmed@kau.edu.sa (H.M.A.A.)
* Correspondence: sefiuadewuyi@gmail.com; Tel.: +234-8070583557

Received: 3 April 2020; Accepted: 5 May 2020; Published: 9 May 2020

Abstract: The comminution of ores consumes a high portion of energy. Therefore, different pretreatment methods of ores prior to their comminution are considered to reduce this energy. However, the results of pretreatment methods and their technological development are scattered in literature. Hence, this paper aims at collating the different ore pretreatment methods with their applications and results from published articles, conference proceedings, and verified reports. It was found that pretreatment methods include thermal (via oven, microwave, or radiofrequency), chemical additive, electric, magnetic, ultrasonic, and bio-milling. Results showed that the chemical pretreatment method has been used at an industrial scale since 1930, mainly in cement production. The microwave pretreatment results showed positive improvements at pilot scale mining applications in 2017. The results of ore pretreatment using electric and ultrasonic methods showed up to 24% and 66% improvement in energy consumption, respectively. The former and the latter have been piloted for gold and carbonate ore, respectively. Findings also showed that magnetic, radiofrequency, and bio-milling methods have not led to significant reductions in comminution energy. Based on energy reduction, safety, costs, stage of application, and downstream benefits, microwave and electrical pretreatment methods may be focused for applications in the mining industry.

Keywords: mining operation; ore milling; ore grinding; rock; liberation

---

1. Introduction

The mineral industry consumes significant energy from the national energy demand. The energy consumed in mining as a percentage of national energy consumption in countries such as Australia, USA, Canada, China, Saudi Arabia, and South Africa are 11 [1], 12 [2], 8 [3], 10 [4], 8 [5], and 17.4 [6], respectively. The sources of the energy (for the USA) and the cost trend are presented in Figure 1. In the mineral industry, comminution accounts for the largest share of energy consumption [7,8]. Comminution involves two-unit operations—crushing and grinding. Grinding is always a great concern because it accounts for about 50–70% of the total energy consumption [7,8]. Researchers have elaborately clarified why energy in grinding is so high. The main reason is that as material becomes smaller, more energy is required to create new surfaces. Unfortunately, most of the applied energy is usually lost as noise, heat, and electromechanical during the grinding operation; only about 1% is used for the materials' grinding [9–12]. Hence, there is a need to look for means of rationalizing energy consumption in the grinding process (Figure 2a) [12,13]. Comminution energy depends on minerals to be comminuted as per their type, amount, and properties. Moreover, it relies on the used machine and its operation parameters and skills of operating manpower [14]. The trend of comminution energy (Canada mining industry) between 1990 and 2015 is as presented in Figure 2b. The mean and standard deviation of the energy consumed within these years are 70.71 petajoules and 7.14, respectively. The comminution energy increased by 6.97 petajoules within the period, with a possible increase in future due to declined ore grades and the increase in demand for minerals [14].

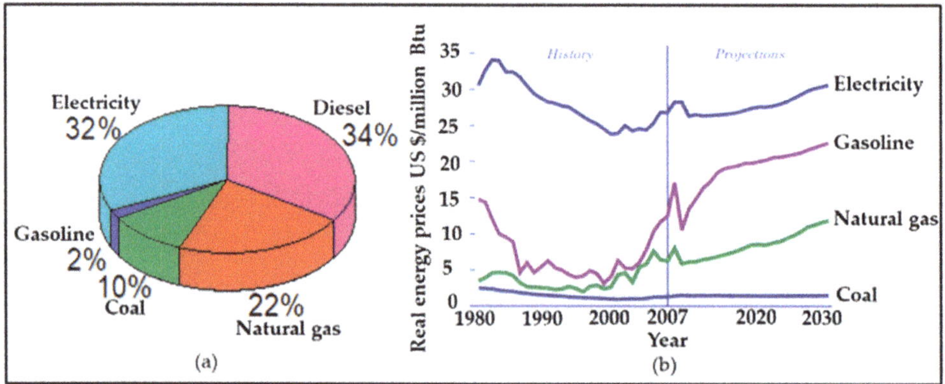

Figure 1. (**a**) Mining operations energy sources (data sources: [13,15]). (**b**) Energy costs trend [16].

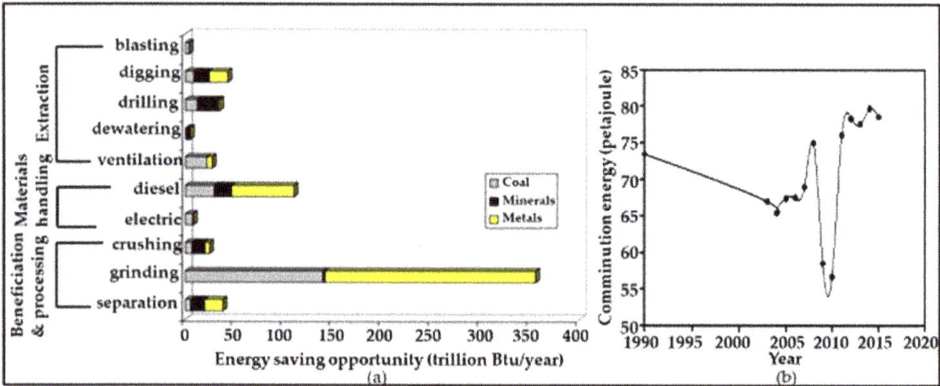

Figure 2. (**a**) Energy-saving opportunity for energy intensive process in mining [13]. (**b**) Comminution energy trend (Canada) between 1990 and 2015 (calculated based on 50% of total mining energy) (data source: [17]).

To ameliorate the situation, early studies have been tailored towards the optimization of grinding circuits. Some of the approaches are the optimization of the mill load, filling ratio [18], ore to media ratio, media size distribution [19], mill length to diameter ratio, and adding mill riffles [20]. In the last few decades, the pretreatment of ore prior to comminution has been proposed [21]. Pretreatment denotes an independent operation performed on ores prior to grinding. The main objective of pretreating ore is to create intergranular cracks so that the grinding operation can become easier, which can lead to a reduction in the grinding energy [22–26]. Among the pretreatment methods are thermal (via furnace, microwave, or radio frequency), chemical additives, electric, magnetic, ultrasonic, radio frequency, and bio-milling. These methods can be categorized as presented in Figure 3. Some of the pretreatment methods have showed positive results towards the reduction in ore's grinding energy requirement but are yet to be commercialized. Previously, some review works had been performed that focused on a single method—chemical additive [27], thermal via furnace [28], thermal via microwave [29], and electric pretreatments [30]. Somani et al. (2017) provided a combined review of thermal, ultrasonic, and electrical pretreatments [21]. In this work, a holistic review of well-known pretreatment methods is attempted in order to give a concise overview of the development in comminution energy reduction and provide for future research needs.

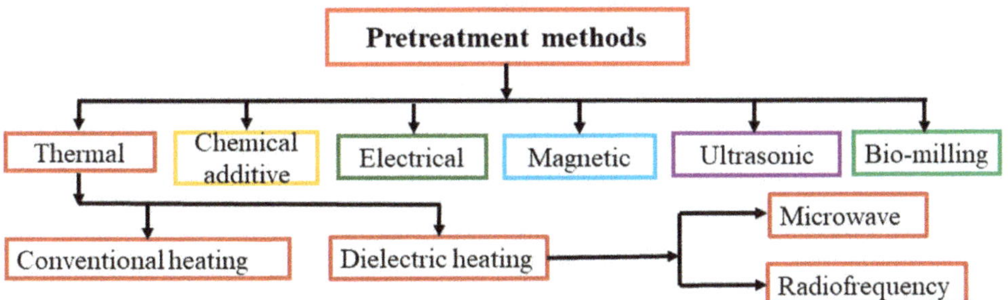

Figure 3. Classification of pretreatment methods.

2. Thermal Pretreatment

Thermal pretreatment is a method that applies heat energy through conduction, convection, or radiation to create granular or inter-granular cracks in rocks [21]. When the temperature of the material changes, the physical properties of the material are altered, which leads to the displacement of some grains leading to fractures. Basically, there are two means of applying heat energy to rocks—through conventional (furnace) and electromagnetic (radiofrequency and microwave) or dielectric heating.

2.1. Thermal Pretreatment via Furnace

The use of a furnace to heat rock is usually referred to as conventional heating. When rock is subjected to heating, stresses are developed within the rock matrix that leads to cracking as a result of thermal expansion and contraction. Thermal pretreatment was introduced in mineral comminution with the prime objective to increase liberation and reduce the energy used in the process [31]. Since rock is an aggregate of minerals, the heating of rocks causes crystals to expand in different orientations depending on the mineral constituents. The expansion of crystals may cause the internal cracking of rocks, which can reduce the competence of rocks before comminution. Inter-crystalline may open up around 200 °C upward, which can increase the existing discontinuity and create new ones in granite [32], gypsum, and celestite [33]; however, this may not be the case for all rock types because ore texture, crystallinity, and size has a significant effect on ore response to thermal pretreatment [34,35]. In addition, a crystal shape may contribute to the crack pattern when grains are displaced due to the expansion of the minerals. The extent of the cracking of rock in different directions depends on the thermal confined stress at any direction. This approach has been used in different studies to create cracks in rocks since early work in the 20th century [36,37]. Omran et al. (2015) demonstrated that, as the temperature of the furnace increases (400 °C and above, at 1 h residence time), the cracks developing in the iron ore matrix increase (approximately 10% increase for every 100 °C) and consequently, the particles' liberation is improved (Figure 4) [34]. However, the maximum temperature after which increasing the temperature has no effect on the particles' liberation from the iron ore still needs to be established. The thermal pretreatment of rock is of interest in rock drilling and excavation, underground storage, nuclear waste, deep petroleum boring, tunneling, dam, reservoir [38], geothermal [39], archeological study [40], building construction [41], and ore comminution and beneficiation [42].

Effect of Thermal Pretreatment (via Furnace) on Ore Grindability

A review of early work on conventional heating as related to rock grindability was discussed by Fitzgibbon and Veasey (1990) [28]. It was reported that the grinding resistance (grindability) of rock can be lowered using a thermal pretreatment approach. Thermal alteration in rocks has different effects, such as structural damage, phase-change, decomposition, and desorption [43]. The anisotropic nature of minerals causes thermal stress concentrations at points and grain boundaries, which may lead to the fracturing of individual mineral grains when heat is applied to rocks [43].

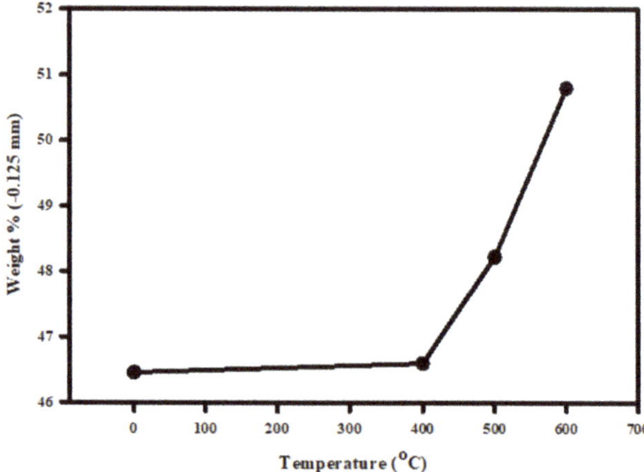

Figure 4. Effect of thermal treatment via furnace on iron ore grinding (100 g, 2.45 GHz, and 1 h) (data source: [34]).

The effect of thermal pretreatment on the grindability of celestite and gypsum was studied [33]. The rock sample of size fractions −1.168 + 0.6 mm (500 g) was ground using a ceramic ball (20 and 25 mm diameters, 588 g). The grinding time ranged between 5 and 30 min at 5 min intervals. The work indexes of untreated celestite and gypsum calculated using the Hardgrove method were 6.76 kWh/ton and 5.18 kWh/ton, respectively [33]. It was reported that there was no significant effect of heat pretreatment on celestite's grindability within 200 °C, while that of gypsum decreased by 22.8% within the same temperature range. This result was expected because gypsum, being one of the hydrate minerals, can decompose easily due to the removal of water molecules when exposed to temperatures within 10–250 °C. When dehydration occurs, gypsum ($CaSO_4 \cdot 2H_2O$) may transformed to plaster of Paris (hemihydrate mineral: $CaSO_4 \cdot O \cdot 5H_2O$), which usually leads to structural failure [33,44]. This shows that the existence of water molecules in the rock structure may help ore response to thermal treatment. Conversely, celestite required higher temperatures up to 1140 °C before effective changes could be observed [33,45].

Recently, the impact of the thermal pretreatment of manganese ore (selected from Qom mining site, Iran) was investigated [44]. Different size fractions (−1.7 + 1.18 mm, −1.18 + 0.6 mm, −0.6 + 0.3 mm, and −0.3 + 0.15 mm) were separately investigated to determine the appropriate size range in which thermal pretreatment can cause a significant effect on the breakage characteristic of manganese ore. A sample was thermally treated in a furnace for 60 min at 750 °C. The treated sample was ground in a ball mill (diameter; 20 cm, height; 25 cm) using ball charges of different sizes of 8.869 kg. An untreated sample of the same mass was ground under the same grinding conditions and the results were compared. The specific rate of breakage approach was adopted in the study, using the first order kinetic model. The slope of the semi-logarithm curve for mass retained against grinding time was used to estimate the breakage characteristic of the ore. The results of the breakage characteristics of treated (thermal treatment) and non-treated manganese ore were compared. An improvement in grinding rate of 37% was obtained for the size fraction −0.30 + 0.15 mm. This result may be attributed to the fact that heat better penetrated to the lower size range than the higher one when treated under the same conditions. The results from different studies related to the pretreatment of ore via furnace are presented in Table 1.

Table 1. Summary of thermal pretreatment via furnace on the grindability of minerals/ores.

Mineral/Ore	Mass (kg)	Size Fraction (mm)	Treatment Temperature (°C)	Soaking Time (min)	Cooling	Improvement in Grindability	Reference
Celestite	0.5	−1.168 + 0.6	200	60	Air	0	[33]
Cassiterite	-	-	650	-	Water	45	[46]
Gypsum	0.5	−1.168 + 0.6	200	60	Air	22.8	[33]
Manganese		−0.3 + 0.15	750	60	Air	37	[42]
Quartzite	0.65	−4 + 0.25	650	65	Water	18	[24]
Quartzite	0.65	−4 + 0.25	650	65	Alkali	32	[24]
Quartzite	0.65	−4 + 0.25	650	65	Acid	28	[24]
Quartzite	0.65	−4 + 0.25	650	65	Salt solution	20	[24]
Hematite	0.1	-	600	60	-	4.2	[34]

Downstream Benefits, Economic Assessment and Industrial Applications of Thermal Pretreatment (via Furnace)

Despite improvements in grindability that may reach up to 45% for some ores (Table 1), thermal pretreatment via furnace has neither been piloted nor adopted in the mining industry. The following challenges have been associated with the method: 1) non-uniformity in rock heating; 2) surface heating; 3) not environmentally friendly—it releases gases to the environment; 4) safety issues related to high temperature; and 5) high energy consumption. Despite all these challenges, thermal pretreatment through furnace is still being pursued, not only to improve the grindability of ores but also to improve downstream operations. Dash et al. (2019) demonstrated that thermal treatment can improve the magnetic separation of low-grade hematite ore [47]. The representative samples (200 g, 10 mm) were treated in a laboratory furnace and the samples were water quenched after the treatment times were reached. It was then ground to −75 µm using a ball mill. The analysis of samples after magnetic separation (wet high intensity magnetic separation (WHIMS); solid % = 25) showed that the iron yield can be improved within a range of 15–20% when hematite is treated between 500 and 800 °C [47]. Early studies show that the thermal treatment of tin ore can save 93 W/t of the ore processed, however 117 kWh/t will be consumed by the furnace at 100% efficiency [48]. This shows that the method may not be economically viable. The issue of the high energy consumption of the furnace is the major challenge; however, further investigation shows that an improvement in liberation usually leads to a high recovery, which may offset the energy consumed during the pretreatment. In fact, a 1% increase in recovery has been argued to be enough to cover the expended energy on the pretreatment via furnace [28]. Even when the improvement in liberation is insignificant, thermal pretreatment can still lead to a better recovery, especially for iron ores due to increases in their magnetic properties [47]. Nevertheless, further research is still needed to investigate the downstream benefits of the thermal pretreatment (via furnace) which will lead to thorough economic assessments and possible applications in the mining industry.

2.2. Thermal Pretreatment via Electromagnetic Waves (Dielectric Heating)

The transfer of energy without the need of a specific material medium of propagation is termed electromagnetic (EM) waves. EM waves are categorized based on wavelength (λ) and frequency (f) (Figure 5), since all EM waves travel at the same speed ($c = f\lambda$).

Among the EM spectrum, radio waves and microwaves (MW) are being used for heating applications in some industries, such as food, geological exploration, pharmaceutical, plastic, construction, medical therapy, and mining. Radio wave heating is usually referred to as radiofrequency (RF) heating [49]. Both RF and MW techniques are referred to as dielectric heating because their heating mechanism relies on the dielectric properties of the material to be heated. Materials with significant dielectric properties respond to dielectric heating. In mining, dielectric heating has gained the attention of researchers because of its peculiar advantages over conventional heating, such as the production of clean energy, fast processing, heating uniformity, and better heat penetration to the ore matrix.

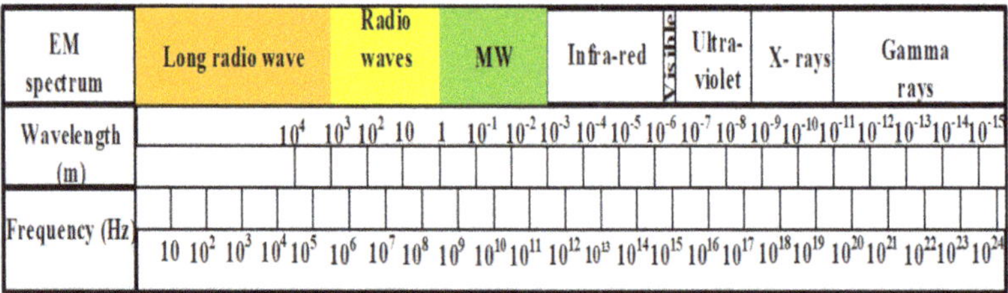

Figure 5. Electromagnetic spectrum.

In contrast to heating through furnaces, which employ heat transferred through convection, conduction, and radiation from the material's surface, dielectric heating involves energy conversion from electromagnetic into heat energy. When material is subjected to an electromagnetic field, electric and magnetic polarization may arise within the frequency of the electromagnetic device used in the process, which may lead to heating. This heating occurs in different mechanisms: dipolar polarization, conduction, and interfacial polarization. In dipolar polarization, the atomic dipole reorientation occurs, but there exists a phase difference between the dipoles and the electric field orientation, which leads to the heating of the material. When a material consisting of a significant electrical conductor is subjected to an electromagnetic field, ions or electrons move which leads to electric polarization. This causes the heating of the material due to its electrical resistance. In such cases, the heating mechanism is through conduction (like that of conventional heating), and a high microwave power and long residence time will be required to make a significant effect on the rock's strength reduction. In rocks with heating behavior such as Goethite (hydrated iron ore), the iron mineral heats, but the bulk ore must be heated to a point at which the hydroxy ion of the water molecules is released. For interfacial polarization, a sample consisting of conducting and non-conducting material with a significant dielectric property is subjected to an electromagnetic field such that both dipolar polarization and the conduction heating mechanism occur. Therefore, when ore is subjected to an electromagnetic filed, both the atomic dipole and electric polarization may occur, leading to microcracks. The extent of the microcracks depend on whether there is an existing fracture and the arrangement of crystals and even the crystal shape, which may help to increase the intergranular cracking.

The main difference between the microwave and radiofrequency means of dielectric heating is their frequency range, which determines the device's configurations. Microwave radiation has a higher frequency compared to the radiofrequency (Figure 5). However, within the frequency range of radio waves, 10–100 MHz is mostly employed for a heating purpose [49]. The typical device set up for the two approaches is as presented in Figure 6. Both methods have been explored for the dielectric heating of ore with the sole aim to reduce its strength and consequently, the comminution energy can be minimized.

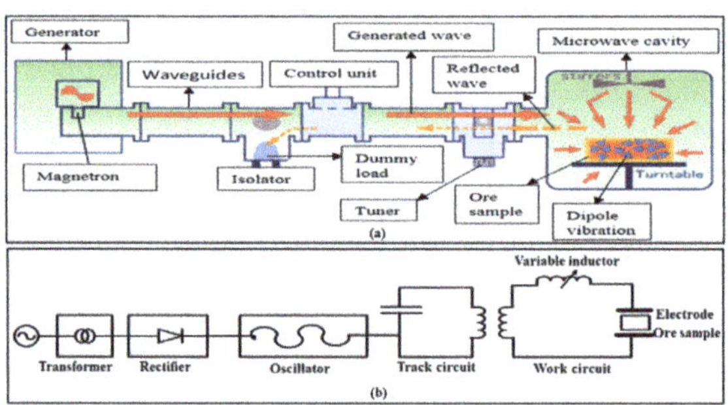

Figure 6. (**a**) Ore pretreatment in a microwave oven (modified after [50]). (**b**) Ore pretreatment in a radiofrequency device [51].

2.2.1. Effect of Microwave Pretreatment on Ore Grindability

Research towards the microwave treatment of minerals was firstly performed in 1975 [52]. The research basically focused on the minerals' responses to microwave treatment, with an emphasis on their dielectric properties. Findings suggested that the approach can be used to heat the ore with minimal processing time compared to conventional heating. It was later demonstrated that ores have varying degrees of response to microwave treatment—hyperactive, active, and difficult to heat [22]. Magnetite, pyrrhotite, and chalcopyrite were grouped as hyperactive minerals because they responded well to the microwave radiation treatment. Chalcocite, galena, and pyrite were active, while albite, marble, quartz, and other gangue minerals were difficult to heat (Figure 7; difficult to heat minerals were arranged downward in decreasing order of their response to the microwave treatment) [53].

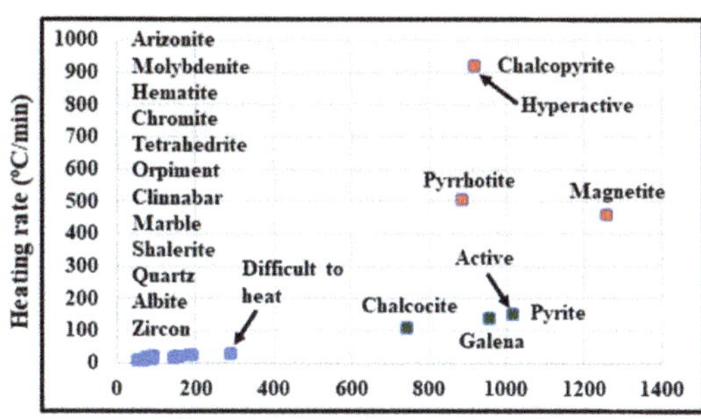

Figure 7. Heating response of minerals in a microwave (data source: [53]).

Afterwards, an attempt was made to investigate the microwave pretreatment's effect on iron ore grindability [7]. The next notable research focused on the effect of variation in the mineralogy of ores and the subsequent impact on their grindability when pretreated in the microwave [54]. Since then, many studies in this direction have been conducted with a focus on ore size [55], grain size [56], texture [57], microwave parameters (exposure time and microwave power), and the mode of cooling system after the microwave treatment [54]. The influence of microwave radiation on the material is usually more pronounced on larger particle sizes than on fine particles [58]. Coarser particles exhibited more fractures than fine particles when the same microwave power and residence time were applied [34,35,54,59]. Omran et al. (2014) demonstrated that iron ore with a larger grain size responded better to microwave radiation than a lower one treated under the same microwave conditions (Figure 8) [55,56].

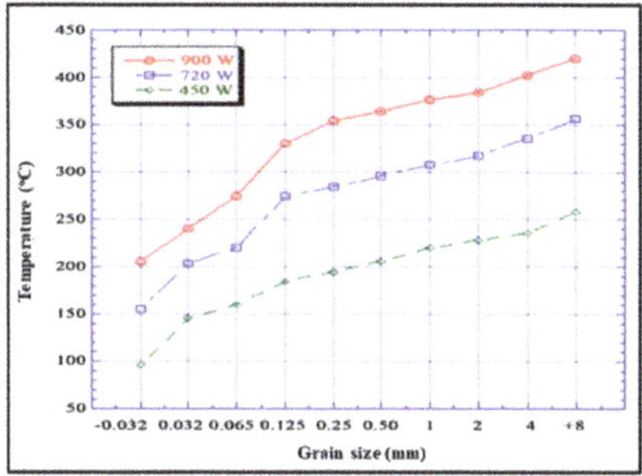

Figure 8. Effect of the grain size and microwave power on the iron ore temperature after the microwave (2.45 GHz) treatment (reproduced with permission: [55]).

As the size of the ore particles becomes smaller, the material hardness and resistance to grinding usually increases, which reduces the probability of creating flaws in the particles. Hence, a higher energy will be required to grind such particles [60]. However, when particles are pretreated in the microwave, the mineral liberation may reach the desired size quickly. To achieve that, the particle size after which microwave treatment has no improvement on the rock strength's reduction must be established, which is limited in the literature; hence, there is a need for further studies [58]. However, a simulation study on coal suggested that at the diameter 50 mm and height within 60 to 100 mm, coal's response to microwave radiation (2.45 GHz) was at optimum conditions [58]. The findings also indicated that the higher the height of the sample, the lower the temperature, but the better the electric field and temperature distribution [58]. Furthermore, at a large size, the hardness of the rock could be reduced when exposed to microwave radiation. This was demonstrated by Sikong and Bunsin (2009) using granite samples selected from Thailand [61]. The prepared representative samples (16 × 16 × 30 mm^3) were labeled thus: the dry sample, air cooled after the microwave treatment, was D-D; D-W was the dry sample water quenched after the microwave treatment; W-D was the wet sample (water soaking for 60 min) air cooled after the microwave treatment; and W-W was the wet sample (water soaking for 60 min) water quenched after the microwave treatment. The hardness of the samples before and after the microwave treatment (2.45 GHz, 600 W) was determined, and the results are presented in Figure 9a. The authors concluded that the microwave treatment of granite samples at lower exposure times has a beneficial effect on the reduction in rock hardness, and that soaking the granite samples in water hampered the reduction in rock hardness.

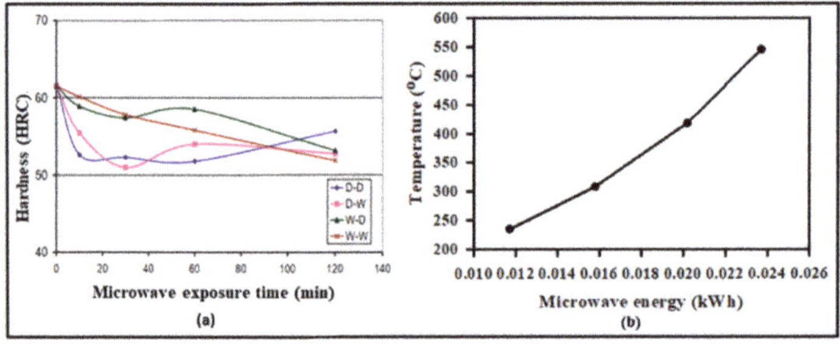

Figure 9. (**a**) Effect of the microwave exposure time on the hardness of granite (reproduced with permission) [61]. (**b**) Response of high phosphorus oolitic iron ore to microwave energy (data source: [34]).

For the influence of ore texture on the ore's response to microwave pretreatment, Batchelor et al. (2015) used lead–zinc, nickel, and copper ores to demonstrate that a high reduction in strength can be achieved in ores with a consistent texture after exposure to microwave radiation [57]. Microwave applied energy has a major influence on ore response to microwave radiation. This was studied using a high phosphorus oolitic iron (HPOI) ore selected from Egypt [34]. In the study, a 100 g representative sample was subjected to microwave radiation (f = 2.45 GHz), and the final bulk sample's temperature and the microwave's energy consumption were measured using a thermocouple and energy meter (CLM 1000), respectively. The results indicated that the higher the applied energy, the better the final temperature reached by the samples (Figure 9b). A scanning electron microscopy (SEM) analysis of untreated and microwave-treated samples indicated that the higher the applied energy, the better the intergranular cracking in the microwave-treated ore. At a microwave (900 W) exposure time of 80 s, transgranular cracking occurred within the oolite, and part of the sample melted at 90 s exposure time. It was concluded that microwave pretreatment can improve the liberation of oolite from other minerals in the ore; consequently, grinding energy can be saved [34].

Omran et al. (2015) compared microwave (900 W, 60 s) and conventional heating (furnace—600 °C, 1 h) pretreatments using HPOI ore. The same grinding operation was performed for the untreated and pretreated samples. The results showed that under size products (0.125 mm) increased—from 46.6% to 59.76% and 50.80%, and the equivalent to approximately 80% and 30% intergranual cracks developed in the samples for the microwave and furnace-treated samples, respectively. Under the treatment conditions, the energy consumption for the microwave was 0.0237 kWh, while that of the furnace was 5.33 kWh [34]. This suggests that the furnace consumed about 224.9 times the energy of the microwave, indicating that the microwave may be more economical than the furnace as an ore pretreatment method. Some of the studies tailored towards energy reduction in comminution, using the microwave pretreatment method as presented in Table 2.

Table 2. Summary of the thermal pretreatment via microwave (MW) (frequency = 2.45 GHz) on the grindability of minerals.

Ore/Mineral	Mass (kg)	Size Fraction (mm)	MW Power (Kw)	Time (min)	Improvement in Grindability (%)	Reference
Magnetite	0.35	−3.36	3.0	3.5	21.4	[7]
Hematite	0.35	−3.36	3.0	3.5	23.7	[7]
Tactonite	0.35	−3.36	3.0	3.5	18.2	[7]
Carbonatite *	0.50	−22.5	2.6	1.5	85.0	[54]
Ilmenite *	0.50	−22.5	2.6	1.5	92.0	[54]
Gold *	0.50	−22.5	2.6	4	0	[54]
Copper *	0.50	−22.5	2.6	1.5	68.0	[54]
Copper-zinc	0.50	-	2.6	1.5	50.0	[62]
Copper-zinc *	0.50	-	-	1.5	65.0	[62]
Copper *	0.50	-	2.6	1.5	70.0	[62]
Lead–zinc *	3.00	−19 + 2	4.0	5.0	30.0	[63]
Ultramafic nickel	0.10	−1 + 0.425	1.2	15	3.6	[64]
Iron	0.50	−19.05 + 12.7	0.9	2.0	50.0	[65]
Lignite	0.10	−4.45 + 0.154	0.9	0.5	81.0	[66]

* Water cooling after microwave treatment.

Downstream Benefits, Economic Assessment and Industrial Applications of Microwave Pretreatment

Microwave pretreatment has been employed by many researchers at laboratory scales, as earlier discussed. Since 1991, when the first laboratory study of the method was performed with promising results for reducing comminution energy [7], the main issues are at the technological level of finding a microwave oven suitable for large-scale ore treatment and the economic feasibility of the method. There is no divergence of opinion that the improvement in grindability alone could not be used to adjudge the economic viability of the microwave pretreatment method. Walkiewicz et al. (1991) discussed that other benefits that can make microwave pretreatment economically viable include the reduction in the tear and wear of the mill, the mill liner, and the milling medium; and a possible increase in throughput with the reduction in recycled ore. Apart from these, the method can increase the grade and recovery of targeted minerals or elements of interest in some rocks [67]. This had been studied using Ilmenite ore [68]. Untreated and microwave-treated (at different power levels—1.3 kW and 2.6 kW) representative samples (200 g, −16 mm) were crushed to 100% passing 220 μm. A two-stage high-intensity wet magnetic (first, 0.045 T was used to remove magnetite; second, 1 T was used to remove ilmenite) separation of Titanium (Ti) from the ore was performed for each of the samples. The result showed that the grade of Ti increased from approximately 1.8% to 3.5%, equivalent to a 7.2% increase in recovery for the microwave-treated samples at 2.6 kW. Indeed, for samples treated at 1.3 kW, the Ti grade increased from approximately 1.8% to 4.4%, equivalent to a 12.8% improvement in recovery [68]. The heap leaching of fine disseminated minerals usually consumes time and leads to a low recovery [69,70]. Therefore, the pretreatment of ore before heap leaching was suggested to improve the process [69]. This was demonstrated using sulphide ore [69]. The results showed that the microwave treatment (f = 2Hz, pulse time = 100 μs, power = 5.6 kW, time = 30 s) of the samples (+9.2 −12.5 mm, 6 kg) prior to heap leaching (800 mL solution; 14 g/L sulphuric acid +3.75 g/L ferric sulphate, at 25 °C) improved the yield in the range of 7% to 12% [69]. Cai et al. (2018) studied the combined effects of the microwave pretreatment, acid leaching, and magnetic separation of high phosphorus oolitic hematite (HPOH) [71]. The first phase of the study indicated that the microwave pretreatment (2.5 kW, 45 s, water quenched after treatment) of the representative HPOH samples reduced their work index from 15.25 kWh/t to 10.11 kWh/t. The next phase was used to study the effect of microwave pretreatment on magnetic separation, while in the last stage, the combined effect of microwave pretreatment and acid leaching (concentrated hydrochloric acid, 1:1; solid:liquid, 45 min) on the magnetic separation (magnetic intensity = 900 kA/m, pulse frequency = 45 MHz) was investigated. In all cases, the improvements in hematite liberation, iron grade, recovery, and dephosphorization were analyzed, and the results were as presented in Table 3.

Table 3. Improvement in hematite liberation, iron grade, recovery, and dephosphorization [71].

Method	Liberation (−0.038) (%)	Grade (%)	Recovery (%)	Dephosphorization (%)
Microwave	30.11	5.65	17.99	3.27
Microwave + Acid leaching	54.80	14.26	34.62	43.49

The effect of microwave pretreatment on the downstream process has been studied on the bioleaching of massive zinc sulphide ore [70]. Different coarse particle sizes were microwave pretreated (frequency = 2.45 GHz, time = 1 s, power = 5.50–5.92 kW), and a continuous column (10 L, 140 mm diameter, and 500 mm height) leaching operation was performed for 350 days. The mic-organism in the leaching process was L. ferriphilum, which acts as a catalysis in the oxidation of Fe (II) to Fe (III). Bioleaching of the same particle size and similar ore material using the same approach has been conducted earlier [72]. The findings showed that the microwave pretreatment of the ore improved the efficiency of the bioleaching operation, as presented in Table 4.

Microwave pretreatment indeed has downstream benefits that can make it economically viable [73]. Most of its downstream benefits are related to the increased grade and recovery, especially for fine mineral particles of interest disseminated in gangue. Extracting such mineral through microwave-assisted leaching has been shown to improve the mineral grade and recovery significantly. The detail of some of the early work using this approach can be found in the literature [73].

Table 4. Zinc leaching efficiency for non-treated and microwave-treated samples [70].

Size Fraction (mm)	Zinc Leaching Efficiency (%)		
	Non-Treated [74]	Non-Treated [72]	Microwave Treated [72]
−5 + 4.475	79.4	73.7	93.1
−16 +9.5	68.7	65.6	81.3
−25 + 19	59.1	58.7	72.0

An economic analysis of the microwave processing of arsenopyrite gold ore (200 t/day) was performed by EMR Microwave Technology Corp. (Fredericton, NB, Canada) in 1997. The results of their findings suggested that microwave processing of the ore was economically viable in both capital and operating costs [73]. The obtained results encouraged the EMR Microwave Technology Corp. (Canada) to conduct a pilot scale study using refractory gold ore, which was probably the first of its kind [67]. In the study, a fluidized bed reactor was developed and coupled with a microwave to produce gold concentrate. The use of the developed technique caused the conversion of pyrite to hematite and elemental sulfur, which led to gold liberation from the ore matrix. The economic analysis of the process as compared to the pressure oxidation and roasting methods is presented in Table 5.

Table 5. Economic comparation of microwave reactor with pressure oxidation and roasting method for the treatment of refractory gold (200 t/day concentrate from 2000 t/day operation) [67].

Method	Operating Costs (US$/t)	Capital Costs (US$ Million)
Pressure oxidation	33.58	26.50
Roasting	13.74	6.90
Microwave reactor	8.60	3.84

The microwave-assisted floatation of copper carbonate ore has also been suggested to be economically viable [74,75]. Comparative batch floatation tests were conducted on untreated and microwave-treated ore (5–12 kW, 0.1–0.5 s), and the results suggested an improvement of 6–15% copper recovery. The best scenario of the economic study of the process indicated a less than two years payback period for the microwave-assisted floatation method [74,75]. The parameters to scale up the technology were also suggested. It was concluded that with a microwave power density of approximately 10^9 Wm^{-3}, and a microwave cavity capable of treating 100–1000 t/h of ore at approximately 0.1 residence time, the process would be commercially viable [74,75].

Recently, a pilot scale study of the microwave pretreatment method has also been conducted, with a possible reduction in comminution energy of up to 9%. It was demonstrated that, apart from the improvement in grindability, the method has the advantages of improving recovery, reducing the wear and tear of ball charges, and increasing the life span of the comminution equipment [25]. Despite the promising results of the microwave pretreatment method, it has not been adopted either in the cement nor the mining industry. The challenges that impair progress in the adoption of the microwave pretreatment in mining and cement industries are the needs for more understanding of microwave interaction with materials, multi-disciplinary research, expertise in microwave engineering, and more pilot scale demonstration of the method for the most demanding ores [76]. Additionally, a design of a high-power large-scale industrial microwave oven is highly demanded. A microwave model study suggests that to achieve the fast processing of ore within 0.002 to 0.2 s, a power density of 1×10^{10} W/m^3 to 1×10^{12} W/m^3 is required [77]. Despite the amount of expended efforts by researchers to demonstrate the applicability of the microwave pretreatment method in reducing the comminution energy, a study has not been conducted to envisage the overall energy that can be saved by considering the full cycle of mineral production. This could be as a result of limited information as to the estimation of the mineral production cycle in terms of electrical energy consumption. A report by the Mining Association of Canada and Natural Resources could be a good source of guidance for estimating the total energy saved in mining activities when microwave pretreatment is incorporated into the mineral production cycle [78].

2.2.2. Effect of Radiofrequency Pretreatment on Ore Grindability

Radio frequency dielectric heating has not drawn the attention of researchers like the microwave has, even though significant research had been conducted for its application in the food industry. Little is known about the possibility of the method to improve the grindability of ore. The first attempt toward this objective was performed using dolerite, marble, and sandstone [51]. The findings indicated that the product size distribution of RF-treated samples improved compared to the as-received samples for the dolerite and sandstone, which may suggest an improved grindability, while that of marble remained the same. Nevertheless, larger particle sizes increased for the studied samples, indicating that the RF method is not appropriate to reduce the comminution energy of the studied samples. There is a need to investigate this technique using other ores [79].

3. Chemical Additive Pretreatment

Chemical additives pretreatment is one of the oldest and probably the most convenient means of pretreating ore before grinding. Generally, there are two types of chemicals used, mostly for cement production. The first type is used to modify the surface structure of the particle to improve grinding, while the second type is used as a cement strength enhancer [80]. Any chemical material that can cause an improvement in ore size reduction when mixed with the feed is termed a chemical additive. It is usually referred to as a grinding aid/additive, since it is applicable to ore at the grinding stage. The grinding aid can be applied in both dry and wet grinding operations. Dry grinding, unlike the wet one, does not require the addition of liquid during the grinding operation. Dry grinding is mostly performed in cement production, while both dry and wet grinding are applicable in other mineral production. Grinding aid is usually active on the ore surface. It reacts with ore molecules to cause local stresses and allows fragmentation at the grain boundaries during grinding. This may cause an improvement in the size distribution and reduce the agglomeration of particles [27].

Effect of Grinding Aids on Ore Grindability

Considerably more research has been carried out on the use of grinding aids to improve the grindability of material in dry grinding operations than in wet grinding. This has occurred as a result of the amount of cement clinker needed to be ground daily in tumbling mills. The grinding of cement clinker to the desired fine size is one the greatest concerns in the cement industry and accounts for huge operation costs in term of energy consumption (Figure 10a). The use of grinding aids in the cement industry to improve the grindability and product throughput of the cement clinker dates to 1930 [81]. Most grinding aids are organic liquids, such as tri-ethanol amine, glycerol, alcohols, propylene glycol, organosilicones, diethylene glycol, and resins, etc. [27,81]. Grinding aids are added to the clinker during the final grinding stage (Figure 10b) to reduce the comminution energy. The summary of the findings from the literature on the use of some grinding additives for the improvement of cement clinker's grindability is as presented in Table 6.

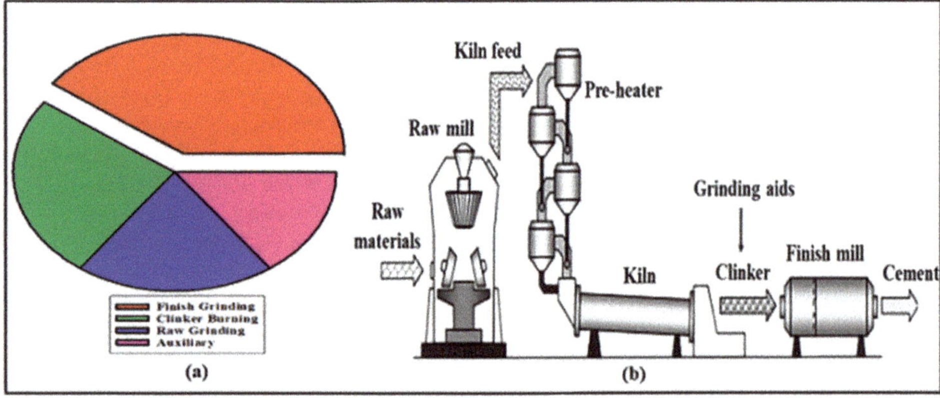

Figure 10. (a) Energy distribution in cement production equipment (data source: [82]). (b) Schematic of dry cement production (modified after [82]).

Table 6. Improvement in grindability of cement clinker using grinding aids.

Grinding Aid	Weight (%)	Improvement in Grinding (%)	Reference
Tri-ethanol amine	0.1	22–29	[27]
Tri-ethanol amine	0.06	16	[83]
Propylene glycol	0.05	25–50	[27]
Organosilicones	0.05	70	[27]
Triisopropanolamine	0.015	26	[84]
Polycarboxylate ether + tri-ethanol amine (1:2)	0.4	22	[81]
Ethylene glycol	0.05	7	[85]

Apart from cement clinker fine grinding where chemical additives have shown a beneficial effect on the grindability of material mostly in dry grinding operations, they have also been used for ore comminution in the mining industry. The use of CaO (200 g/t) as an additive for the grinding of magnetite ore (40% Fe, quartz as gangue) has been demonstrated, and has caused the differential grinding of quartz from Fe content [86].

As per the reduction in comminution energy in the wet grinding operation of minerals, sodium silicate, aero801, and sodium oleate have been demonstrated for celestite [87]. At concentrations of 100 g/t and 1000 g/t, sodium silicate and sodium oleate improved the grinding of celestite. At a 10 g/t concentration, both sodium silicate and sodium oleate caused adverse effects (cumulative 80%, passing size is larger than referenced celestite ground at the same operating conditions) on the grinding of celestite at all grinding periods. Aero801 improved the grindability of celestite at 10 g/t, 100 g/t, and 1000 g/t; however, 100 g/t was reported as the appropriate concentration for the grinding of celestite [87].

Downstream Benefits, Economic Assessment, and Industrial Applications of Chemical Additive

Toprack et al. (2014) conducted an industrial scale study to investigate the appropriate chemicals that showed significant improvements in cement production [80]. In the study, six chemicals were investigated for the improvement in production and 28-day strength enhancement of cement (Table 7). The raw materials considered for the cement production were gypsum, limestone, clinker, and fly ash. An existing closed grinding circuit that consisted of a two-compartment ball mill and dynamic air classifier was used for the study. The dynamic air classifier allowed fine particles to move to the silo, while coarse particles returned to the mill for further grinding. The energy consumptions of the reference and the chemically treated samples were compared, as presented on Table 7. In terms of energy saving, all the six chemical additives used in the study reduced the comminution energy significantly. However, the organic and inorganic modified amines, hydroxylamine, and the mixture of polycarboxylate and amines had adverse effects on the 28-day cement strength [80]. An economic analysis of the whole system showed that cement production using any of the chemical additives used in the study was more profitable than cement production without the chemical additive (Table 7).

Table 7. Effect of chemical additives on the comminution energy and cement strength enhancement [80].

Chemical	Dose (g/t)	Energy Saving (%)	Strength Enhancement (%)	Total Cost Saving (Euro/t)
Organic and inorganic modified amines	300	14.54	−3.15	0.15
Alkanolamines	345	17.34	2.23	0.20
Amine acetate (aqueous solution)	570	13.54	4.45	0.13
Hydroxylamine	808	17.01	−3.89	0.30
Mixture of polycarboxylate and amines	330	16.33	−4.82	0.30
triethanolamine	331	14.37	3.53	0.24

4. Electrical Pretreatment

Electrical pretreatment is among the most targeted technologies that has been studied and reported in literature, after microwave pretreatment [21]. Its principle is based on the passage of a high-voltage electrical pulse (HVEP) into the rock matrix to cause fragmentation. The variation in the electrical conductivity of rock causes the expansion and explosion of rock grains when a high voltage is passed into the rock matrix. The non-conductive part of the rock resists the current flow, which leads to a structural change due to HVEP. The expansivity of the mineral grain varies, and therefore micro cracks can be generated in different degrees. A high pressure is also built up within the rock matrix, such that the tensile strength of the rock is exceeded (electrical disintegration (ED) method). This pressure occurs as a result of the change of state (from solid to gas) of some particles within the rock when the electric current passes through the rock lumps [88]. These amount to the deformation and weakening of the rock due to the high temperature (about 10^4 K) generated by the charge displacement current [89,90]. Different technical terms are found in the literature to represent electrical pretreatment; however, there are slight variations in the procedures or parameters used in creating micro cracks. Some of the technical names are as listed in Table 8.

Table 8. Technical terms used for the electrical pretreatment [30].

Technical Term	Electrode Channel	Voltage Changing Time (ns)
Electrical disintegration (ED)	Rock	<500
Electrical pulse disaggregation (EPD)	Water	<500
Electrodynamic disintegration (EDD)	Water	<500
Electrohydraulic disintegration (EHD)	Water	>500

Electrical pretreatment equipment consists of a high voltage (HV) power source, a sample chamber, and an HV pulse generator that has an arrangement of capacitors with a rectifier. The arrangement of the capacitors depends on the expected capacitance that gives the required voltage. The rock sample to be tested is usually put in water because it has a high dielectric strength and creates a plasma which prevents electrical discharge outside the rock. The rise in voltage is the same for all techniques except electrohydraulic disintegration (EHD), which may result in a lower energy efficiency [30]. For ED, an HV pulse is directly passed into a rock lump that has been immersed in water through the electrode, which makes this procedure quite different from other approaches that require the dipping of the electrode into water in order to generate a shock wave [91]. Electrical pulse disaggregation (EPD), electrodynamic disintegration (EDD), and EHD require water, but more energy is needed for the EHD method and the deformation is generally due to exceeding the compressive strength of the rock [92]. There are divergent opinions on the classification of electrical pretreatment. Some researchers are of the opinion that electrical fragmentation is divided into two categories—one that requires water for breakage and the other without water [93,94]. In this regard, EPD, EDD, and EHD belong to the same group, while ED is the second type. Another view is that it is quite difficult to distinguish between the methods because the electrode gap that is usually associated with EPD, EDD, and EHD may not occur due to rock shape variation [91]. In this case, the classification is based on the voltage rising time.

Effect of Electrical Pretreatment on Ore Comminution

The investigation of EPD and ED to be used in mineral processing was started in the early 1970s and research continued until 2002, when an EPD-suitable device (CNT EPD Spark-2) was designed by the research team of CNT Mineral Consulting Inc. (Ottawa, ON, Canada) [95]. The machine has been used to liberate undamaged diamond crystal from the host rock and emerald from quartzite. The good thing about the machine is that the original shape of the crystal is retained, unlike in conventional crushers that can deform the crystal or break it into fine particles. The ED technique was used to disintegrate granite, copper, kimberlite, and nickel sulfide rocks [89]. The feature associated with ED is that the disintegration occurs at the grain boundary without causing unnecessary fine products and liberating valuable minerals [89]. This can reduce the amount of ore to be crushed in a conventional crusher. This method is even more appropriate to be referred to as secondary blasting or pre-crushing, since it is more suitable for larger rock sizes (boulders).

A comparison of ED with a roll-crusher was performed using coal feeds of different specific gravities (1.35–1.45) and size distributions (4.0–5.6 and 5.6–8.0) [94]. The cathode and anode electrodes of the ED device are stainless steel (with a 2 mm sieve size) and brass disks, respectively. The coal samples (Nantun, China) were crushed and sieved to obtain different size distributions, as earlier stated. Representative samples from the two size distributions were mixed (1:1) to get a 200 g feed sample. A total of 100 g of the sample was fed into the sample chamber, such that there existed five layers with 200 g each. The initial voltage supplied through the cathode was 16 kV, and the value was increased up to 56 kV before the sample disintegration occurred. The voltage and current waveforms were studied using the oscilloscope. The ED test was repeated 60 times to arrive at good conclusions. A representative sample prepared as that of the ED test was crushed using a roll-crusher, and a size distribution analysis of both test methods was performed. The results of optical images showed that rough and smooth surfaces were generated for the ED and the roll-crusher, respectively. In addition, mineral matter was exposed in the case of the ED test products, which indicates that the disintegration occurred at the grain boundaries [94].

The EPD technique was used to liberate minerals from copper (New South Wales, Australia), gold–copper, and lead–zinc (Queensland, Australia) ores [96]. A sample size in the range of 12–45 mm (3600 kg) was collected from mine sites and each ore type was divided into two (one half for the EPD test while the other half was for the conventional crusher test). The products from the two tests were used to carry out a standard bond rod mill test. The closing screen aperture considered for the test was 1.18 mm. The results showed that the percentage changes in the work indices (improvement) between the EPD and conventional crusher for the copper, gold–copper, and lead–zinc ores were 18%, 24%, and 6%, respectively [96]. Similar research was carried out using a platinum group metal ore (South Africa) and samples from Australia, as earlier mentioned. Coarser products were generated in the EPD method, with less fine materials and valuable minerals liberated than that of the conventional crusher [97].

The HVEP technique was used to investigate the liberation of magnetite ore using a −2 mm (200 g) representative feed sample. The sample treated with HVEP and the untreated one were ground under the same grinding conditions using a rod mill. An improvement of 13.19% in the liberation of iron minerals was achieved using HVEP when compared to the untreated sample [98].

Recently, a high-voltage electric pulse crusher (HVEPC) was designed and used for the crushing of phosphate ore [99]. The size fractions of the phosphate ore used in the study ranged between −75 and 50 mm. The bond crushability index of the phosphate ore reduced by 10.6% (compared with the conventional crusher) when the HV pulse-specific energy ranged between 3 and 5 kWh/t. The effects of capacitance, voltage, and PSD of the sample on crushing using the HVEPC were also investigated. It was found that an increase in the capacitance and voltage lead to an improvement in the crushing of the phosphate ore at size ranges of −19 + 12.5, −12.5 + 6.35, and −6.35 + 3.35 mm. The summary of some of the laboratory experiment successes of electrical pretreatment are presented in Table 9.

Table 9. Summary of the electrical pretreatment.

Ore/Mineral	Size (mm)	Electrode Gap (mm)	Voltage (kV)	HVP Specific Energy (kWh/t)	Improvement in Grindability (%)	Reference
Copper	−12.5	10–40	90–200	3	18.0	[96]
Gold–copper	−12.5	10–40	90–200	3	24.0	[96]
Lead–zinc	−12.5	10–40	90–200	3	6.0	[96]
Copper–gold	−12.5	10–40	90–200	3	0.0	[96]
Magnetite	−2.0	3	30	-	13.2	[98]
Phosphate	−75 + 50	-	40	5	10.6	[99]

Downstream Benefits, Economic Assessment, and Industrial Applications of Electrical Pretreatment

Recently, a pilot scale HV pulses (HVP or EPD) testing machine (Figure 11) developed by SELFRAG AG (Kerzers, Switzerland) was used for the investigation of the particle weakening behavior of ores (gold–copper ore, New South Wales, Australia; iron oxide copper–gold (IOCG) ore, South Australia; and hematite ore) [100]. The machine had a setting system that allowed capacitance and voltage regulation. The pulse energy (ranged 50–200 kV) could be kept constant when varying the voltage or capacitance of the machine. The machine could process ore up to 10 th^{-1} (3–10 th^{-1}), depending on the pulse energy, PSD, and density of the ore. The PSDs of the tested ores were 22.4–26.5, 31.5–37.5, and 45–53 mm. It was found that the higher the specific energy of the HVP machine, the better the breakage characteristics measured using the fineness indicator (t_{10}), which connotes a cumulative percentage passing size equivalent to one tenth of the original size before the pretreatment and pre-weakening assessments. It was reported that the HVP pretreatment caused a reduction in the competency of gold–copper and IOCG ores by 81.7% and 131.8% respectively, while that of hematite increased by 40.7% [100]. However, an economic evaluation of this method was not performed in the study, which calls for further research. Nevertheless, the HVP machine produced at Julius Kruttschnitt Mineral Research Centre (JKMRC, Indooroopilly, Queensland, Australia) in 2009 consumed considerable energy (1–3 kWh/t) during the ore pretreatment [96]. Safety due to high-voltage generation for rock weakening has usually being associated with electrical pretreatment; however, this has been put into consideration in the JKMRC machine, and electromagnetic shielding against high electric voltages has been introduced [101]. The electrical method has been applauded for its reduction in the ore to be comminuted; this can assist with rejecting gangue after treatment (pre-concentration) [102,103], and for the reduction in fine particles in the final product after comminution [104]. The former has been suggested to be performed at the mining site so that the haulage cost can be reduced and the rejected can be used as back filling [105]. With that approach, the energy cost can be lowered not only for haulage but also for comminution and processing. The inclusion of the HVP machine in the mining cycle also has environmental benefits, as tailing can be reduced since some wastes would have been rejected from the mining site. A simulation of this approach suggested that 5 kWh/t can be saved for 2000 t/h copper–gold operations [105]. Parker et al. (2015) discussed that electrical pretreatment (electro-comminution) has the potential to improve mineral liberation, which may increase the recovery using the floatation method. In the study, the authors compared the surface chemistry of the untreated and electrical treated samples and found that the latter improved the surface chemistry of the ore as well as the liberation of chalcopyrite in the coarse size range [106].

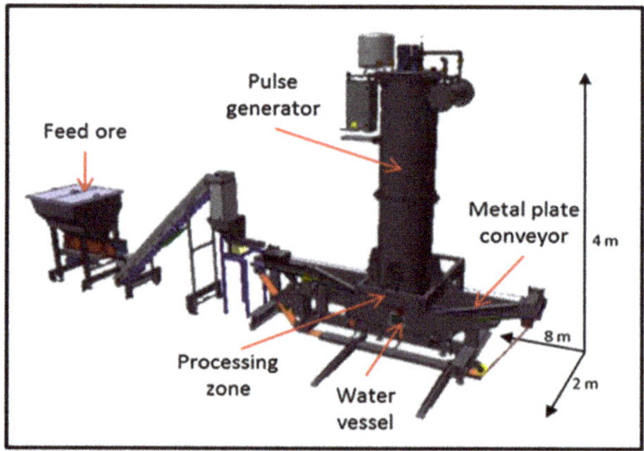

Figure 11. High voltage pulse pilot scale testing machine at Kerzers, Switzerland (reproduced with permission) [100].

5. Magnetic Pretreatment

Magnetic pretreatment is quite a new method investigated for the reduction in the comminution of iron-rich minerals, especially magnetite. The idea behind the method is that the magnetic pulses generated through a magnetic field may probably change the magnetic dipoles of magnetic minerals and hence introduce microcracks in the rock. This method was proposed in 2016 by J. Yu and his research team [107]. A generator was developed that can create magnetic pulses through a dielectric pipe where coils were mounted. The magnetic field and frequency of the pulses generated were 5.0 T and 0.05 Hz, respectively. The magnetic pulse generator (MPG) was used for the treatment of magnetite ore collected from Dagushan city, Liaoning, China. The representative sample was placed in a plastic container and subjected to magnetic pulse treatment in the MPG for 2.5, 5.0, and 7.5 min using separate samples. The magnetic field intensity of the MPG was kept constant throughout the treatment periods. The pretreated samples and the untreated one (500 g each) were ground for 3 min using a laboratory cone call mill. The energy consumption (measured with a DTZ 119 smart power meter) for grinding untreated and treated samples reflects a 10 W difference, representing 1% energy saving in the grinding process. The PSD analysis performed showed no significant change in the size distribution of products. The improvement in the bond ball mill work index of the treated sample was 0.68% (test sieve size = 74 µm). To sum it up, MPG did not sufficiently weaken the magnetite ore and there was no significant improvement in the work index. [107]. The magnetic method may provide promising results in future research.

6. Ultrasonic Pretreatment

The ultrasonic pretreatment of ore uses a three-dimensional acoustic wave (sound wave above 20 kHz) that passes through the ore matrix to create fractures. The velocity of this wave varies as it passes through the ore media, due to the change in grain size or mineralogy, which causes a kind of reflection and refraction at the grain boundaries. The tensile stress caused by this phenomenon creates microcracks or extends the existing cracks in the ore matrix. The use of ultrasonic waves for the fragmentation of particles was reported to have firstly been attempted by Gärtner in 1953 [52]. In 1981, the Energy and Mineral Research (EMR) company with the support of the US Department of Energy (DOE) constructed an ultrasonic device consisting of a rotating roller that had the capacity to treat 4.5–13.6 kg/h of ore, with an energy requirement of 3 kWh/t for 80% of the product finer than 75 μm. At that time, the energy requirement to achieve the same size as that of using a hammer mill was 20 kWh/t. The major disadvantage of the device, when compared to the conventional machine, was that it had a very low comminution capacity. The DOE, US, in 1988 commenced an investigation into the construction of their own ultrasonic device using the previous idea from 1981. The constructed device was used for the treatment of coal, and resulted in little liberation of the samples. Indeed, the device consumed more energy than that of conventional comminution mills [108]. However, this method has been reported to have improved the grindability of copper ores by up to 32% [108]. Gaete-Garretón et Al. (2000) provided a solution to the limitation (small capacity) of the previous ultrasonic devices by incorporating ultrasonic treatment into a high-pressure roller mill with a capacity of 20 t/day [109]. The developed device and ball mill were used separately for the grinding of prepared copper ore feeds. It was reported that 66% energy was saved when compared with the ball mill. A similar idea was developed called the Ultrasonic High-Pressure Roller Press (UHPRP), which was used in the investigation of the grinding characteristic of copper ore [110]. Findings show that 6% energy was saved when compared with samples without ultrasonic treatment [110]. The combination of microwave and ultrasonic pretreatment was investigated for the disintegration of iron from phosphorus gangue. The findings showed that the disintegration and removal of fine particles from the samples were higher for microwave-treated samples compared to the ultrasonically treated one. The combination of the two methods indicated that a 20% improvement in disintegration was achieved during the process [111]. An ultrasonic device has been piloted for the treatment of carbonate ore which improved its liberation and downstream processes [112]. The economic and industrial applications of this method are limited in the literature; hence, further research is still needed.

7. Bio-Milling Pretreatment

Bio-milling is the process of size reduction in particles using living organisms. Since microorganisms can cause the weathering of rock, researchers have therefore suggested that their use for ore comminution could improve the size reduction process. This has found direct application in the milling of particulate matter in the nano range. The use of fungal biomass for milling chemically synthesized $BiMnO_3$ had been reported [113]. It was discussed that $BiMnO_3$ (150–200 nm) was reduced to <10 nm without any effect on either the crystallinity or phases of the material [113]. Research using this approach for ore comminution energy reduction may find relevance in the future.

8. Ranking of Methods of Ore Pretreatment

To look for an immediate way forward to reduce comminution energy, Canada Mining Innovation Council (CMIC) ranked both conventional and some emerging technologies based on the responses and analysis of a questionnaire filled in by professionals in the mining sector [114]. Consideration was given to energy reduction, costs, safety, stage of the technology, and downstream benefits. For all these parameters considered for rating the pretreatment methods, 1–5 were used as rating numbers in an ascending order of benefit, except cost. For the cost implication of the methods, the rating five means that the cost is very low, meaning that one is the costliest rating number [114]. Based on the available literature [80,100,107], there are developments in the stages of technologies, which was part of the ranking matrix used by the CMIC; hence, there is a need for updating the ranking of emerging technologies so that research can be channeled toward the proven methods (Table 10).

Table 10. Ranking of ore pretreatment methods (* cement production) [114].

Methods	Energy Reduction	Cost	Safety	Scale of Application	Downstream Benefits	Total
Conventional heating	2	1	2	2	2	9
Microwave	3	2	3	4	3	15
Radiofrequency	-	2	3	1	1	7
Chemical *	2	3	3	4	4	16
Electrical	3	2	3	4	4	16
Magnetic	-	-	3	1	-	4
Ultrasonic	1	1	2	3	3	10
Bio-milling	-	-	2	1	-	3

9. Conclusions

Comminution is an important operation to liberate minerals/ores. It accounts for about 50% to 70% of the total electrical energy required in mining activities. Many methods have been explored to reduce this energy demand. These methods include thermal (via furnace, microwave, and radiofrequency techniques), chemical, electrical, magnetic, ultrasonic, and bio-milling. Thermal pretreatment via furnace showed that improvements in the grindability of some ore can be achieved by up to 45%. Nevertheless, the high energy demand of furnaces, their non-uniform heating, safety issues, and environmental pollution due to their release of process gasses are major concerns. The development of a solar convergence device that can produce high thermal energy to heat ore can be investigated, since solar energy is environmentally friendly and may be cheaper in the long run. Microwave dielectric heating is the most pursued of all methods, with promising results at both laboratory and pilot scale studies. Findings from the former suggested that a 3–92% improvement in the grindability of some ores can be achieved, while the latter shows a maximum of 9% reduction in the grindability of copper ore. In contrast to the microwave, radiofrequency dielectric heating has not produced promising results for some selected rocks. The chemical method has been demonstrated to be appropriate to aid comminution, especially in the cement industry, with an improvement in the grindability of clinker in the range of 7–70%. However, the cost of the chemicals used in the process is still a challenge. The results of studies using electrical and ultrasonic methods showed improvements in grindability of up to 24% and 66%, respectively. The former has been piloted for gold, copper, and iron-related ores, while the latter has been used for carbonate rocks with promising results. For magnetic and bio-milling methods, progress has not been made, as suitable approaches for ore comminution energy reduction have not been found, but these methods may find relevance in the future. To sum up, microwave and electrical pretreatments should be given preferential attention based on their stage of technology, energy reduction, cost, safety, and downstream benefits.

Author Contributions: Conceptualization, S.O.A. and H.A.M.A.; methodology, S.O.A.; formal analysis, S.O.A.; investigation, S.O.A.; resources, S.O.A., H.A.M.A.; data curation, S.O.A.; writing—original draft preparation, S.O.A.; writing—review and editing, S.O.A., H.A.M.A.; visualization, H.A.M.A, H.M.A.A.; supervision, H.A.M.A, H.M.A.A.; funding acquisition, H.M.A.A., H.A.M.A. All authors have read and agreed to the published version of the manuscript.

Funding: This project was supported by the Deanship of Scientific Research (DSR), King Abdulaziz University, Jeddah under grant No. (DG-026-306-1441). The authors, therefore, gratefully acknowledge the DSR technical and financial support.

Acknowledgments: The authors also acknowledge the contributions of anonymous reviewers that improved this work.

Conflicts of Interest: The authors declare no conflict of interest.

References

1. Ball, A.; Billing, J.; McCluskey, C.; Pham, P.; Pittman, O.; Lawson, S.; Ahmad, S.; Starr, A.; Rousseau, J.; Lambert, N. *Australian Energy Update, commonwealth of Australia 2018*; Department of the Environment and Energy: Canberra, Australia, 2018; pp. 1–39.
2. EIA Energy Use in Industry-US. Available online: https://www.eia.gov/energyexplained/use-of-energy/industry.php (accessed on 26 April 2020).
3. NRC Energy Efficiency Trends in Canada 1990 to 2013. Available online: https://www.nrcan.gc.ca/energy/publications/19030 (accessed on 26 April 2020).
4. Sandklef, K. *Energy in China: Coping with Increasing Demand*; FOI-Swedish Defence Research Agency: Stockholm, Sweden, 2004; pp. 1–46.
5. SEEC. National Energy Conservation Campaign. Available online: https://www.seec.gov.sa (accessed on 12 February 2019).
6. Mogodi, P. Analysis of Energy Consumption, Economy and Management at new Denmark Colliery. Ph.D. Thesis, University of the Witwatersrand, Johannesburg, South Africa, 2012.
7. Walkiewicz, J.W.; Clark, A.E.; McGill, S.L. Microwave-assisted grinding. *IEEE Trans. Ind. Appl.* **1991**, *27*, 239–243. [CrossRef]
8. Musa, F.; Morrison, R. A more sustainable approach to assessing comminution efficiency. *Miner. Eng.* **2009**, *22*, 593–601. [CrossRef]
9. Magdalinović, N. Truths and misconceptions in the ore comminution. *Min. Metall. Eng. Bor* **2016**, *1*, 37–46. [CrossRef]
10. Austin, A.G.; Luckie, P.T.; Klimpel, R.R. *Process Engineering of Size Reduction: Ball Milling*; Society of Mining Engineers of the American Institute of Mining, Metallurgical and Petroleum Engineers: New York, NY, USA, 1984.
11. Sadrai, S. High Velocity Impact Fragmentation and the Energy Efficiency of Comminution. Ph.D. Thesis, University of British Columbia, Vancouver, BC, Canada, 2007.
12. Toifl, M.; Hartlieb, P.; Meisels, R.; Antretter, T.; Kuchar, F. Numerical study of the influence of irradiation parameters on the microwave-induced stresses in granite. *Miner. Eng.* **2017**, *103*, 78–92. [CrossRef]
13. U.S. DOE. *Mining Inustry Energy Bandwidth Study*; US Department of Energy: Washington, DC, USA, 2007.
14. Jack, J.; Alex, S. Energy Consumption in Mining Comminution. *Procedia CIRP* **2016**, *48*, 140–145.
15. Bogunovic, D.; Kecojevic, V.; Lund, V.; Heger, M.; Mongeon, P. Analysis of energy consumption in surface coal mining. *Soc. Mini. Metall. Explor.* **2009**, *326*, 79–87.
16. Worrell, E.; Blinde, P.; Neelis, M.; Blomen, E.; Masanet, E. *Energy Efficiency Improvement and Cost Saving Opportunities for the U.S. Iron and Steel Industry*; University of California: Berkeley, CA, USA, 2010; pp. 1–160.
17. Wang, T. Canada's Mining Industry Energy Consumption 1990–2015. Available online: https://www.statista.com/statistics/964258/energy-consumption-mining-industry-canada/ (accessed on 26 April 2020).
18. Yin, Z.; Peng, Y.; Zhu, Z.; Yu, Z.; Li, T. Impact Load Behavior between Different Charge and Lifter in a Laboratory-Scale Mill. *Materials* **2017**, *10*, 882. [CrossRef]
19. Sahoo, A.; Roy, G.K. Correlations for the grindability of the ball mill as a measure of its performance. *Asia-Pac. J. Chem. Eng.* **2008**, *3*, 230–235. [CrossRef]
20. Kanda, Y.; Kotake, N. Comminution Energy and Evaluation in Fine Grinding. In *Handbook of Powder Technology*; Elsevier B.V.: Amsterdam, The Netherlands, 2007; pp. 529–550.
21. Somani, A.; Nandi, T.K.; Pal, S.K.; Majumder, A.K. Pre-treatment of rocks prior to comminution—A critical review of present practices. *Int. J. Min. Sci. Technol.* **2019**, *53*, 195–211. [CrossRef]
22. Chen, T.T.; Dutrizac, J.E.; Haque, K.E.; Wyslouzil, W.; Kashyap, S. The relative transparency of minerals to microwave radiation. *Can. Metall. Q.* **1984**, *23*, 349–351. [CrossRef]
23. Chunpeng, L.; Yousheng, X.; Yixin, H. Application of microwave radiation to extractive metallurgy. *Chin. J. Mater. Sci. Technol.* **1990**, *6*, 121–124.
24. PococK, J.; Veasey, T.J.; Tavares, L.M.; King, R.P. The effect of heating and quenching on grinding characteristics of quartzite. *Powder Technol.* **1998**, *95*, 137–142. [CrossRef]
25. Batchelor, A.R.; Buttress, A.J.; Jones, D.A.; Katrib, J.; Way, D.; Chenje, T.; Stoll, D.; Dodds, C.; Kingman, S.W. Towards large scale microwave treatment of ores: Part 2—Metallurgical testing. *Miner. Eng.* **2016**, *111*, 5–24. [CrossRef]

26. Sahoo, B.K.; De, S.; Meikap, B.C. Improvement of grinding characteristics of Indian coal by microwave pre-treatment. *Fuel Process. Technol.* **2011**, *92*, 1920–1928. [CrossRef]
27. Fuerstenau, D.W. grinding aids. *KONA* **1995**, *13*, 5–18. [CrossRef]
28. Fitzgibbon, K.E.; Veasey, T.J. Thermally Assisted Liberation—A review. *Miner. Eng.* **1990**, *3*, 181–185. [CrossRef]
29. Kingman, S.W. Recent developments in microwave processing of minerals. *Int. Mater. Rev.* **2006**, *51*, 1–12. [CrossRef]
30. Shi, F.; Manlapig, E.; Zuo, W. Progress and Challenges in Electrical Comminution by High-Voltage Pulses. *Chem. Eng. Technol.* **2014**, *37*, 765–769. [CrossRef]
31. Veasey, T.J.; Wills, B.A. Review of Methods of Improving Mineral Liberation. *Miner. Eng.* **1991**, *4*, 747–752. [CrossRef]
32. Homand-Etiennet, F.; Houpertt, R. Thermally Induced Microcracking in Granites: Characterization and Analysis. *Int. J. Rock Mech. Min. Sci. Geomech. Abstr.* **1989**, *26*, 125–134. [CrossRef]
33. Sener, S.; Bilgen, S.; Ozbayoglu, G. Effect of heat treatment on grindabilities of celestite and gypsum and separation of heated mixture by differential grinding. *Miner. Eng.* **2004**, *17*, 473–475. [CrossRef]
34. Omran, M.; Fabritius, T.; Mattila, R. Thermally assisted liberation of high phosphorus oolitic iron ore: A comparison between microwave and conventional furnaces. *Powder Technol.* **2015**, *269*, 7–14. [CrossRef]
35. Ali, A.Y. Understanding the Effects of Mineralogy, Ore Texture and Microwave Power Delivery on Microwave Treatment of Ores. Ph.D. Thesis, University of Stellenbosch, Stellenbosch, South Africa, 2010.
36. Holman, B.W. Heat treatment as an agent in rock breaking. *Trans. IMM* **1926**, *26*, 219.
37. Yates, A. Effect of heating and quenching Cornish tin ores before crushing. *Trans. IMM* **1918**, *28*, 1918–1919.
38. Kumari, W.G.P.; Ranjith, P.G.; Perera, M.S.A.; Chen, B.K.; Abdulagatov, I.M. Temperature-dependent mechanical behaviour of Australian Strathbogie granite with different cooling treatments. *Eng. Geol.* **2017**, *229*, 31–44. [CrossRef]
39. Kumari, W.G.P.; Ranjith, P.G.; Perera, M.S.A.; Chen, B.K. Experimental investigation of quenching effect on mechanical, microstructural and flow characteristics of reservoir rocks: Thermal stimulation method for geothermal energy extraction. *J. Pet. Sci. Eng.* **2018**, *162*, 419–433. [CrossRef]
40. Salomon, H.; Vignaud, C.; Lahlil, S.; Menguy, N. Solutrean and Magdalenian ferruginous rocks heat-treatment: Accidental and/or deliberate action ? *J. Archaeol. Sci.* **2015**, *55*, 100–112. [CrossRef]
41. Heap, M.J.; Lavallée, Y.; Laumann, A.; Hess, K.U.; Meredith, P.G.; Dingwell, D.B.; Huismann, S.; Weise, F. The influence of thermal-stressing (up to 1000 °C) on the physical, mechanical, and chemical properties of siliceous-aggregate, high-strength concrete. *Constr. Build. Mater.* **2013**, *42*, 248–265. [CrossRef]
42. Heshami, M.; Ahmadi, R. Effect of thermal treatment on specific rate of breakage of manganese ore. *J. Min. Environ.* **2018**, *9*, 339–348.
43. Huotari, T.; Kukkonen, I. *Thermal Expansion Properties of Rocks: Literature Survey and Estimation of Thermal Expansion Coefficient for Olkiluoto Mica Gneiss*; Posiva Oy: Olkiluoto, Finland, 2004; Volume 4, p. 62.
44. Axenenko, O.; Thorpe, G. Modeling of dehydration and stress analysis of gypsum plasterboards exposed to fire. *Comput. Mater. Sci.* **1996**, *6*, 281–294. [CrossRef]
45. Torres, J.; Mendez, J.; Sukiennik, M. Transformation enthalpy of the alkali-earths sulfates ($SrSO_4$, $CaSO_4$, $MgSO_4$, $BaSO_4$). *Thermochim. Acta* **1999**, *334*, 57–66. [CrossRef]
46. Scheding, W.M.; AJ, S.; Binns, D.; Parker, R.H.; Wills, B.A. The effect of thermal pretreatment on grinding characteristics. *Cambame Sch. Mines J.* **1881**, *81*, 43.
47. Dash, N.; Rath, S.S.; Angadi, S.I. Thermally assisted magnetic separation and characterization studies of a low-grade hematite ore. *Powder Technol.* **2019**, *346*, 70–77. [CrossRef]
48. Wonnacott, G.; Wills, B.A. Optimisation of Thermally Assisted Liberation of a Tin Ore with the Aid of Computer Simulation. *Miner. Eng.* **1990**, *3*, 187–198. [CrossRef]
49. Parodi, F. Dielectric Heating—An Overview|ScienceDirect Topics. Available online: https://www.sciencedirect.com/topics/chemical-engineering/dielectric-heating (accessed on 20 January 2010).
50. Micro Denshi, C.L. Basics of Microwave. Available online: https://www.microdenshi.co.jp/en/microwave/index.html (accessed on 25 December 2019).
51. Swart, A.J. *Evaluating the Effects of Radio-Frequency Treatment on Rock Samples: Implications for Rock Comminution*; Panagiotaras, D.D., Ed.; InTech: Shanghai, China, 2012; pp. 457–484.

52. Wong, D. Microwave Dielectric Constants of Metal Oxides at High Temperature. Ph.D. Thesis, University of Alberta, Edmonton, AB, Canada, 1975.

53. Walkiewicz, J.W.; Kazonich, G.; McGill, S.L. Microwave heating characteristics of minerals and compounds. *Miner. Metall. Process.* **1988**, *39*, 39–42.

54. Kingman, S.W.; Vorster, W.; Rowson, N.A. The Influence of Mineralogy on Microwave Assisted Grinding. *Miner. Eng.* **2000**, *13*, 313–327. [CrossRef]

55. Omran, M.; Fabritius, T.; Abdel-Khalek, N.; El-Aref, M.; Elmanawi, A.E.-H.; Nasr, M.; Elmahdy, A. Microwave Assisted Liberation of High Phosphorus Oolitic Iron Ore. *J. Miner. Mater. Charact. Eng.* **2014**, *2*, 414–427. [CrossRef]

56. Like, Q.; Jun, D.; Liqun, Y. Meso-mechanics simulation analysis of microwave-assisted mineral liberation. *Frat. Integrità Strutt.* **2015**, *9*, 543–553.

57. Batchelor, A.R.; Jones, D.A.; Plint, S.; Kingman, S.W. Deriving the ideal ore texture for microwave treatment of metalliferous ores. *Miner. Eng.* **2015**, *84*, 116–129. [CrossRef]

58. Lin, B.; Li, H.; Chen, Z.; Zheng, C.; Hong, Y.; Wang, Z. Sensitivity analysis on the microwave heating of coal: A coupled electromagnetic and heat transfer model. *Appl. Therm. Eng.* **2017**, *126*, 949–962. [CrossRef]

59. Ali, A.Y.; Bradshaw, S.M. Quantifying damage around grain boundaries in microwave treated ores. *Chem. Eng. Process.* **2009**, *48*, 1566–1573. [CrossRef]

60. Jankovic, A. Variables affecting the fine grinding of minerals using stirred mills. *Miner. Eng.* **2003**, *16*, 337–345. [CrossRef]

61. Sikong, L.; Bunsin, T. Mechanical property and cutting rate of microwave treated granite rock. *Songklanakarin J. Sci. Technol.* **2009**, *31*, 447–452.

62. Vorster, W.; Rowson, N.A.; Kingman, S.W. The effect of microwave radiation upon the processing of Neves Corvo copper ore. *Int. J. Miner. Process.* **2001**, *63*, 29–44. [CrossRef]

63. Kumar, A.; Kamath, B.P.; Ramarao, V.V.; Mohanty, D.B. Microwave Energy Aided Mineral Comminution. In Proceedings of the International Seminar on Mineral Processing Technology, Chennai, India, 8–10 March 2006; pp. 398–404.

64. Bobicki, E.R.; Liu, Q.; Xu, Z. Microwave Treatment of Ultramafic Nickel Ores: Heating Behavior, Mineralogy, and Comminution Effects. *Minerals* **2018**, *8*, 524. [CrossRef]

65. Kumar, P.; Sahoo, B.K.; De, S.; Kar, D.D.; Chakraborty, S.; Meikap, B.C. Iron ore grindability improvement by microwave pre-treatment. *J. Ind. Eng. Chem.* **2010**, *16*, 805–812. [CrossRef]

66. Zhu, J.; Liu, J.; Yuan, S.; Cheng, J.; Liu, Y.; Wang, Z.; Zhou, J.; Cen, K. Effect of microwave irradiation on the grinding characteristics of Ximeng lignite. *Fuel Process. Technol.* **2016**, *147*, 2–11. [CrossRef]

67. Kingman, S.W.; Rowson, N.A. Microwave Treatment of Minerals—A Review. *Miner. Eng.* **1998**, *11*, 1081–1087. [CrossRef]

68. Kingman, S.W.; Corfield, G.M.; Rowson, N.A. Effects of microwave radiation upon the mineralogy and magnetic processing of a massive Norwegian Ilmenite ore. *Magn. Electr. Sep.* **1998**, *9*, 131–148. [CrossRef]

69. Schmuhl, R.; Smit, J.T.; Marsh, J.H. The influence of microwave pre-treatment of the leach behaviour of disseminated sulphide ore. *Hydrometallurgy* **2011**, *108*, 157–164. [CrossRef]

70. Charikinya, E.; Bradshaw, S.M. An experimental study of the effect of microwave treatment on long term bioleaching of coarse, massive zinc sulphide ore particles. *Hydrometallurgy* **2017**, *173*, 106–114. [CrossRef]

71. Cai, X.; Qian, G.; Zhang, B.; Chen, Q.; Hu, C. Selective liberation of high-phosphorous oolitic hematite assisted by microwave processing and acid leaching. *Minerals* **2018**, *8*, 245. [CrossRef]

72. Ghorbani, Y.; Petersen, J.; Harrison, S.T.L.; Tupikina, O.V.; Becker, M.; Mainza, A.N.; Franzidis, J.P. An experimental study of the long-term bioleaching of large sphalerite ore particles in a circulating fluid fixed-bed reactor. *Hydrometallurgy* **2012**, *129*, 161–171. [CrossRef]

73. Al-Harahsheh, M.; Kingman, S.W. Microwave-assisted leaching—A review. *Hydrometallurgy* **2004**, *73*, 189–203. [CrossRef]

74. Sahyoun, C.; Rowson, N.A.; Kingman, S.W.; Groves, L.; Bradshaw, S.M. The Influence of Microwave Pre-Treatment on Copper Flotation. *Trans. S. Afr. Inst. Min. Metall.* **2005**, *105*, 7–13.

75. Bradshaw, S.; Louw, W.; van der Merwe, C.; Reader, H.; Kingman, S.; Celuch, M.; Kijewska, W. Techno-economic considerations in the commercial microwave processing of mineral ores. *J. Microw. Power Electromagn. Energy* **2007**, *40*, 228–240. [CrossRef]

76. Kingman, S.W. Microwave Processing of Materials. Ph.D Thesis, Chemical Engineering, Stellenbosch University, Stellenbosch, South Africa, 2018.
77. Jones, D.A.; Kingman, S.W.; Whittles, D.N.; Lowndes, I.S. The influence of microwave energy delivery method on strength reduction in ore samples. *Chem. Eng. Process.* **2007**, *46*, 291–299. [CrossRef]
78. CIPEC. *Benchmarking the Energy Consumption of Canadian Open-pit Mines*; CIPEC: Ottawa, ON, Canada, 2005; pp. 1–56.
79. Swart, A.J.; Mendonidis, P. Evaluating the effect of radio-frequency pre-treatment on granite rock samples for comminution purposes. *Int. J. Miner. Process.* **2013**, *120*, 1–7. [CrossRef]
80. Toprak, N.A.; Altun, O.; Aydogan, N.; Benzer, H. The influences and selection of grinding chemicals in cement grinding circuits. *Constr. Build. Mater.* **2014**, *68*, 199–205. [CrossRef]
81. Mishra, R.K.; Weibel, M.; Müller, T.; Heinz, H.; Flatt, R.J. Energy-effective grinding of inorganic solids using organic additives. *Chimia* **2017**, *71*, 451–460. [CrossRef]
82. Gled Gennadievich Mejeoumov. Improved Cement Quality and Grinding Efficiency by Means of Closed Mill Circuit Modeling. Ph.D Thesis, Civil Engineering, Texas A&M University, College Station, TX, USA, 2007.
83. Allahverdi, A.; Babasafari, Z. Effectiveness of triethanolamine on grindability and properties of portland cement in laboratory ball and vibrating disk mills. *Ceram. Silik.* **2014**, *58*, 89–94.
84. Katsioti, M.; Tsakiridis, P.E.; Giannatos, P.; Tsibouki, Z.; Marinos, J. Characterization of various cement grinding aids and their impact on grindability and cement performance. *Constr. Build. Mater.* **2009**, *23*, 1954–1959. [CrossRef]
85. Hashem, F.S.; Hekal, E.E.; Wahab, M.A. El The influence of Triethanol amine and ethylene glycol on the grindability, setting and hydration characteristics of Portland cement. *Int. J. Petrochem. Sci. Eng.* **2019**, *4*, 81–88.
86. Rao, R.B.; Narasimhan, K.S.; Rao, T.C. Effect of additives on grinding of magnetite ore. *Min. Metall. Explor.* **1991**, *8*, 144–151. [CrossRef]
87. Cebeci, Y.; Bayat, O. Effect of flotation reagents on the wet grinding of celestite concentrate. *Indian J. Chem. Technol.* **2004**, *11*, 382–387.
88. Andres, U. Parameters of Disintegration of Rock by Electrical Pulses. *Powder Technol.* **1989**, *58*, 265–269. [CrossRef]
89. Andres, U.; Timoshkin, I.; Soloviev, M. Energy consumption and liberation of minerals in explosive electrical breakdown of ores. *Miner. Process. Extr. Metall.* **2001**, *110*, 149–157. [CrossRef]
90. Singh, V.; Dixit, P.; Venugopal, R.; Venkatesh, K.B. Ore pretreatment methods for grinding: Journey and prospects. *Miner. Process. Extr. Metall. Rev.* **2018**, *40*, 1–15. [CrossRef]
91. Shi, F.; Zuo, W.; Manlapig, E. Characterisation of pre-weakening effect on ores by high voltage electrical pulses based on single-particle tests. *Miner. Eng.* **2013**, *50*, 69–76. [CrossRef]
92. Fujita, T.; Yoshimi, I.; Shibayama, A.; Miyazaki, T.; Abe, K.; Sato, M.; Yen, W.T.; Svoboda, J. Crushing and Liberation of Materials by Electrical Disintegration. *Eur. J. Miner. Process. Environ. Prot.* **2001**, *1*, 113–122.
93. Cho, S.H.; Mohanty, B.; Ito, M.; Nakamiya, Y.; Owada, S.; Kubota, S.; Ogata, Y.; Tsubayama, A.; Yokota, M.; Kaneko, K. Dynamic fragmentation of rock by high-voltage pulses. In Proceedings of the 41st U.S. Symposium on Rock Mechanics (USRMS): 50 Years of Rock Mechanics—Landmarks and Future Challenges, Golden, CO, USA, 17–21 June 2006; pp. 1–9.
94. Ito, M.; Owada, S.; Nishimura, T.; Ota, T. Experimental study of coal liberation: Electrical disintegration versus roll-crusher comminution. *Int. J. Miner. Process.* **2009**, *92*, 7–14. [CrossRef]
95. Cabri, L.J.; Rudashevsky, N.S.; Rudashevsky, V.N.; Oberthür, T. Electric-pulse disaggregation (EPD), Hydroseparation (HS) and their use in combination for mineral processing and advanced characterization of ores. In Proceedings of the 40th Annual Meeting of the Canadian Mineral Processors, Ottawa, ON, Canada, 22–24 January 2008.
96. Wang, E.; Shi, F.; Manlapig, E. Pre-weakening of mineral ores by high voltage pulses. *Miner. Eng.* **2011**, *24*, 455–462. [CrossRef]
97. Wang, E.; Shi, F.; Manlapig, E. Mineral liberation by high voltage pulses and conventional comminution with same specific energy levels. *Miner. Eng.* **2012**, *27*, 28–36. [CrossRef]
98. Peng Gao, S.Y.; Yuexin Han, Y.L.; Chen, H. Experimental Study on the Effect of Pretreatment with High-Voltage Electrical Pulses on Mineral Liberation and Separation of Magnetite Ore. *Minerals* **2017**, *7*, 153.

99. Razavian, S.M.; Rezai, B.; Irannajad, M. Investigation on pre-weakening and crushing of phosphate ore using high voltage electric pulses. *Adv. Powder Technol.* **2014**, *25*, 1672–1678. [CrossRef]

100. Zuo, W.; Shi, F.; van der Wielen, P.K.; Weh, A. Ore particle breakage behaviour in a pilot scale high voltage pulse machine. *Miner. Eng.* **2015**, *84*, 64–73. [CrossRef]

101. van der Wielen, P.K. Application of High Voltage Breakage to a Range of Rock Types of Varying Physical Properties. Ph.D Thesis, Earth Resources, University of Exeter, Exeter, UK, 2013.

102. Bowman, D.J.; Bearman, R.A. Coarse waste rejection through size based separation. *Miner. Eng.* **2014**, *62*, 102–110. [CrossRef]

103. Zuo, W.; Shi, F.; Manlapig, E. Pre-concentration of copper ores by high voltage pulses. Part 1: Principle and major findings. *Miner. Eng.* **2015**, *79*, 306–314. [CrossRef]

104. Sperner, B.; Jonckheere, R.; Pfänder, J.A. Testing the influence of high-voltage mineral liberation on grain size, shape and yield, and on fission track and ^{40}Ar/^{39}Ar dating. *Chem. Geol.* **2014**, *371*, 83–95. [CrossRef]

105. Shi, F.; Zuo, W.; Manlapig, E. Pre-concentration of copper ores by high voltage pulses. Part 2: Opportunities and challenges. *Miner. Eng.* **2015**, *79*, 315–323. [CrossRef]

106. Parker, T.; Shi, F.; Evans, C.; Powell, M. The effects of electrical comminution on the mineral liberation and surface chemistry of a porphyry copper ore. *Miner. Eng.* **2015**, *82*, 101–106. [CrossRef]

107. Yu, J.; Han, Y.; Li, Y.; Gao, P. Effect of magnetic pulse pretreatment on grindability of a magnetite ore and its implication on magnetic separation. *J. Cent. South Univ.* **2016**, *23*, 3108–3114. [CrossRef]

108. Menacho, J.; Yerkovic, C.; Gaete Garreton, L. Explorating the ultrasonic comminution of copper ores. *Miner. Eng.* **1993**, *6*, 607–617.

109. Gaete-Garretón, L.F.; Vargas-Hermández, Y.P.; Velasquez-Lambert, C. Application of ultrasound in comminution. *Ultrasonics* **2000**, *38*, 345–352. [CrossRef]

110. Dodds, J.A.; Gaete-Garreton, L.; Vargas-Hernandez, Y.; Chamayou, A.; Valderama-Reyes, W.; Montoya-Vitini, F. Development of an ultrasonic high-pressure roller press. *Chem. Eng. Sci.* **2003**, *58*, 4317–4322.

111. Omran, M.; Fabritius, T.; Elmahdy, A.M.; Abdel-Khalek, N.A.; Gornostayev, S. Improvement of phosphorus removal from iron ore using combined microwave pretreatment and ultrasonic treatment. *Sep. Purif. Technol.* **2015**, *156*, 724–737. [CrossRef]

112. Leibtag, S. Ultrasonic Dispersing and Wet-Milling of Calcium Carbonate (CaCO$_3$). Available online: http://blog.sonomechanics.com/blog/ultrasonic-dispersing-and-wet-milling-of-calcium-carbonate (accessed on 27 April 2020).

113. Mazumder, B.; Uddin, I.; Khan, S.; Ravi, V.; Selvraj, K.; Poddar, P.; Ahmad, A. Bio-milling technique for the size reduction of chemically synthesized BiMnO$_3$ nanoplates. *J. Mater. Chem.* **2007**, *17*, 3910–3914. [CrossRef]

114. Lee, K.; Rosario, P. *Canada Mining Innovation Council (CMIC)—Comminution Technology Appraisal Study*; Hatch: Mississauga, ON, Canada, 2016; pp. 1–56.

Article

Quantitative Microstructural Analysis and X-ray Computed Tomography of Ores and Rocks—Comparison of Results

Oleg Popov [1], Irina Talovina [2], Holger Lieberwirth [1,*] and Asiia Duriagina [2]

[1] Institute of Mineral Processing Machines, TU Bergakademie Freiberg, Lampadiusstr. 4, 09599 Freiberg, Germany; Oleg.Popov@iam.tu-freiberg.de

[2] Department of Historical and Dynamic Geology, Saint-Petersburg Mining University, 22 line, 2, 199106 St-Petersburg, Russia; i.talovina@gmail.com (I.T.); gayfutdinovaam@yandex.ru (A.D.)

* Correspondence: holger.lieberwirth@iam.tu-freiberg.de; Tel.: +49-3731-39-2528

Received: 9 January 2020; Accepted: 24 January 2020; Published: 31 January 2020

Abstract: Profound knowledge of the structure and texture of rocks and ores as well as the behavior of the materials under external loads is essential to further improvements in size reduction processes, particularly in terms of liberation size. New analytical methods such as computer tomography (CT) were adopted to improve the understanding of material characteristics in rocks and ores relevant to mineral processing, particular the crushing and grinding and the modelling/simulation thereof. Results obtained on the texture and structure of identical samples of rather different rocks and ores (copper ore, granodiorite, kimberlite) are compared by CT with quantitative results from traditional optical microscopy obtained by quantitative microstructural analysis (QMA). While the two approaches show a good agreement of the results in many areas, the measurements with the two different methods also exhibit remarkable differences in other areas, which are discussed further. In conclusion, both methods have their specific advantages starting from sample preparation to the accuracy of information obtained concerning certain parameters of mode and fabric. While sample preparation is faster with CT and information on special distribution of metal minerals is more reliable, the information on mode, grain size and clustering seem to be more precise with QMA. Based on the results, it can be concluded that both methods are comparable in many areas, but in in the field of spatial distribution, they are merely complementary.

Keywords: quantitative microstructural analysis; X-ray computed tomography; selective comminution; texture; structure; mineral processing; crushing; grinding

---

1. Introduction

An increasing demand for raw materials worldwide meets a general trend towards the exploitation of lower grade new ore deposits [1–3]. Thus, ever larger quantities of ore have to be processed to produce the required amounts of concentrates. Not only rising energy costs, per processed ton of ore as well as in absolute terms, but also the accompanying CO_2 emissions and the related social license to proceed are reasons for the resource industries to look for smarter solutions, particularly for the energy-intensive comminution processes.

In a study comprising numerous Australian copper mines, it was found that on average, 36 percent of the energy utilized is consumed solely by comminution processes [4]. Comminution processes for copper production alone consume about 1.223 MWh/t Cu [4]. Milling, in particular fine grinding in the area of liberation size, consumes thereby the largest mass specific amounts of energy. Investigation of the distribution of the valuable components in the ore, their grain size as well as the behavior of the rock and ore under certain load is therefore essential, to select the right type, magnitude and frequency of load and thus, respectively, the right milling machine working at appropriate operating parameters [5,6].

The influence of rock and ore properties on the processing characteristics is a well-established fact [7–12]. A profound knowledge of the mineral behavior is required on all comminution steps, be it for the efficient production of preconcentrates by selective comminution on coarse particle size level [13] or the high voltage impulse comminution [14,15], or in the fine grinding. The importance of such an understanding on the microscopic level is emphasized by a number of publications and even global initiatives such as the Coalition for Energy Efficient Comminution (CEEC) [16,17]. It is also essential for the modelling and simulation of comminution processes by numerical methods, such as with the Discrete Element Method (DEM). Modelling of comminution processes with those methods provides fascinating new insights into comminution processes [18–21]. For those methods the calibration of the models for prognostic purposes poses a tremendous challenge if the differences of the mode and fabric of the various ores and rocks cannot be quantified [22–24].

The increased focus on the ore microstructure led to a number of efforts introducing new technologies into the analysis of ore microstructures. In recent years, new technologies such as the scanning electron microscope (SEM)-based Mineral Liberation Analysis (MLA) or Quantitative Evaluation of Minerals by Scanning Electron Microscopy (QEMSCAN) made their way into the mineralogist's laboratories [25,26]. More recently, also Computer Tomography (CT) was applied for mineralogical analyses [27–30]. CT is a nondestructive technique that allows for the visualization of internal structures of objects based on their different X-ray density. Originally, the method was used mainly for medical purposes, but since the 1990s it has also been used in other applications, first in petroleum geology [31–33] and later in process mineralogy [34] and process technology [27,29,30,35].

The investigation of structure and texture of ores is a pretty new but fast growing application area of CT [36,37]. While a number of publications support the usefulness of applying CT for analyzing mineral structures is still missing, a comparison with results acquired by traditional methods, e.g., optical microscopy. This may be caused by the missing quantitative criteria for such a comparison. The Quantitative Microstructural Analysis (QMA), a method based on optical microscopy and proven in numerous industrial applications, shall be used in this paper to compare the results obtained by the new method CT with the proven QMA. Optical microscopy has been used for decades to investigate mineral raw materials using thin or polished sections. Under polarized light, an experienced mineralogist directly identifies the various mineral varieties of the grains. Yet, optical microscopy has its limitations with regard to the image and grain size resolution, and the mineral identification, finally limited by the wave length of the light. QMA will be briefly introduced in the following chapter. A more detailed description can be found for instance in [9,10].

The research presented will therefore present a comparison of structural and textural data, obtained for identical samples of three rather different materials using both methods. This allows to evaluate the capacity of CT for obtaining relevant material data of rocks and ores on microstructural level for modelling, simulation, machine and process design in crushing and milling.

2. Methods and Materials

2.1. QMA—Microstructural Analysis Based on Optical Microscopy

Descriptions of texture and structure of rocks are of great value as they can be used in the interpretation of rock formations. Microstructural descriptions are traditionally obtained by geologists using thin or polished sections for petrographic analysis. The traditional method has limitations in describing the full 3D rock structure with quantitative measures and in drawing conclusions and deriving interpretations of the microstructural information.

Quantitative Microstructural Analysis (QMA) is a collective term for a number of methods for analysing the geometry and mechanical properties of microstructural constituents. The objective is usually to investigate the relationship between the microstructure on the one hand and the properties of the rock or its history of origin on the other. Microstructural characteristics are often used for the classification of rocks. Methods for microstructural assessment have long been part of rock research, rock testing and production monitoring. The widening range of possible applications (petrography, mineralogy, metallography, investigation of the structure of building materials, etc.) and the development of theoretical principles (stochastic geometry, digital image processing) have led to QMA becoming a separate field of science over recent years [9,10].

The basis of the QMA of a rock sample is normally a microscope analysis of thin or polished sections. As rock-forming minerals and rock microstructures mostly have a complex three-dimensional structure, the information that can be derived from one-dimensional cut surfaces is often insufficient for the spatial quantitative characterization of the minerals and microstructures of geological materials. A three-dimensional information or the reconstruction of minerals and rock microstructures of geological materials is required. The results of mathematical-petrographic rock characterization are a precondition for predicting the relationships between rock characteristics and relevant product properties or system characteristics (e.g., wear, energy consumption) with the help of mathematical statistical modelling (e.g., multiple regression and correlation analysis) [8,38].

All the strength properties of rocks are directly dependent on the type and degree of mutual bonding of the mineral grains in the microstructure. The greater the specific surface on which the mineral grains are in contact with one another, the more mechanically competent is the rock and the more resistant to the effects of comminution processes.

The degree of grain bonding depends on grain shape and grain size. The cohesiveness of the microstructure rises as grain size decreases and specific surface correspondingly increases. With otherwise identical properties, fine-grained rocks therefore always manifest greater strength than medium and coarse-grained types. The more irregular the grain shape, the greater the grain surface area and the more intense the mutual intergrowth of adjacent individual grains [11].

The QMA usually starts with a proper extraction of a defined number of rock samples from the deposit. According to the QMA approach developed at the Institute of Mineral Processing Machines (Freiberg, Germany), model three polished sections perpendicular to each other are prepared from each rock specimen. These thin sections are evaluated with the help of a polarization microscope, assessed by means of various stereological methods (Figure 1) and classified in a library of thin sections. A mathematical model is used to characterize the rocks quantitatively, based on the calculation of relevant structural and textural characteristic data [13,23].

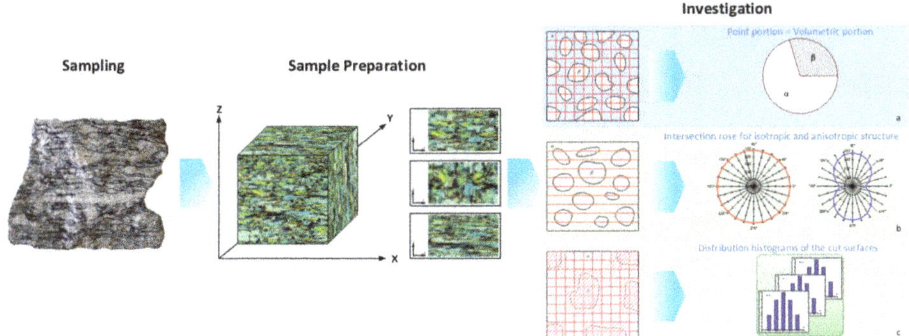

Figure 1. Process flow for QMA rock analysis.

Rock characteristics are mathematically derived in the form of characteristic values (see Table 1). These are first determined completely independently from a specific application, but can be interpreted for use with specific applications in comminution or other processes. In the following, the most important characteristic values gained by QMA (according to [8]) are explained in greater detail:

Table 1. Rock characteristics as established by QMA under the optical microscope.

Raw Material		Rock Type: Granodiorite			Phase Related Features			Raw Material Features
		Deposit: Kindisch Location: Saxony, Germany						
		Properties	Symbol	Unit	Quartz	Feldspar	Mica	Σ Microbodies
Mode	Content	Volumetric Portion	ε_V	%	34	48	18	100
Fabric	Texture	Size	Mean diameter	$d_{50,3}$	mm	0.732	0.928	0.647	0.810

<table>
<tr><td rowspan="2">Raw Material</td><td colspan="3" align="center">Rock Type: Granodiorite</td><td colspan="3" align="center">Phase Related Features</td><td rowspan="2">Raw Material Features</td></tr>
<tr><td colspan="3" align="center">Deposit: Kindisch   Location: Saxony, Germany</td></tr>
<tr><td colspan="3" align="center">Properties</td><td>Symbol</td><td>Unit</td><td>Quartz</td><td>Feldspar</td><td>Mica</td><td>Σ Microbodies</td></tr>
<tr><td rowspan="13">Mode

Fabric</td><td colspan="2">Content</td><td>Volumetric Portion</td><td>ε_V</td><td>%</td><td>34</td><td>48</td><td>18</td><td>100</td></tr>
<tr><td rowspan="6">Texture</td><td rowspan="2">Size</td><td>Mean diameter</td><td>$d_{50,3}$</td><td>mm</td><td>0.732</td><td>0.928</td><td>0.647</td><td>0.810</td></tr>
<tr><td>Scatter parameter</td><td>σ_{ln}</td><td>-</td><td>0.410</td><td>0.384</td><td>0.414</td><td>0.398</td></tr>
<tr><td>Grain surface</td><td>Specific surface</td><td>S_V</td><td>mm^2/mm^3</td><td>7.480</td><td>5.180</td><td>11.200</td><td>7.050</td></tr>
<tr><td rowspan="2">Shape</td><td>Elongation</td><td>E</td><td>-</td><td>1.274</td><td>1.330</td><td>1.968</td><td>1.425</td></tr>
<tr><td>Flatness</td><td>F</td><td>-</td><td>1.267</td><td>1.248</td><td>1.059</td><td>1.221</td></tr>
<tr><td>Roughness</td><td>Rougjness degree</td><td>K_R</td><td>%</td><td>30</td><td>13</td><td>38</td><td>23</td></tr>
<tr><td rowspan="6">Structure</td><td rowspan="3">Orientation</td><td>Degree of linear orientation</td><td>K_{lin}</td><td>%</td><td>16</td><td>18</td><td>42</td><td>10</td></tr>
<tr><td>Degree of areal orientation</td><td>K_{fl}</td><td>%</td><td>12</td><td>12</td><td>3</td><td>11</td></tr>
<tr><td>Degree of isotropic orientation</td><td>K_{is}</td><td>%</td><td>72</td><td>70</td><td>55</td><td>79</td></tr>
<tr><td>Distribution</td><td>Degree of clustering</td><td>C</td><td>%</td><td>36</td><td>41</td><td>13</td><td>34</td></tr>
<tr><td>Space filling</td><td>Space filling degree</td><td>ε_{VF}</td><td>%</td><td>-</td><td>-</td><td>-</td><td>100</td></tr>
</table>

Mode (Volume Percentage of the Mineral Phases)

The volume percentage of a mineral phase is defined as the quotient of the sum of the volume of the individual mineral grains in the sample and the sample volume. It can be most easily determined with the help of the point counting method [39].

Grain Size

The individual microbodies are polydispersed in the rock microstructure so that the size of the microbodies has to be approximated with a size distribution function. The size distribution function can be frequently approximated by a logarithmic normal distribution (LND), which is uniquely characterized by a median $d_{50.3}$ and a scatter parameter σ_{ln} [8].

Grain Shape

The shape of a microbody corresponds to its outer appearance, which is determined by its grain shape and roughness. The grain shape is a geometric particle characteristic, i.e., a characteristic that makes allowance for all three dimensions [40]. The grain shape of a microbody can be regarded approximately as an ellipsoid with the main axes a, b and c, where $a \geq b \geq c$. The relationship of the main axes to each other describe the grain shape of the microbody. The elongation E describes the "needle shape" of a particle and the flatness F of the particle its "platy shape".

Roughness

The shape of severely curved, serrated and similar complex microbodies can be analysed statistically and characterized by the roughness. This parameter can be recorded statistically in thin sections and characterized with indices. The roughness K_R is defined as the ratio of the difference between the "real" surface area $S_{V(R)}$ and the "ideal" surface area $S_{V(I)}$ relative to the "real" surface area $S_{V(R)}$ of the individual phases. The calculation of the "ideal" surface area $S_{V(I)}$ is based on the microbody size distribution with consideration of the microbody shape and the phase volume percentage. The "real" surface area $S_{V(R)}$ is calculated from the three-dimensional rose of intersections. The outer shape of the microbodies largely determines the character of the microstructure [8].

Orientation

From three 2D-roses of intersections orthogonal to each other, with an approximation, the parameters of a spatial rose of intersections and their orientation angle in the space are calculated. A spatial rose of intersections of an oriented microstructure is the superposition of the roses of intersection of ideal boundary surface systems. From the parameters of the spatial rose of intersections, the orientation degrees (K_{is}, K_{lin} and K_{fl}) of a spatial arrangement (percentages of linear- and areal-oriented as well as nonoriented boundary surfaces) can be derived.

Distribution

Microbodies of a phase can be evenly distributed in the space, but they can also form clusters in which the microbodies of a phase share boundary surfaces. The degree of clustering C is defined as the quotient of the sum of the boundary surfaces between the grains of the same mineral group and the total boundary surface of this mineral group. The degree of clustering can be calculated with the help of linear analysis [8].

Space Filling

Rocks are practically never completely compact, i.e., the aggregate space is not completely filled by solid constituents, but there are cavities (pores) between them, which normally contain gas (e.g., air) and/or aqueous solutions. The space filling degree ε_{VF} can vary within wide boundaries for different rocks. The size, distribution and type of pores also show great differences. To assess many properties of the rocks, it is necessary to examine the character of the space filling more closely.

2.2. X-ray Computed Tomography (CT)

CT evolved as an effective supplement in the complex studies of petrophysical properties of rocks. It includes the definition of porosity, fracturing, cavities and their distribution by diameter, volume, and sphericity degree. Moreover, microtomography is a useful tool for the measurement of mineral phases with equal X-ray density throughout the whole sample. The main advantages of the method include also the wide visualization potentials with comparably low measurement time.

Principles of CT are described in detail in numerous papers [27,29,35]. The method consists in essence in weakening the power of the X-ray beam when passing through a certain volume of the material. During scanning in X-ray beam, the sample is rotated around a vertical axis by 180° with a certain step. The result of the scan is fixed by the shadow projection (using Al or Brass filters behind the sample and before the detector). Shadow projections are graphic files where each pixel contains information about the amount of X-ray absorption by the object at a given point. This value is expressed in 256 grey color shades (the darker, the higher the absorption). The processing of the obtained results consists in the mathematical transformation of the shadow vertical projections into a series of horizontal sections of the object using special software CTAn (e.g., Version 1.95, Bruker, Kontich, Belgium). After reconstruction, the most important step of the analysis is the volumetric visualization.

During the binarization of images, e.g., with the help of the CTVox software (Version 2.14, Bruker, Kontich, Belgium) it is possible to create three-dimensional images of the sample for further visual studies and morphometric analysis (Figure 2).

In this paper, the CT studies of typical ore/rock samples were carried out by using a microtomograph "SkyScan-1172" (Bruker, Kontich, Belgium) with resolutions from 19.58 to 32.32 μm, equipped with the certified programs Skyscan1172 μCT, NRecon, DataViewer, CTVox, CTAn, CTVol, SkyScan_MTS, Gidropora. The shooting parameter set for scanning the samples is shown in Table 2.

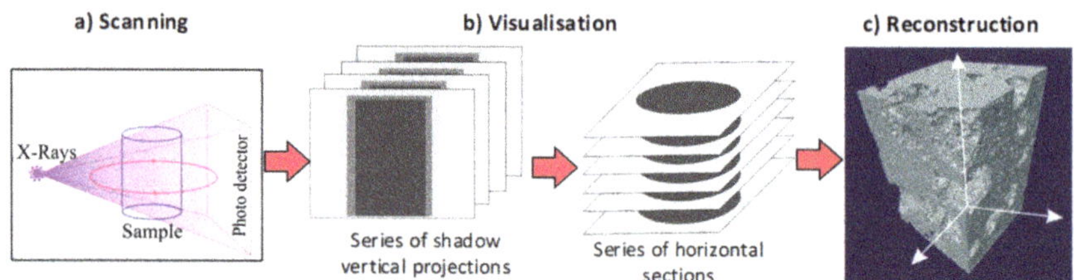

Figure 2. Basic principles of CT: (**a**) scanning of the sample in X-ray beam, (**b**) mathematical transformation of the shadow vertical projections into a series of horizontal sections of the object, (**c**) creation of three-dimensional images of the sample.

Table 2. The shooting parameters set for scanning the samples.

Parameters	Unit	Copper Sandstone	Granodiorite	Kimberlite
Accelerating voltage	kV	100	125	130
Current rate	mA	80	61	61
Resolution	μm	19.58	32.32	19.58
Filter	mm	Al 0.1	Brass 0.25	Brass 0.25
Rotation step	grade		0.100	

2.3. Sample Materials

A rock can be described based on the content of its mineral components and its fabric (Figure 3). It consists of the individual mineral phases, but often also contains defects (e.g., pores, cracks), which form the continuous phases (gas or liquid phase) and glasses as well as cryptocrystalline masses which are combined to a nondifferentiated phase. Each single constituent of a phase—also termed microbody or grain—takes up a certain volume that is defined by boundary surfaces.

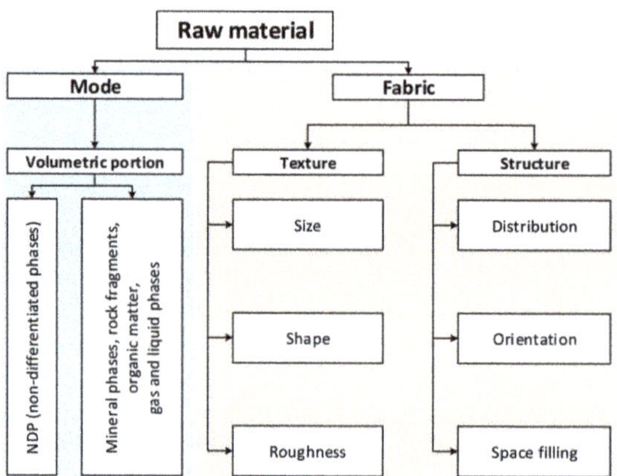

Figure 3. Characterization of raw materials.

To introduce the characteristics of the rock fabric in technology assessment, the texture must be expressed in figures. Based on the application of stereological methods, the rock is quantitatively described by means of its

- mode (volume percentage of the mineral phases),
- characteristic values of the texture (e.g., grain size and shape, roughness),
- characteristic values of the structure (e.g., orientation, distribution and space filling).

The term "rock fabric", comprising texture and structure, can be understood as the quantity of all geometric data of a piece of rock. The characteristics of the rock fabric, quantified with the help of QMA methods, are based on the formal characterization of the rocks by geologists in order to facilitate understanding between geologists and engineers with their respective terminology.

The texture of a rock is the term used to describe the nature of its composition by individual components. It is dictated by the form of the individual mineral components and their mutual geometric relationships. Accordingly, the grain shapes and the grain sizes of the mineral components are essential elements of the rock texture.

The structure of a rock depends on the spatial arrangement of its components. In this case, the components are less the individual grains and far more complexes of the same minerals or other identical structural elements. In this context, the orientation of the components as the directional texture, the distribution of the same as the distribution texture and relative density (compact and porous rocks) are taken into consideration. Rocks with strata of varying mineral composition therefore have an inhomogeneous distribution texture; lava with gas bubbles is characterized by their incomplete space filling.

For the comparison of QMA and CT three different samples were investigated. Their characteristics, described by mode and fabric, but inherently representing also mechanical properties relevant for comminution, ore content vary widely, to cover a range of materials from metal and non-metal mining as well as the aggregate industry. The need to crush and, to a certain extent at least, grind the material to produce a sellable product is the bracket that connects those industries. The investigated samples contain:

- copper ore (green sandstone ore) from Legnica (Poland),
- granodiorite from Kindisch (Saxony, Germany),
- kimberlite from Letseng (Lesotho).

3. Results

In the following, the results of the QMA and the CT analyses of the three different sample materials are presented.

3.1. Copper Ore

3.1.1. QMA Results (Copper Ore)

Mode

Copper ore from Legnica consists of 17% copper minerals in a quartz (64%) and calcite matrix (19%) (see Figure 4).

Figure 4. Copper ore (green sandstone ore): (**a**) hand specimen, (**b**) microphotograph under reflected light.

Grain Size

The quartz shows a mean grain size of 265 μm and the copper minerals with 277 μm are of similar size. The standard deviation σ_{ln} of the diameter distribution of the copper minerals shows a scatter parameter of 0.504. Figure 5 shows the grain size distribution of both minerals as identified by QMA.

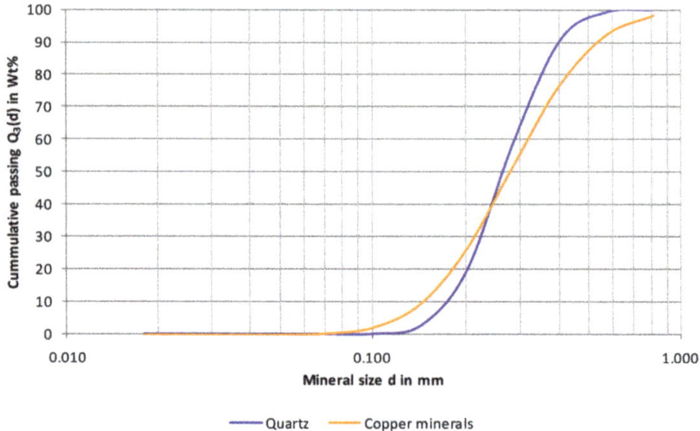

Figure 5. Mineral grain size distribution of the copper ore according to QMA.

Grain Shape

The shape of the copper minerals of the copper ore is predominantly platy-like. This is confirmed by the values for their elongation E and flatness F (E = 1.224; F = 1.156). In comparison, the shape of the quartz is predominantly needle-like. This is confirmed by respective elongation and flatness values (E = 1.458; F = 1.162).

Roughness

The roughness K_R of the copper minerals is around 62% and is therefore in a very high range compared to minerals of other rock fabrics.

Specific Surface Area

The specific surface area S_V of the copper minerals is in the medium range and is around 53 mm^2/mm^3 per mineral grain. The specific surface area and the roughness characterize the "intergrowth" of the minerals with each other. This intergrowth is an important information since it allows to draw conclusions on the breakage behaviour of a material, the specific energy consumption, wear of comminution machine parts etc.

Degree of Clustering

The degree of clustering C shows whether a mineral grain forms an agglomeration in the rock with another mineral grain of the same type. The quartz shows a 6% cluster formation, that is, 6% of all quartz grain surfaces lie next to another quartz grain surface.

Space Filling

The very good space filling means that moisture can only adhere to the surface of particles, as there are no pores in the ore sample.

3.1.2. CT Results (Copper Ore)

Mode

Copper ore consists, according to CT, of 5.2% of ore minerals, quartz grains (53.1%) and a matrix of crystallized calcite (40.9%). The distribution of copper minerals in the entire rock volume is shown in Figure 6. It can be seen that it is possible to clearly distinguish dense copper-containing minerals from quartz and calcite on the tomogram, whereas the latter two minerals are quite difficult to distinguish from each other, because of their close X-ray density.

Figure 6. The distribution of copper minerals (green) in the quartz-calcite matrix (gray) in sandstone.

Grain Size

Using the CTVox and CTan programs, it is possible to analyse in detail not only the sizes of the grains composing the rock, but also their interrelations, which is especially important for the purposes of process mineralogy. Copper minerals are fairly evenly distributed throughout the rock volume, sometimes forming individual grains with an average size of 70 μm and fine-grain aggregates 260 μm in size (Figure 7a,b), sometimes up to 2.0–4.5 mm (Figure 7c).

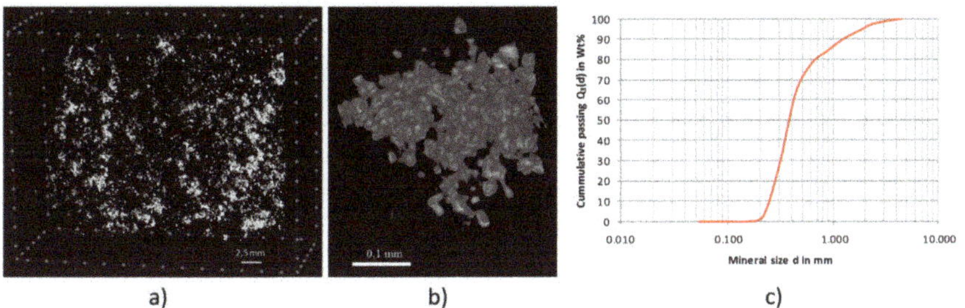

Figure 7. (**a,b**) Aggregates of copper minerals in the volume of sandstone and (**c**) aggregate distribution according to the maximum diameter.

Grain Shape

Copper minerals have a quite isometric shape, the coefficient of sphericity, calculated as the ratio of the short to the long axis, is 0.78 (when the shape of the grain approaches isometric, this coefficient reaches 1, Figure 8a).

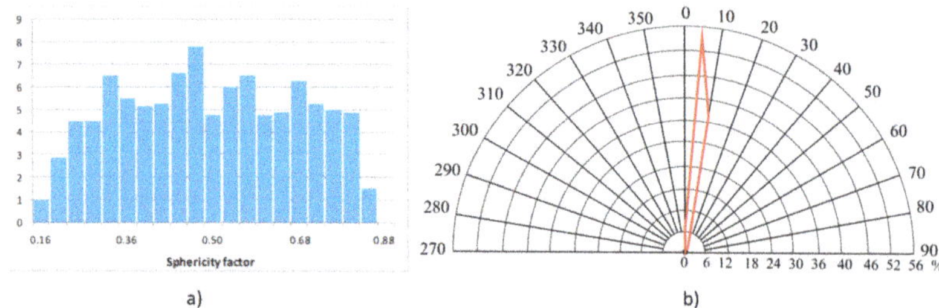

Figure 8. (**a**) Distribution of aggregates of copper minerals according to the sphericity coefficient and (**b**) the orientation of the longest axis, grade

Roughness

Specific surface of copper minerals, calculated as the ratio of the total surface of minerals to their volume and representing the degree of tortuosity of the boundaries (roughness), is equal to 31.86 mm^2/mm^3.

Orientation

The orientation of copper grains and aggregates is also marked, expressed quantitatively as the deviation of the longest axis of grains from the vertical in grades (Figure 8b).

Porosity

One of the most important characteristics of rocks for comminution, linking the strength of the rock with a defect in its structure, is porosity. It determines the interpretation of the strength properties and behavior of the rock during comminution [30]. The total porosity of the sandstone sample is 1.4%, the open porosity is 0.68%. The pore connectivity is low. In general, the pores of the subcapillary size are less than 0.2 μm, the predominant pore size is 25–50 μm, but there are also single pores with a diameter of more than 100 μm. The pore sphericity factor is 0.62. The void density, calculated as the ratio of the number of pores per unit volume of rock, is 9.10 mm^{-3}.

3.2. Granodiorite

3.2.1. QMA Results (Granodiorite)

Mode

Granodiorite from Kindisch consists of 34% quartz, 48% feldspar and 18% mica (see Figure 9).

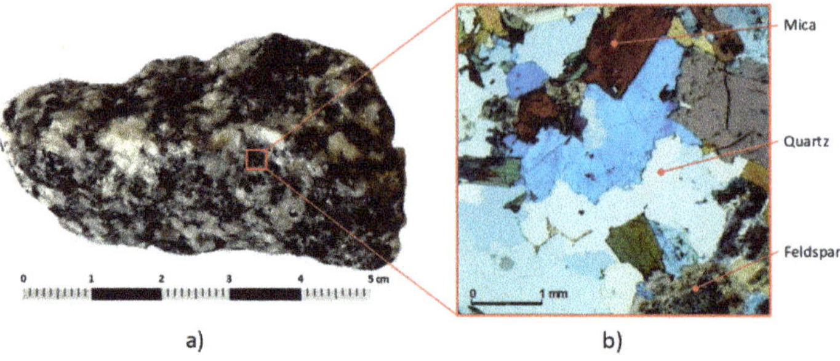

Figure 9. Granodiorite (**a**) hand specimen, (**b**) microphotograph under transmitted light with lambda/4 plate.

The effects of these minerals on the mechanical characteristics of the rock and, hence, its comminution behaviour and abrasivity are quite different. The mineral quartz is a positive development in terms of strength, weathering and temperature resistance. Due to its high Mohs hardness, it increases the abrasivity of a rock in comminution. The feldspars consist of polysynthetic twinned plagioclase and alkali feldspars. The most singular property of mica is its physical structure. As a sheet mineral it can be split into strong, flexible films having good high-temperature resistance. Its mechanical properties, however, vary substantially depending on the direction of loads relative to the direction of the films. While the grain strength is rather high with compression loads rectangular to the film planes the material strength with regard to compression loads acting in the planes is usually comparably low.

Grain Size

The mean grain size of all minerals is rather different. The quartz shows a mean grain size of 732 μm, the feldspar 928 μm, the mica 647 μm. The overall mean grain size of the rock is 810 μm (Figure 10). The diameter distribution of the mica shows a scatter parameter of 0.414.

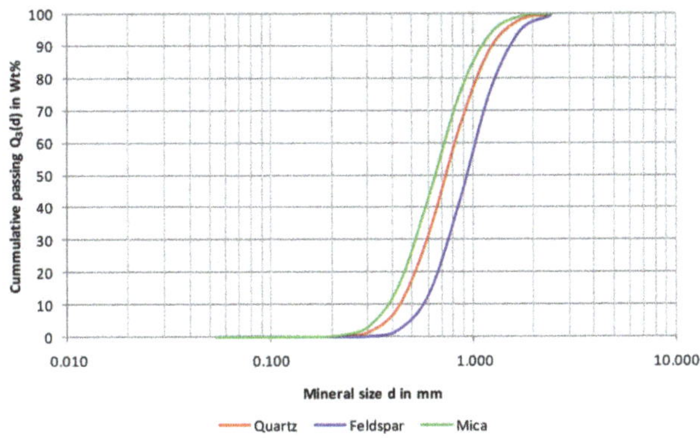

Figure 10. Mineral grain size distribution of granodiorite.

Grain Shape

The grain shape of the mica looks like needle-shaped. This is confirmed by the values of their elongation and flatness (E = 1.968; F = 1.059). In comparison, the overall grain shape of all minerals of granodiorite is mainly needle-platy, with elongation and flatness values (E = 1.425; F = 1.221).

Roughness

The roughness K_R of the mica minerals is approximately 38% and is therefore in very high range compared to minerals of other rock fabrics.

Specific Surface Area

The specific surface area S_V of the mica minerals is approximately 11 mm^2/mm^3 per mineral grain.

Degree of Clustering

The mica shows a 13% cluster formation, which means 13% of all quartz grain surfaces lie next to another mica grain surface.

Isotropic Orientation

The isotropic orientation degree K_{is} is about 79%. These values give an indication that the production of cubic products is only possible by increased machine expenditures (e.g., multi-stage comminution and preferably by impact loading) and matched operating mode of the processing plant (e.g., high stress intensity).

3.2.2. CT Results (Granodiorite)

Mode

The granodiorite contains 89.3% of quartz-feldspar mass and 7.7% of mica according to the CT results. Mica is fairly evenly distributed throughout the rock volume, forming slightly oriented flattened grains of 0.594 mm in size (Figure 11a). There are also individual aggregates of 1.8–3.0 mm (Figure 11b).

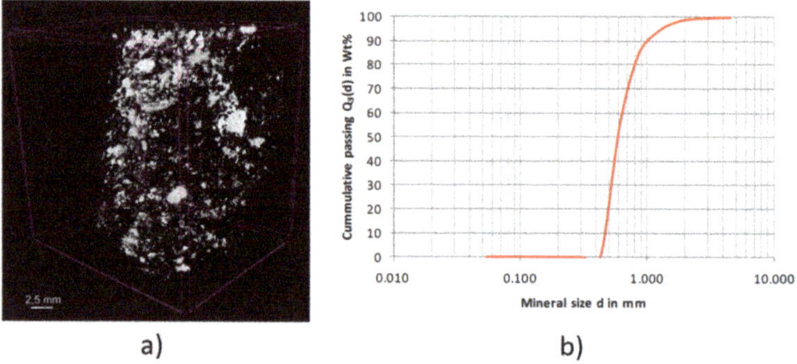

a)   b)

Figure 11. (**a**) Mica in the volume of granodiorite and (**b**) their distribution according to the maximum diameter.

Grain Shape

Mica grains have a flattened morphology (K_{sph} = 0.48) (Figure 12a).

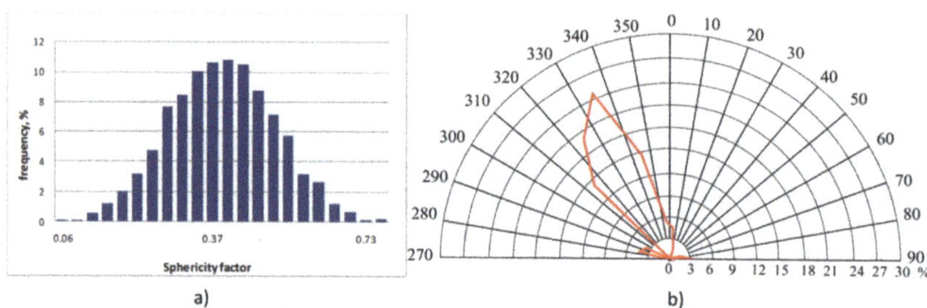

a)   b)

Figure 12. (**a**) Distribution of mica according to the sphericity factor and (**b**) the orientation of the longest axis.

Specific Surface Area

The specific surface area, i.e., the degree of tortuosity of the boundaries is 20.2 mm^2/mm^3.

Isotropic Orientation

There is also a strong orientation of the mica grains (Figure 12b).

Porosity

The total porosity of the granodiorite sample is 3.05%, the opened porosity is 0.91%. Basically, pores are of subcapillary and capillary size (15–40 μm), there are single pores with a diameter of more than 100 µm. The predominant shape of pores is close to spherical (sphericity factor is 0.80). The void density is 2.06 mm^{-3}.

3.3. Kimberlite

3.3.1. QMA Results (Kimberlite)

Mode

Kimberlite from Letseng consists of so-called phenocrysts (21% olivine, 20% pyroxene) and nondifferentiable phase (59%). This nondifferentiable phase (NDP) consists mostly of an irregular mixture of serpentine, chlorite, pyroxene, olivine and ore minerals (Figure 13).

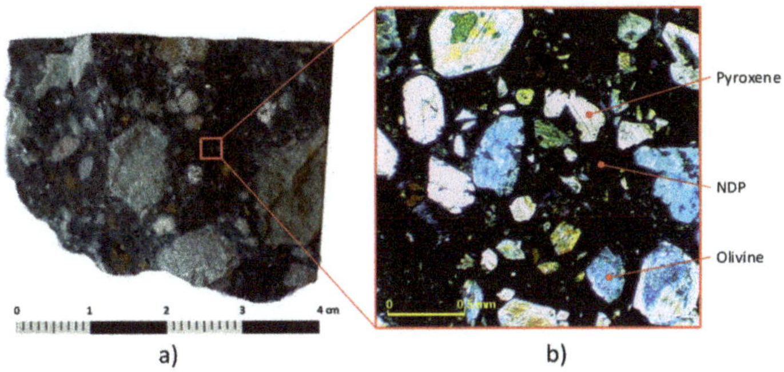

Figure 13. Kimberlite: (**a**) hand specimen, (**b**) microphotograph under transmitted light with lambda/4 plate.

Grain Size

The grain size differs between the phenocrysts. The olivine shows a mean grain size of 679 µm, the pyroxene a size of 324 µm, the phenocrysts having a mean grain size of 503 µm overall (Figure 14). The diameter distribution of the phenocrysts shows a scatter parameter of 0.334.

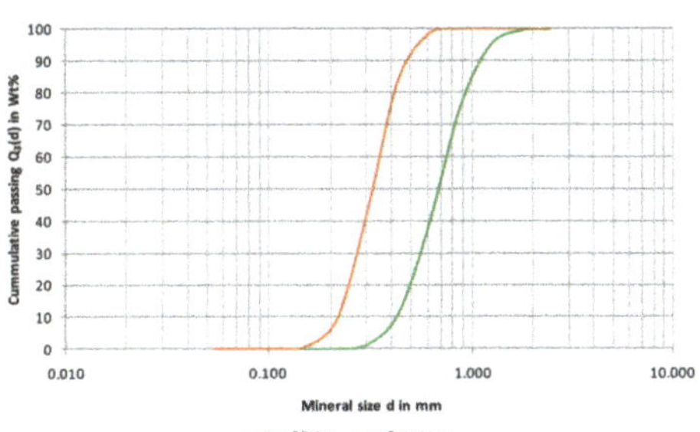

Figure 14. Mineral grain size distribution of phenocrysts (olivine and pyroxene).

Grain Shape

The shape of the phenocrysts is predominantly cubic, confirmed by the values for their elongation and flatness (E = 1.238; F = 1.067).

3.3.2. CT Results (Kimberlite)

Mode

Kimberlite consists of a fine-grained carbonate-serpentine mass (26.4%) with olivine inclusions (together 32.7%), pyroxene (33.5%) and accessory minerals (7.4%).

Grain Shape

Olivine and pyroxene grains are often of irregular shape, the sphericity coefficient for olivine is 0.23, and pyroxene is 0.48 (Figure 15a).

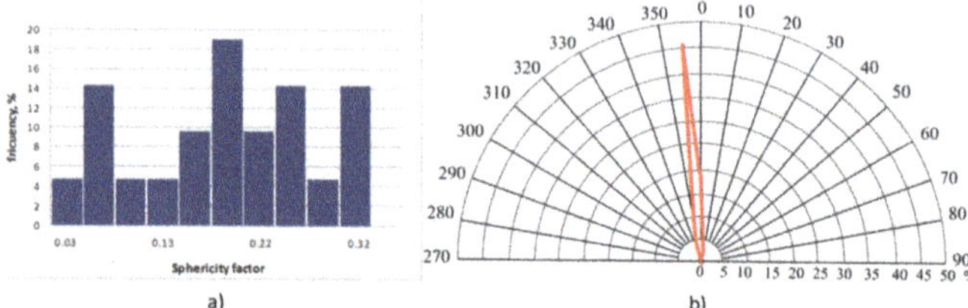

Figure 15. (**a**) Distribution of olivine grains according to the sphericity coefficient and (**b**) the orientation of the longest axis in grade

Specific Surface

The specific surface for both olivine and pyroxene is large (40.17 and 56.10 mm^2/mm^3 respectively), which again indicates a complex morphology of grains.

Orientation

Olivine grains are slightly orientated in the rock volume (Figure 15b).

Clustering

The pyroxene and olivine grains are distributed in the rock very unevenly, they form separate clusters of 1.161–1.595 mm in size respectively (Figure 16a), and single aggregates up to 15 mm in size as well (Figure 16b).

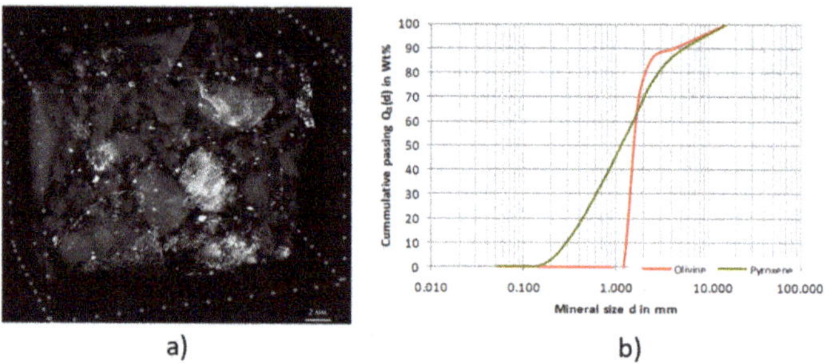

Figure 16. (**a**) Olivine and pyroxene grains in the volume of kimberlite and (**b**) their distribution according to the maximum diameter.

Porosity

A sample of kimberlite is characterized by a minimal porosity: the total porosity is 0.13% and the opened porosity is 0.02%. In general, the pores are of 15–50 μm in size and have irregular shape (sphericity factor is 0.66). The void density is 12.81 mm^{-3}.

4. Discussion

Analysis of the data obtained in the study of three different rock types using CT and QMA allows to reveal interesting textural and structural features. Thus, for example, both methods give sufficiently close measurements for such a parameter like the sizes of grains and mineral aggregates, especially if they sharply differ from the surrounding minerals in X-ray density. In this regard, the maximum correspondence of the data (0.647 and 0.594 mm) is observed for mica grains enclosed in a quartz-feldspar matrix of granodiorite, which is particularly well observed in Figure 17.

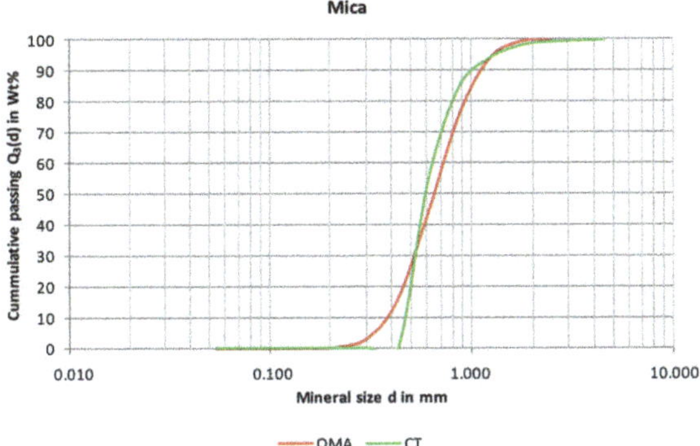

Figure 17. Grain size distributions of mica in granodiorite.

Thus, the CT method was proven to be an efficient tool to analyse materials with high X-ray density such as ore minerals (native elements, sulphides and metal oxides) in a matrix consisting of low X-ray density materials such as silicates (quartz, feldspar, olivine, pyroxenes, amphiboles, etc.) or carbonate matrix. The CT often analyzes not individual grains of minerals, but their aggregates. So, there may be slightly overestimated mean diameters of grains and their aggregates, which occurred with copper minerals in copper ore (see Figure 18). The calculated sizes of these minerals according to the CT data exceed the QMA data (0.381 and 0.277 mm, respectively), since one grain is taken as the aggregate of grains of copper minerals.

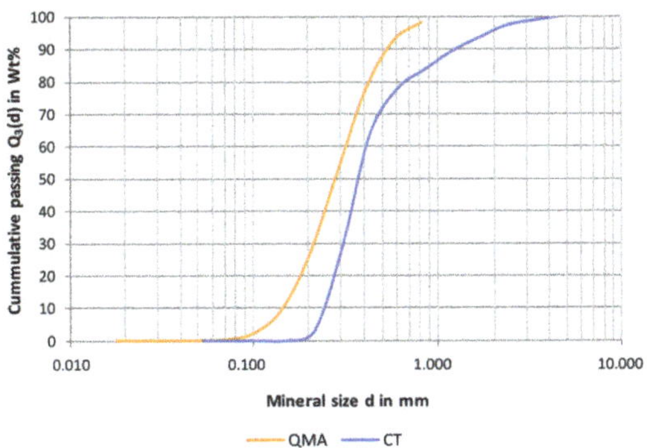

Figure 18. Grain size distributions of copper minerals in copper ore.

Similar distortions can occur for other morphometric parameters, such as, for example, the specific surface area of grains (see Table 3). In such cases, tomographic studies must be accompanied by additional petrographic and electron microscopic studies, but in this case, one of the main advantage of CT—the speed of the method—is lost.

Table 3. Comparison of the specific surface of minerals obtained with QMA and CT.

Parameter	Copper Ore (Copper Minerals)		Granodiorite (Mica)		Kimberlite (Olivine)	
	QMA	CT	QMA	CT	QMA	CT
Specific surface S_V mm^2/mm^3	52.84	31.86	11.20	20.20	6.23	40.17

Some differences in estimated parameters can also be explained by the peculiarity of sample preparation for CT and QMA: by the CT method the total volume of a sample is scanned while for QMA three orthogonally oriented sections are investigated. Accordingly, the sample volume is different in both cases. For example, several large phenocrysts of olivine in the very heterogeneous kimberlite sample influenced significantly the values of the average morphometric parameters of the whole sample (see Figure 19).

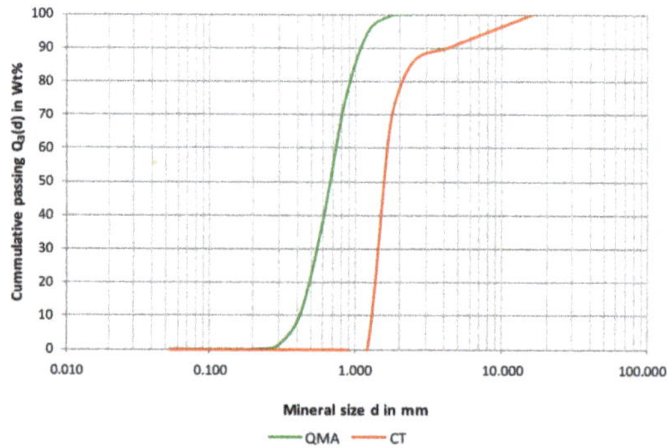

Figure 19. Grain size distributions of olivine in kimberlite.

5. Conclusions

In the paper, a comparative analysis of textural-structural rock characteristic using QMA and CT methods is carried out. The undoubted advantage of the CT method is the possibility of 2D and 3D data visualization with the help of specialized programs such as CTVox and CTan, which help to analyse not only the grain and agglomerate sizes, but also their distribution in the total rock volume and relationships with each other. Although, possibilities of CT are not enough for measuring separate grains while their similar density. In these cases, the QMA results are much more representative.

The sample preparation for each method differs as well. Thus, CT is a non-destructive method and it is possible to analyse hand specimens. Whereas QMA requires more difficult and lengthy preparations (three orthogonally orientated thin sections or polished sections should be prepared).

So, it can be concluded that it is possible to use CT as a quick simple nondestructive method, as well as for more serious and difficult measurements, it is better to use QMA method. The analysis of the obtained data indicates a quite good repeatability of measurements, and consequently, the possibility of using the CT method in addition to other methods such as QMA for the purpose of studying the behaviour of rocks and ores during their beneficiation, in particular for the better understanding the important process of comminution.

Author Contributions: Conceptualization, H.L.; methodology, O.P. and I.T.; software, A.D.; validation, A.D.; O.P.; formal analysis, I.T.; H.L.; investigation, A.D.; O.P.; data curation, A.D.; O.P.; writing—original draft preparation, A.D.; O.P.; writing—review and editing, I.T.; H.L.; visualization, A.D.; O.P.; supervision, I.T.; H.L.; project administration, I.T.; H.L.; funding acquisition, I.T.; H.L. All authors have read and agreed to the published version of the manuscript.

Funding: The CT Study was performed as part of the Joint German-Russian Project No 20-55-12002 of the Russian Foundation for Basic Research.

Conflicts of Interest: The authors declare no conflict of interest.

References

1. Bringezu, S.; Ramaswami, A.; Schandl, H.; O'Brien, M.; Pelton, R. Assessing Global Resource Use: A Systems Approach to Resource Efficiency and Pollution Reduction. 2017. Available online: https://wedocs.unep.org/bitstream/handle/20.500.11822/27432/resource_use.pdf?sequence=1 (accessed on 25 January 2020).

2. Schodde, R. Role of Technology and Innovation for Identifying and Growing Economic Resources. In Proceedings of the 12th Biennial Exploration Managers Conference, Hunter Valley, Australia, 26–29 July 2019.

3. Steinbach, V. E-mobility—The Global Race for High-Tech Metals. In Proceedings of the 100 years IEC, Scientific Colloquium - Solutions for the Carbon Challenge, Berlin, Germany, 10–12 October 2019.

4. Ballantyne, G.R.; Powell, M.S. Benchmarking comminution energy consumption for the processing of copper and gold ores. *Miner. Eng.* **2019**, *65*, 109–114. [CrossRef]

5. Hesse, M. *Selektive Zerkleinerung von Erzen und Industriemineralen bei Prallbeanspruchung*; TU Bergakademie Freiberg: Freiberg, Germany, 2017. (In German)

6. Li, X.F.; Li, X.; Li, H.B.; Zhang, Q.B.; Zhao, J. Dynamic tensile behaviours of heterogeneous rocks: The grain scale fracturing characteristics on strength and fragmentation. *Int. J. Impact Eng.* **2018**, *118*, 98–118. [CrossRef]

7. Aleksandrova, T.N.; Nikolaeva, N.V.; Lvov, V.V.; Romashev, A.O. Ore processing efficiency improvements for precious metals based on process simulations. *Obogashchenie Rud* **2019**, *2*, 8–13. [CrossRef]

8. Popov, O. *Beitrag zur Mathematisch-Petrographischen Gefügecharakterisierung für die Beurteilung der Festgesteine Hinsichtlich Ihrer Aufbereitung und Ihrer Produkteigenschaften*; TU Bergakademie Freiberg: Freiberg, Germany, 2007. (In German)

9. Popov, O.; Lieberwirth, H.; Folgner, T. Quantitative Charakterisierung der Festgesteine zur Prognostizierung des Gesteinseinflusses auf relevante Produkteigenschaften und Systemkenngrößen. Teil 1: Anwendung der quantitativen Gefügeanalyse. *AT Miner. Process.* **2014**, *07–08*, 76–88.

10. Popov, O.; Lieberwirth, H.; Folgner, T. Quantitative Charakterisierung der Festgesteine zur Prognostizierung des Gesteinseinflusses auf relevante Produkteigenschaften und Systemkenngrößen. Teil 2: Ausgewählte Beispiele. *AT Miner. Process.* **2014**, *10*, 54–63.

11. Schreiber, S. *Beitrag zur Quantitativen Gesteinscharakterisierung zur Beurteilung von Gesteinen hinsichtlich Ihrer Festigkeiten*; TU Bergakademie Freiberg: Freiberg, Germany, 2018. (In Germany)

12. Tromans, D. Crack Propagation in Brittle Materials: Relevance to Minerals Comminution. *Int. J. Recent Res. Appl. Stud.* **2012**, *11*, 406–427.

13. Hesse, M.; Popov, O.; Lieberwirth, H. Increasing efficiency by selective comminution. *Miner. Eng.* **2017**, *103–104*, 112–126. [CrossRef]

14. Mezzetti, M.; Popov, O.; Lieberwirth, H.; Anders, E.; Hoske, P. Electro Impulse Technology for Processing Complex Ores. In Proceedings of the ESCC European Symposium on Comminution and Classification, Izmir, Turkey, 11–14 September 2017.

15. Zuo, W.; Shi, F.; Manlapig, E. Modelling of high voltage pulse breakage of ores. *Miner. Eng.* **2015**, *83*, 168–174. [CrossRef]

16. Hilden, M.M.; Powell, M.S. A geometrical texture model for multi-mineral liberation prediction. *Miner. Eng.* **2017**, *111*, 25–35. [CrossRef]

17. Napier-Munn, T.; Drinkwater, D.; Ballantyne, G. The CEEC Roadmap for Eco-Efficient Comminution. CEEC/JKTech Workshop on Eco-Efficient Comminution. CEEC the Future. Available online: https://www.ceecthefuture.org/media/downloads/CEECRoadmap.pdf (accessed on 25 January 2020).

18. Delaney, G.W.; Morrison, R.D.; Sinnott, M.D.; Cummins, S.; Cleary, P.W. DEM modelling of non-spherical particle breakage and flow in an industrial scale cone crusher. *Miner. Eng.* **2015**, *74*, 112–122. [CrossRef]

19.   Johansson, M.; Quist, J.; Evertsson, M.; Hulthén, E. Cone crusher performance evaluation using DEM simulations and laboratory experiments for model validation. *Miner. Eng.* **2017**, *103–104*, 93–101. [CrossRef]

20.   Tavares, L.M.; das Neves, P.B. Microstructure of quarry rocks and relationships to particle breakage and crushing. *Int. J. Miner. Process.* **2008**, *87*, 28–41. [CrossRef]

21.   Weerasekara, N.S.; Powell, M.S.; Cleary, P.W.; Tavares, L.M.; Evertsson, M.; Morrison, R.D.; Quist, J.; Carvalho, R.M. The contribution of DEM to the science of comminution. *Powder Technol.* **2013**, *248*, 3–24. [CrossRef]

22.   Klichowicz, M.; Frühwirt, T.; Lieberwirth, H. New experimental setup for the validation of DEM simulation of brittle crack propagation at grain size level. *Miner. Eng.* **2018**, *128*, 312–323. [CrossRef]

23.   Klichowicz, M.; Lieberwirth, H. Modelling of Realistic Microstructures as Key Factor for Comminution Simulations. In Proceedings of the XXVIII International Mineral Processing Congress (IMPC), Québec, QC, Canada, 11–15 September 2016; pp. 1–10.

24.   Kühnel, L. *Untersuchungen zur Gutbettzerkleinerung in einer hydraulischen Stempelpresse*; TU Bergakademie Freiberg: Freiberg, Germany, 2019. (In German)

25.   Leißner, T.; Hoang, D.H.; Rudolph, M.; Heinig, T.; Bachmann, K.; Gutzmer, J.; Schubert, H.; Peuker, U.A. A mineral liberation study of grain boundary fracture based on measurements of the surface exposure after milling. *Int. J. Miner. Process.* **2016**, *156*, 3–13. [CrossRef]

26.   Rahfeld, A.; Kleeberg, R.; Möckel, R.; Gutzmer, J. Quantitative mineralogical analysis of European Kupferschiefer ore. *Miner. Eng.* **2018**, *115*, 21–32. [CrossRef]

27.   Evans, C.L.; Wightman, E.M.; Yuan, X. Quantifying mineral grain size distributions for process modelling using X-ray micro-tomography. *Miner. Eng.* **2015**, *82*, 78–83. [CrossRef]

28.   Talovina, I.V.; Aleksandrova, T.N.; Popov, O.; Lieberwirth, H. Comparative analysis of rocks structural-textural characteristics studies by computer X-ray microtomography and quantitative microstructural analysis methods. *Obogashchenie Rud* **2017**, *3*, 56–62. [CrossRef]

29.   Vaysberg, L.A.; Kameneva, E.E. Vozmozhnosti kompyuternoy mikrotomografii pri issledovanii fiziko-mekhanicheskikh svoystv gornykh porod (The possibilities of computer microtomography in the study of physical and mechanical properties of rocks). *Gornyy zhurnal* **2014**, *9*, 85–90.

30.   Vaysberg, L.A.; Kameneva, E.E.; Pimenov, Y.G.; Sokolov, D.I. Issledovaniye struktury porovogo prostranstva granito- gneysa metodom rentgenovskoy tomografii. (The study of the structure of the pore space of granite-gneiss by x-ray tomography). *Obogashcheniye Rud* **2013**, *3*, 37–40.

31.   Gasumov, R.A.; Gridin, V.A.; Kopchenkov, V.G.; Galai, B.F.; Dudaev, S.A. Research of Mining and Geological Conditions for Geological Exploration in Pre-Caucasus Region. *Zapiski Gornogo instituta* **2017**, *228*, 654–661. [CrossRef]

32.   Ivanov, M.; Burlin, G.; Kalmykov, E. *Petrophysic Research Methods of Core Material (Terrigenous Deposits)*; Book 1; Moscow University Publishing House: Moscow, Russia, 2008. (In Russian)

33.   Shtyrlyaeva, A.A.; Zhuravlev, A.V.; Gerasimova, A.I. Prospects and problems of computer microtomography using for core samples studies. *NGTP* **2016**, *11*, 1. [CrossRef]

34.   Willson, C.S.; Lu, N.; Likos, W.J. Quantification of Grain, Pore, and Fluid Microstructure of Unsaturated Sand from X-ray Computed Tomography Images. *Geotech. Test. J.* **2012**, *35*, 20120075. [CrossRef]

35.   Ho, S.T.; Hutmacher, D.W. A comparison of micro CT with other techniques used in the characterization of scaffolds. *Biomaterials* **2006**, *27*, 1362–1376. [CrossRef] [PubMed]

36.   Ditscherlein, R.; Leißner, T.; Peuker, U.A. Preparation techniques for micron-sized particulate samples in X-ray microtomography. *Powder Technol.* **2020**, *360*, 989–997. [CrossRef]

37.   Leißner, T.; Diener, A.; Löwer, E.; Ditscherlein, R.; Krüger, K.; Kwade, A.; Peuker, U.A. 3D ex-situ and in-situ X-ray CT process studies in particle technology—A perspective. *Adv. Powder Technol.* **2019**. [CrossRef]

38.   Duryagina, A.; Talovina, I.; Shtyrlyaeva, A.; Popov, O.; Duryagina, A.; Talovina, I.; Shtyrlyaeva, A.; Popov, O. Application of Computer X-ray Microtomography for Study of Technological Properties of Rocks. *Key Eng. Miner.* **2018**, *769*, 220–226. [CrossRef]

39. Glagolev, A.A. Quantitative analysis with the microscope by the point method. *Eng. Min. J.* **1934**, *135*, 339–400.
40. Unland, G.; Folgner, T. Automatische Kornformbestimmung durch photooptische Partikelanalyse. *Die Naturstein-Industrie* **1997**, *5*, 20–28.

Article

Comminution Effects on Mineral-Grade Distribution: The Case of an MVT Lead-Zinc Ore Deposit

Gabriele Baldassarre *, Oliviero Baietto and Paola Marini

Department of Environmental, Land and Infrastructure Engineering (DIATI), Politecnico di Torino, Corso Duca degli Abruzzi, 24, 10129 Turin, Italy; oliviero.baietto@polito.it (O.B.); paola.marini@polito.it (P.M.)
* Correspondence: gabriele_baldassarre@polito.it; Tel.: +39-011-090-7614

Received: 21 July 2020; Accepted: 8 October 2020; Published: 9 October 2020

Abstract: Every mining operation is followed by a beneficiation process aimed at delivering quality materials to the transformation industry. Mainly, in order to separate valuable minerals from gangue in mineral processing, the crushing and grinding of extracted ore are crucial operations for the following separation steps. Comminution is the most energy-consuming operation in mining, and the quality of the results is strictly related to the characteristic of the material under treatment, the type of equipment used in comminution, and the circuit design adopted. A preliminary study was performed in order to understand the crushing behavior under different comminution forces of a high-grade mixed Zn-Pb sulfide ore sample, collected in a Mississippi-Valley Type (MVT) deposit, and the distribution of the target minerals among the products of the process. Ore samples were examined and characterized through thin section observation and SEM analyses for the determination of grain size and texture features, while X-ray powder diffraction (XRPD) quantitative analyses were performed for the definition of target mineral concentrations of comminuted product samples. The selected crushing and grinding circuit comprised lab-scale equipment. For each stage of the process, products below the estimated free-grain size threshold were collected, and particle size analyses were carried out. Comminution products were divided into size distribution classes suitable for further separation operations, and XRPD analyses showed a mineral-grade distribution varying with the dimensions of the products. Characterization of the ore material after crushing and grinding force applications in terms of the distribution of target minerals among different-sized classes was achieved. The important trends highlighted should be considered for further investigation related to an efficient separation.

Keywords: comminution; mineral processing; mixed sulfides; sphalerite; galena

---

1. Introduction

The mining industry is characterized by essential operations for a proper transformation process of raw materials into final products. Generally, the exploitation activity in a mine is followed by a beneficiation process designed to increase the quality of the valuable minerals naturally present in the excavated material. The very early stage of the transformation is represented by the comminution and grinding of extracted ore aimed at obtaining different changes in terms of the dimension and shape of the material that will be further treated [1–3]. Progressively, separation processes are set up in order to concentrate and separate target minerals from gangue.

Comminution is a highly energy-consuming operation, and it is characterized by very low efficiency [1,4–6] and the quality of the final product being strictly related to the physical characteristics of the material under treatment [7,8].

The principal objective of comminution is to liberate valued minerals, which is not an easy and predictable outcome. However, it is important to achieve this, because it affects the efficiency and control of the entire downstream process, especially when the ore has to be delivered to a flotation process [9–14]. The shape, dimension, and distribution of minerals also play a strategic role for the achievement of an effective beneficiation process [15,16].

In this study, a representative sample of a specific mineralized area of the Gorno Mining District was comminuted with lab-scale equipment, and the size distribution related to the selected equipment was studied.

Determination of the mineral composition and distribution of the concentration of target minerals among the different-sized classes of the crushing products was the key focus of this work.

The importance of having a clear comprehension of the redistribution of target minerals in mid-processing products could lead to future effective separation designs. Benefits that could be achieved can enhance a reduction in end-process waste, as well as the control of water use, reagents, and machinery utilization during the beneficiation stage.

2. Materials and Methods

2.1. Sampling

The mixed Pb-Zn sulfide ore samples were collected from the Val Vedra mine in the Gorno Mining District, Lombardy, Bergamo, Italy. The Gorno mineralization is defined as MVT [17]; hosted mainly in Triassic shallow water carbonates [18]; and delimited in N-S, E-W, and NW-SE directions by 3 different tectonic faults [19].

About 30 kg of material was grab-sampled in the Pian Bracca extension area at level 1040 m a.s.l., currently under exploration for resource definition by Alta Zinc Ltd. [20]. In the Pian Bracca extension area, the mineralization is developed close to the low-angle Pian Bracca fault, hosted in a graphitic and carbonatic matrix and strongly tectonized, as described in detail by other authors [19].

2.2. Characterization Methods for Samples and Products

2.2.1. Optical Microscopy

Optical microscopy (OM) was performed on uncovered polished 30-μm thin sections, realized from rock sample slices, using a Leica (Wetzlar, Germany) model Ortholux II POL-MK optical polarized-light microscope equipped with a DeltaPix camera and interfaced with DeltaPix software for image acquisition and processing. This was carried out using normal and polarized transmitted lights on polished thin sections for the determination of the main features of the mineralization and the average grain size of the target minerals.

2.2.2. Scanning Electron Microscopy

Scanning electron microscopy (SEM) analyses were performed using a FEI (Hillsboro, OR, USA) model QUANTA INSPECT 200LV microscope equipped with an Energy Dispersive Analysis X-Ray (EDAX) detector and interfaced with xT Microscope Control and an Edax Inc. (Mahwah, NJ, USA) GENESIS Spectrum version 6.04 software. The measurement settings were high vacuum mode, 5.00 kV HV, and backscattered electron detector (BSED). Analyses were performed on uncovered 30-μm thin sections obtained from rock samples.

2.2.3. X-Ray Powder Diffraction Analysis (XRPD)

X-ray powder diffraction (XRPD) analysis was performed using a Rigaku (Tokyo, Japan) model SmartLab SE diffractometer, with Copper K-alpha Radiation (CuKα) at 40 kV and 30 mA, 5–90° 2θ range, 0.01° step width, 1°/min scan speed equipped with a D/teX Ultra 250 (H), and interfaced with the Rigaku (Tokyo, Japan) software SmartLab Studio II package. Quantitative phase analyses were performed by the whole-powder-pattern fitting (WPPF) Rietveld method [21–23], as implemented in the software. A pseudo-Voigt peak shape function was selected. The refined parameters were the phase scale factor, peak shape parameters, lattice parameters, preferred orientation, and structure coefficients. The ICDD PDF-4 2020 (International Center for Diffraction Data, Powder Diffraction File™) database [24] was used for phase recognition and refining. Powder samples were manually prepared using an agate mortar from oven-dried samples.

2.3. Crushing and Grinding Equipment

Physical treatments were carried out on a lab-scale, in order to try to reproduce the typical comminution forces and stresses that characterize comminution circuits [1] and control the gradual size reduction of the sample. Using the available equipment, the selected comminution and grinding route takes into account the necessity to provide an input material dimension, for each machine, suitable for feeding hoppers dimensions. Output materials dimensional characteristics have been set up in order to allow the material to fit in the subsequent machine's hopper. In addition, the equipment arrangement selected for this kind of ore samples was set up taking into account the possibility of the material to be subsequently tested for processing and separation purposes.

Figure 1 shows the comminution equipment arrangement that was selected in order to test the response in terms of the mineral distribution of the material. A preliminary crushing and grinding test on representative samples of the collected materials was performed. OM observations carried out on the preliminary test outputs showed that, generally, 85% of the grains was free below the 0.425 mm size. Consequently, each machine output was screened at 0.425 mm. At the end of the process, all passing 0.425-mm materials were collected as a composite product.

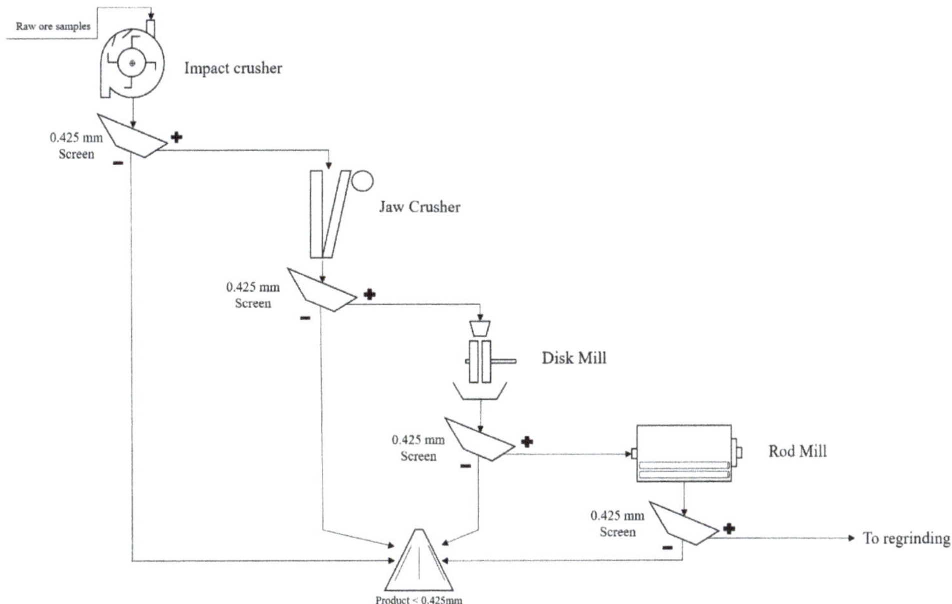

Figure 1. Lab-scale crushing and grinding equipment scheme.

A Hazemag impact crusher was used as the first crushing stage. This machine has a four-lug bolt rotor, powered by an electrical motor. The running speed was 500 rpm, and the gap width between the bolts and impact elements was set at 10 mm. The feeding hopper has 120 × 120 mm aperture, limiting the maximum input material dimension to 50 mm.

A Magutt-10 jaw crusher was used for the second crushing step. It is powered by a 0.75-kW electric engine at 250 rpm. The selected closed-side setting (CSS) was 5 mm, with an open-side setting (OSS) of 15 mm. Feeding hopper is 120 × 60 mm, limiting input material dimension to d <40 mm.

A S.I.M.A. Milano disk mill was used as the first grinding stage. The disk gap selected was 2 mm. Feeding hopper dimensions allow an input material dimension below 5 mm.

A lab-scale rod mill was used as the final stage of grinding. It is equipped with a 193-mm-diameter, 267-mm-long drum, containing nine steel rods with a diameter of 8 mm inside, powered by a 0.4-kW electrical engine. The drum was filled with 1 kg of material at a 30% v/v filling ratio. The grinding time was 4 min at 50 rpm rotation speed.

2.4. Sieve Analysis

Sizing of the material was performed using different combinations of sieves in dry and wet conditions, in order to know the size distribution of the equipment output materials. Dry sieving was selected for the general output materials resulting from each crushing and comminution machine. A portion of the materials was collected and screened by means of nested screens in decreasing sizes, from the top screen to 0.125–0.075 mm, and shaken for 3 min in the lab screen shaker.

Wet sieving was selected for the grain size distribution definition of the composite comminution product passing 0.425 mm. The sieve sequence consisted of 0.425, 0.355, 0.300, 0.25, 0.212, 0.180, 0.125, 0.090, and 0.063 mm. The materials were oven-dried before XRPD analysis.

3. Results

3.1. Optical Microscopy

The observation of thin sections, shown in Figure 2, was carried out with OM in parallel transmitted light (PL) and cross-polarized transmitted light (XPL), with objectives between 4× and 10×. There was an abundant presence of sphalerite minerals surrounded by a calcite matrix. Light-opaque mineralization, such as galena and its altered compounds, were found. It was possible to spot some opaque veins filled with black organic material.

Sphalerite grains were massively present in the samples, showing a yellow-brownish and black coloration transparent in PL and opaque in XPL. Calcite appeared as a white and striated matrix, transparent both in PL and XPL. Opaque minerals, like galena and cerussite, were sparse and usually occurred in the surroundings of sphalerite grains.

Coarser sphalerite grain sizes, ranging in average between 0.400 and 0.425 mm, were defined using digital measurements from the OM images.

3.2. Scanning Electron Microscopy

In order to have elemental information related to the mineralization nature, SEM punctual analyses were performed. The investigation targeted those areas difficult to characterize by OM and the intergranular filled voids present. Figure 3 represents the SEM analysis on uncovered rock slices, which was performed in order to confirm the OM outcomes and the total number of mineral phases present in the sample before XRPD analyses.

SEM images and Energy Dispersive Spectroscopy (EDS) spectra, shown in Figure 3, confirmed a widespread presence of sphalerite [25] in the calcite [26] matrix and highlighted the presence of Zn and Pb alteration compounds, mainly cerussite [27], anglesite [28], and smithsonite [29]. The presence of organic matter in the microfractures was detected [30].

3.3. Crushing and Grinding

In the arranged comminution and grinding scheme, samples entered with maximum size of 50 mm and they were reduced to millimetric and sub-millimetric grains. According to first equipment feeding hopper specifications, a preliminary manual size reduction has been performed with a 5-kg hammer. The purpose of this test was to bring most of the material below 0.425 mm, limiting the size fractions that could negatively affect an eventual flotation separation [31–34]. The process can be divided into two stages:

- crushing using the impact crusher and jaw crusher and
- grinding using a disk mill and rod mill.

The particle size distribution of the broken products from each comminution was analyzed. Granulometric curves for the output materials are shown in Figure 4.

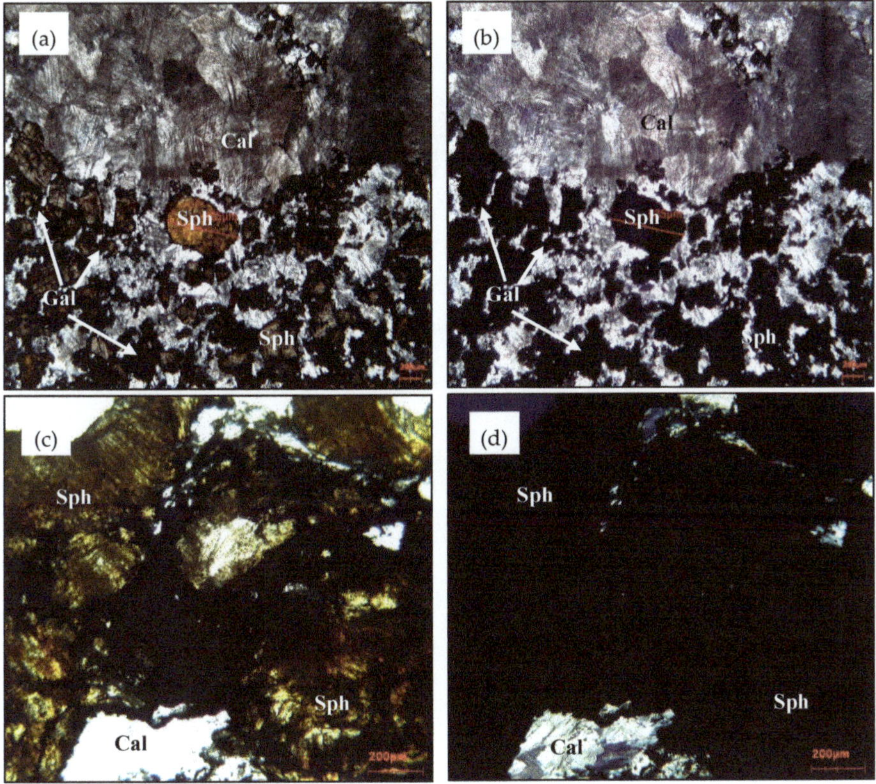

Figure 2. Optical microscopy (OM) thin section images in parallel transmitted light (PL) and cross-polarized transmitted light (XPL) (Sphalerite, Sph; Calcite, Cal; and Galena, Gal): (**a**) thin section n.1, PL and (**b**) XPL and (**c**) thin section n.3, PL and (**d**) XPL.

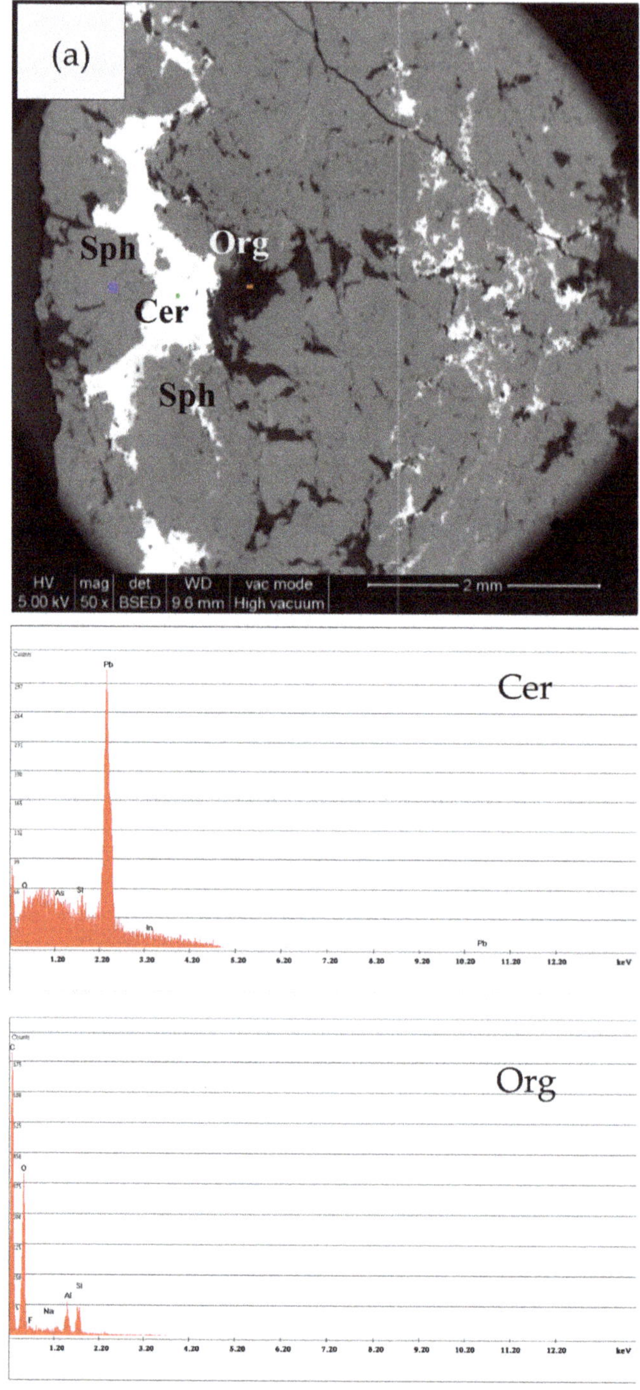

Figure 3. *Cont.*

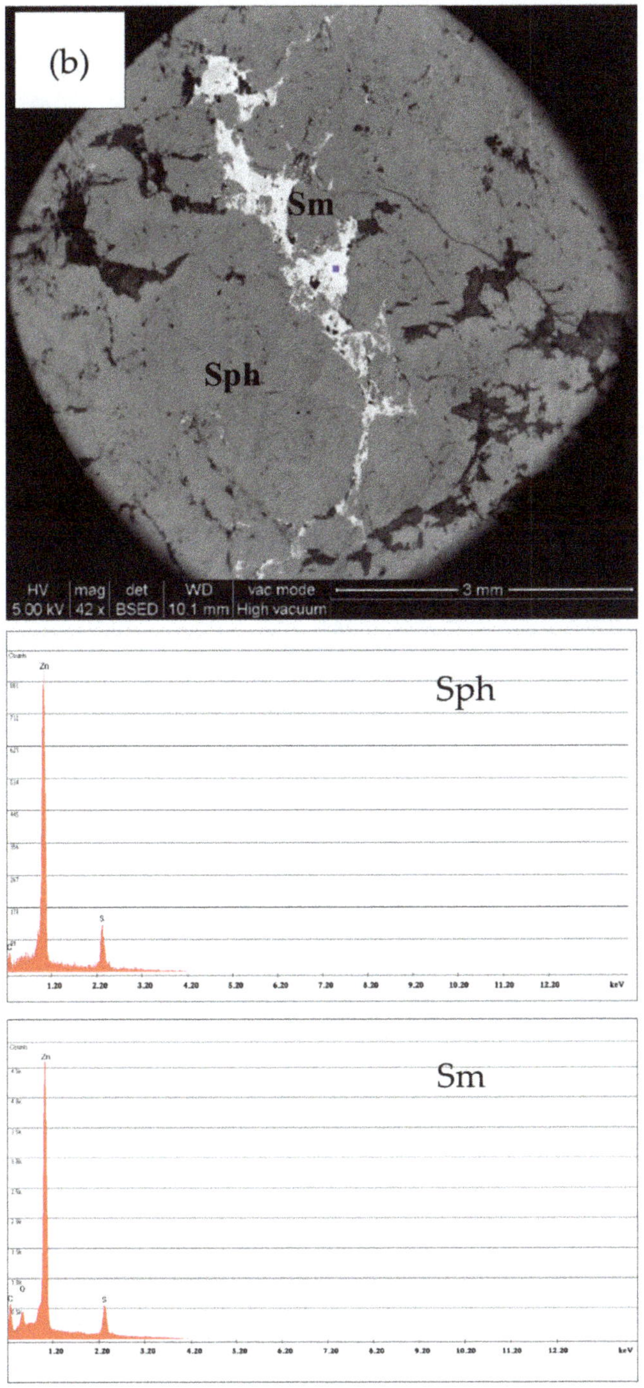

Figure 3. SEM images and spectra of thin sections: (**a**) sphalerite (Sph, blue dot), cerussite (Cer, green dot), organic matter (Org, orange dot), (**b**) sphalerite (Sph), and smithsonite (Sm, blue dot).

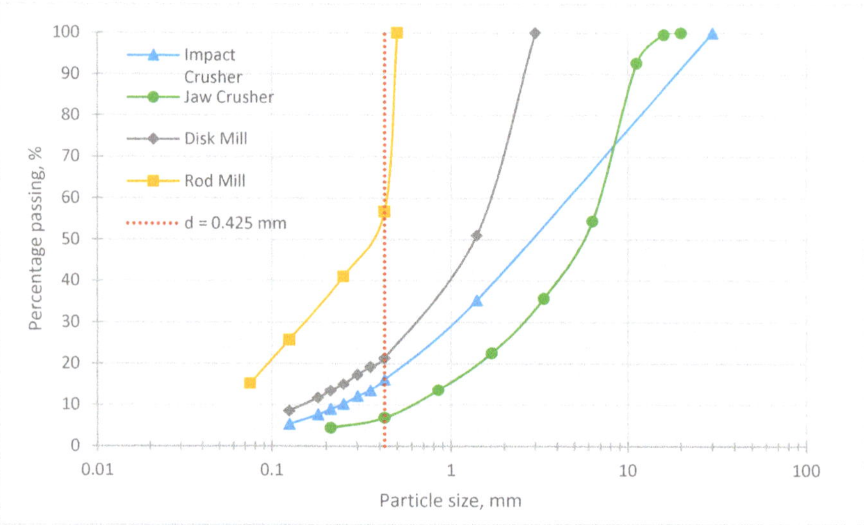

Figure 4. Output product granulometric curves for each equipment.

A mass balance, assuming negligible losses, for 5 kg of materials resulting from crushing and grinding is plotted in Figure 5. The impact crusher, selected as the first stage of crushing, resulted in a dimension reduction with a contained fine production <0.425 mm of 16.2 wt %. The jaw crusher, chosen for the second crushing stage, gave a lower production of fines. In fact, only 6.9 wt % of <0.425 mm was obtained. At the end of the crushing stage, the overall fraction having dimensions below a 0.425 mm-size was 22 wt % of the input materials, while the oversized materials sent to the grinding stages were 78 wt %.

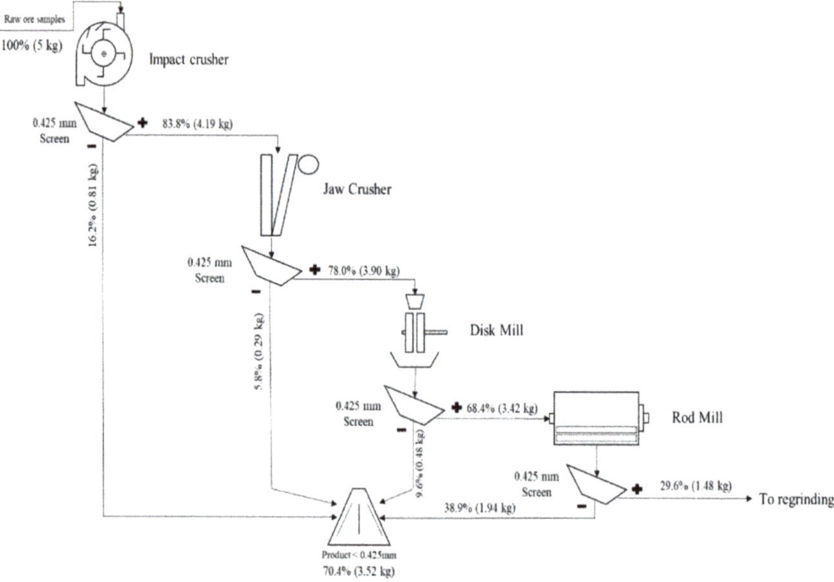

Figure 5. Mass balance resulting from the crushing and grinding stages.

For a further size reduction, a disk mill was selected as the first stage of grinding. It resulted in materials characterized by $d_{50} > 1.4$ mm, with the fraction of the fines ($d < 0.425$ mm) being 12.3 wt % The output material was homogeneously sized. Rod milling was selected for the final grinding stage. It was selected aiming at a drastic reduction of the material below the size of 0.425 mm. The results showed how four min of grinding time produced 56.8 wt % of undersized material. At the end of the secondary crushing test, material having $d < 0.425$ mm was 62.1 wt % of the feed material obtained by primary crushing.

Oversized material, accounting for about 30 wt % of the total output, was not reintroduced in any grinding stage. All the materials passing 0.42–5 mm screens were collected as a composite sample and further analyzed in terms of the granulometric distribution and XRPD.

3.4. Product Material < 0.425-mm Sizing

Comminution products passing the 0.425-mm threshold were more specifically analyzed in terms of granulometric distribution. This was necessary for the assessment of the ore behavior under different comminution forces aiming to deliver good quality material for further separation steps.

Wet sieving methodology was selected. Figure 6 shows the granulometric curve obtained resulting in values of $d_{80} = 0.300$ mm and $d_{50} = 0.160$ mm. These findings show that the presence of fine fractions ($d < 0.063$ mm) is relevant. In fact, material passing 0.063 mm accounted for 27.2 wt % of the total considered sample.

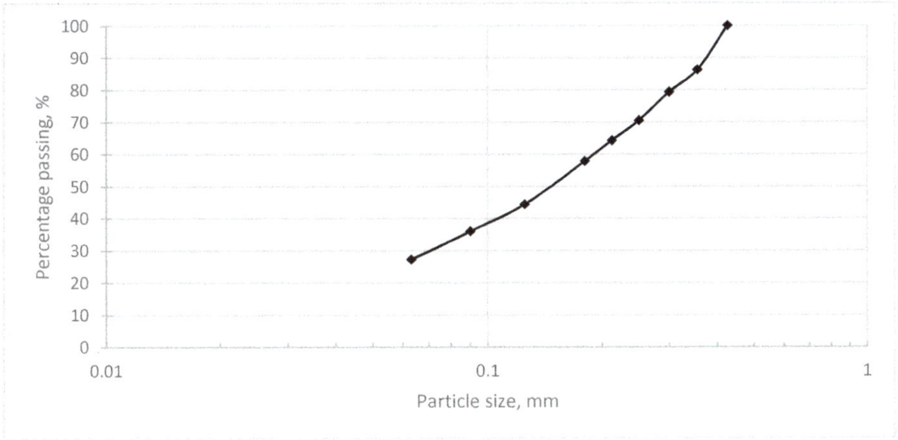

Figure 6. Particle size distribution obtained by means of wet sieving on comminution products < 0.425 mm.

3.5. X-Ray Powder Diffraction Analyses

XRPD analyses, performed on each dimension class obtained by wet sieving from comminution composite output products < 0.425 mm, distinguished five different mineral phases: sphalerite, galena, calcite, anglesite, and cerussite, as reported in Figure 7a–i. Sphalerite's main peaks were observed at 2θ angles of 28,57°, 33.11°, 47.57°, 56.39°, 76.77°, and 88.52° [35]. Calcite's representative peaks were observed at 2θ angles of 29.44° and 48.58° [36]. Galena's peaks were observed at 2θ angles of 26°, 30.09°, 43.06°, and 50.98° [37]. Cerussite's and Anglesite's characteristic peaks were observed at 2θ angles of 24.80°, 25.49°, and 43.48° and 20.84°, 23.36°, and 26.75° respectively [38,39].

Qualitatively, each class showed a constant presence in terms of the mineral phases. Reference spectra [24,35–39] were compared with the experimental ones. These outcomes generally confirmed what was previously observed by OM and SEM.

Quantitative XRPD analyses obtained by Rietveld [21–23] refinement showed residual errors R_{wp} ranging from 5.50% to 6.49% and goodness-of-fit in the range from 2.06 and 2.41, as shown in Table 1.

Table 1. Observed residual errors (R_{wp}) and goodness-of-fit (S) from the quantitative X-ray powder diffraction (XRPD) analysis.

Particle Size (mm)	R_{wp} (%)	S
0.425–0.350	5.90	2.20
0.350–0.300	5.54	2.07
0.300–0.250	5.55	2.07
0.250–0.212	5.55	2.06
0.212–0.180	6.21	2.33
0.180–0.125	5.50	2.05
0.125–0.090	5.86	2.20
0.090–0.063	5.87	2.21
<0.063	6.49	2.41

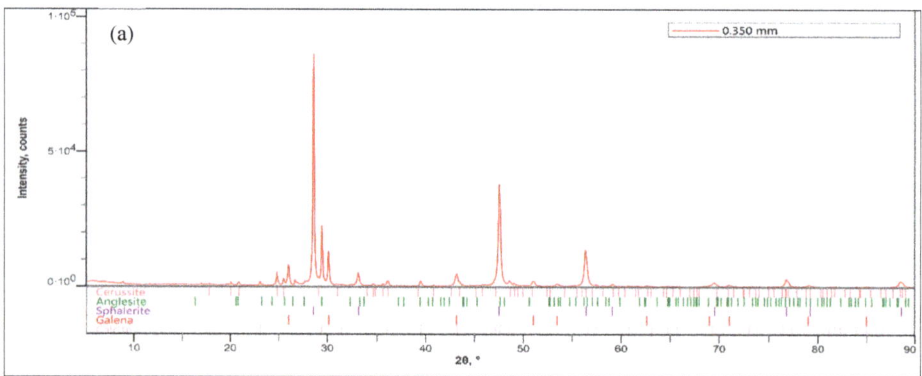

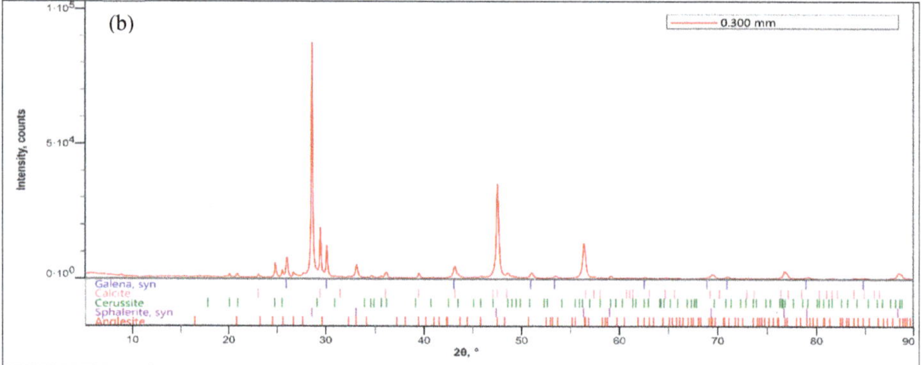

Figure 7. *Cont.*

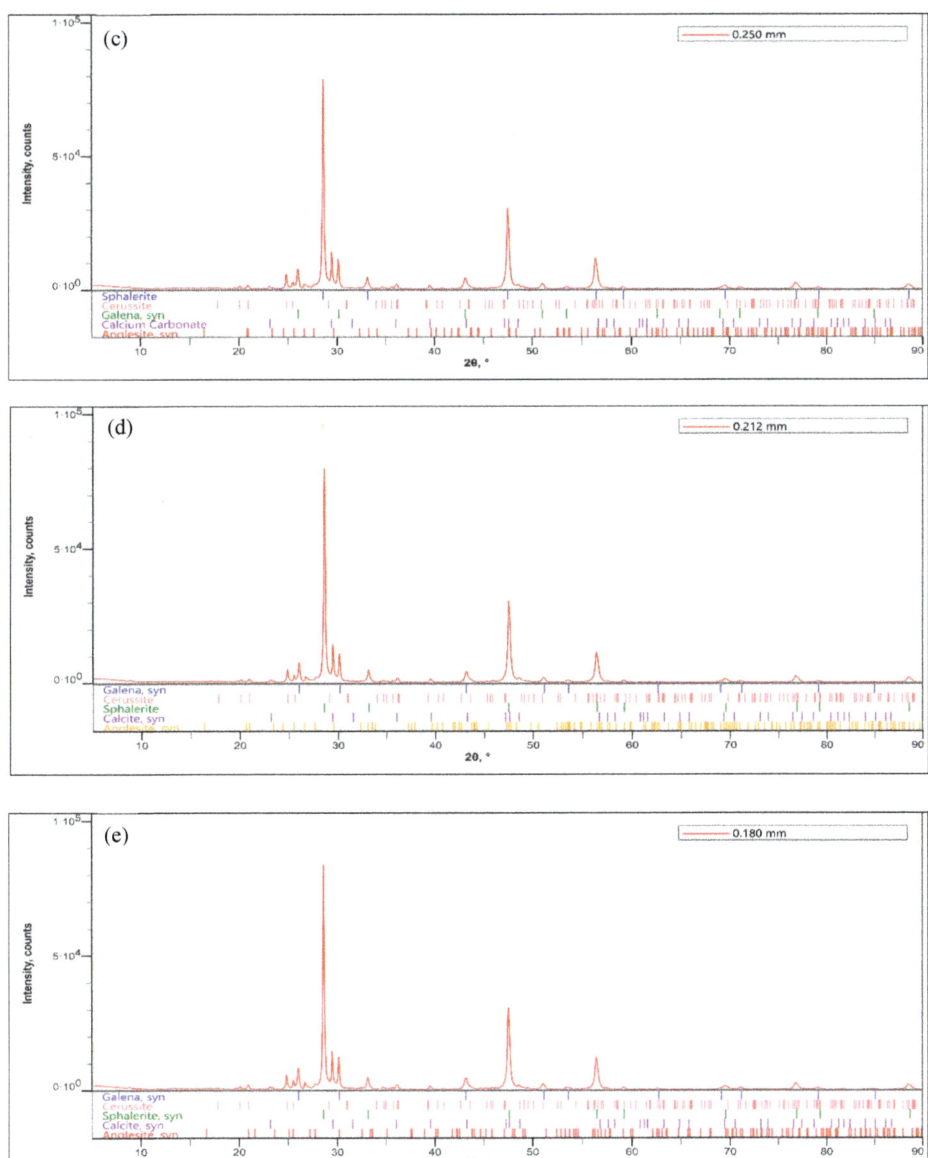

Figure 7. *Cont.*

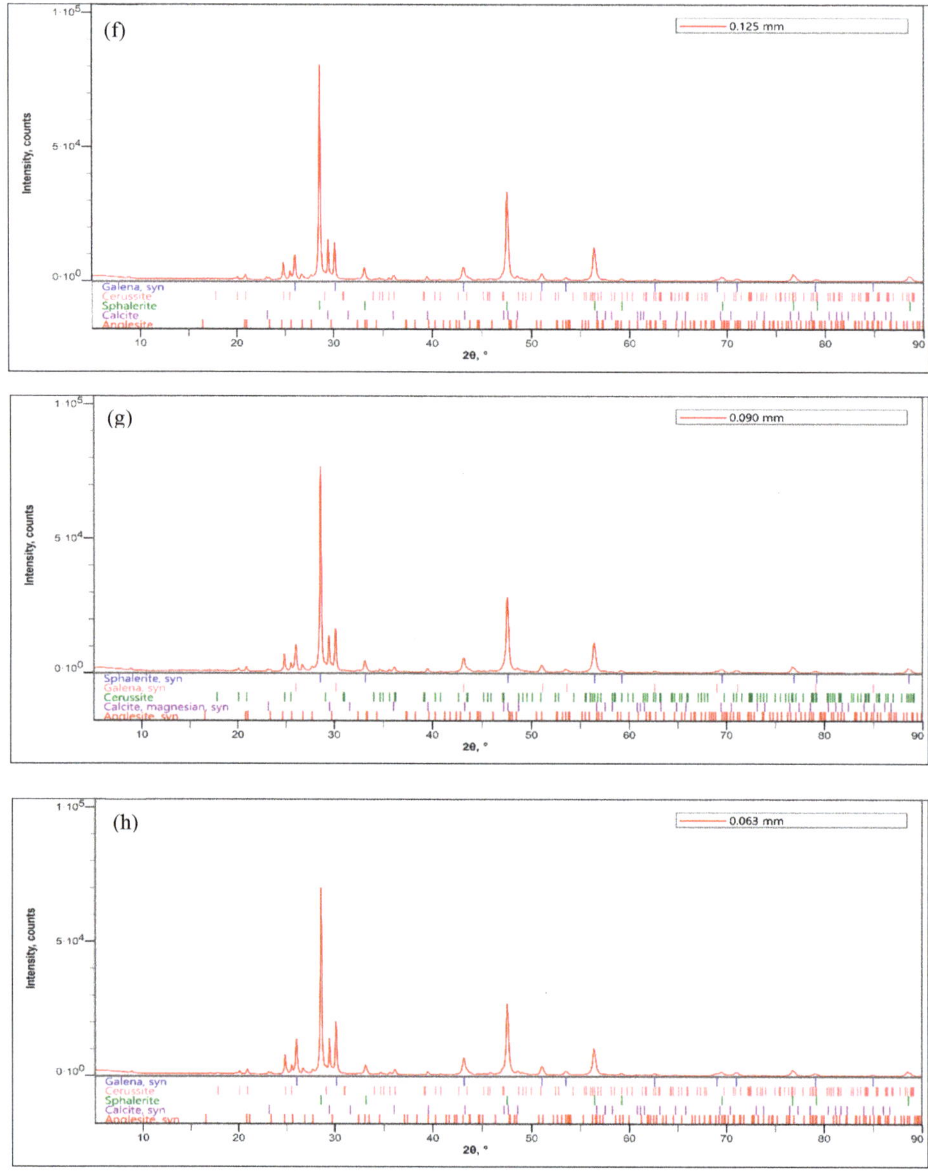

Figure 7. *Cont.*

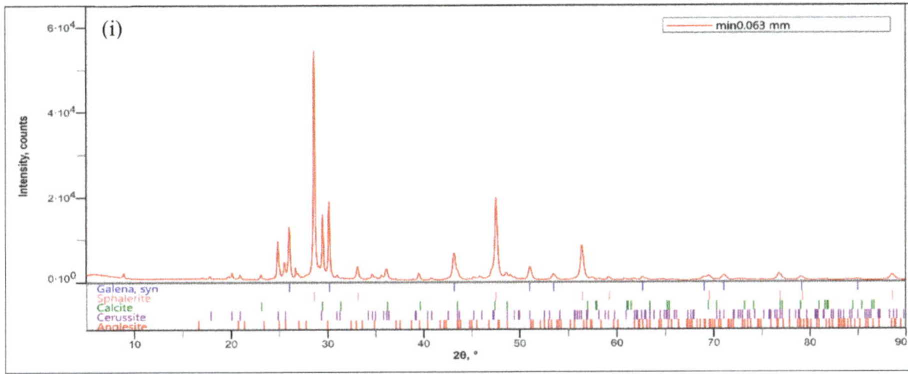

Figure 7. X-ray powder diffraction (XRPD) spectra of different comminution product-sieved classes: (**a**) 0.425–0.350 mm; (**b**) 0.350–0.300 mm; (**c**) 0.300–0.250 mm; (**d**) 0.250–0.212 mm; (**e**) 0.212–0.180 mm; (**f**) 0.180-0.125 mm; (**g**) 0.125-0.090 mm; (**h**) 0.090-0.063 mm; (**i**) <0.063 mm.

According to some authors, a Rietveld refinement showing low R_{wp} and S values can be considered as reliable, with errors in the weight fraction estimation between 0.5 wt % and 1.5 wt % [40,41]. The quantitative results are shown in Table 2. On average, sphalerite was the most abundant phase in the samples, ranging 50–70 wt %, calcite, defined as the gangue mineral, attained 17–27 wt %, while galena ranged 4–9 wt % The other phases could be considered as minor. The accuracy of the quantitative analyses was not assessed in detail.

Table 2. Quantitative XRPD results for each particle size.

Particle Size (mm)	Weight Fraction (%) [1]				
	ZnS	PbS	PbCO$_3$	PbSO$_4$	CaCO$_3$
0.425–0.350	65.59	4.09	2.52	2.93	24.87
0.350–0.300	66.82	4.14	3.04	2.67	23.33
0.300–0.250	69.32	4.28	3.08/	2.50	20.82
0.250–0.212	67.35	4.12	3.11	3.55	21.87
0.212–0.180	70.39	4.60	3.71	4.04	17.26
0.180–0.125	68.22	5.37	3.27	2.79	20.35
0.125–0.090	65.88	6.54	3.97	4.10	19.51
0.090–0.063	62.63	8.41	4.94	3.53	20.49
<0.063	51.50	7.97	6.43	7.20	26.90

[1] Estimation errors must be considered in the range ± 0.5–1.5%, according to [40,41].

Some of the mineral phases present in the samples showed a different concentration among different granulometric classes. The clearest evidence was the ones linked with galena and Pb-related compounds. Their total concentration was around 9.5 wt % in particle sizes 0.425–0.350 mm and reached 21.6 wt % in particle sizes <0.063 mm, constantly increasing their presence with the decrease of their material dimensions. Galena was the most abundant regarding Pb-related compounds and showed a worthwhile concentration for further recovery.

Sphalerite was the most abundant phase in each size, but its presence decreased from 0.180 mm, where it peaked with 70.4 wt %, toward finer classes. The lowest concentration of ZnS was 62.6 wt %, found in the <0.063-mm class. The calcite phase had its lowest presence in class 0.212–0.180 mm with 17.3 wt %, while its highest concentration of 26.9 wt % was measured in the <0.063-mm class.

In general, high values of valuable minerals were found in the different classes of comminution products, underling the necessity of further separation in order to obtain high-quality concentrates from the ore. In the choice of recovery and separation methods of Pb-related minerals—galena, in particular—the increase of their concentrations in the finest classes should be taken into consideration.

4. Discussion

Ore samples were collected in the Pian Bracca extension area in the Gorno Mining District under an exploration operation by Alta Zinc Ltd. In order to have an overview of the characteristics of the sampled ores uncovered, polished thin sections from rock slices were realized and observed. Petrographic observations were necessary for the determination of the shape, composition, and dimensions of grains present in the mineralized mixed sulfide ores collected. The results showed typical characteristics related to MVT deposits [19,42], with relatively high contents of sphalerite minerals embedded in a calcite matrix. Galena was sparsely present in very small crystals, difficult to be distinguished only by OM.

SEM characterization brought additional information, especially concerning alteration products and the filling of the ore micro-fractures. The presence of cerussite, anglesite, and organic matter was detected, confirming the presence of Pb-related minerals and traces of organic matter, arguably linked with the formational geological environment of the site [19]. The observation made on the dimensional characteristics resulted in estimated sizes of valuable mineral grains below the 0.400 mm and 0.450 mm thresholds. On these parameters, a lab-scale crushing and grinding circuit was arranged, aiming to obtain products <0.425 mm. Ground materials passing the 0.425-mm screening were collected as unique products, accounting for 70 wt % of the initial input quantity. The other 30 wt % of the initial input materials needed to be reground and were not considered for further analysis (Figure 5). Passing 0.425 mm, the material grain size distribution was studied: material <0.063 mm accounted for 27 wt % (Table 1).

XRPD quantitative analyses were realized on oven-dried samples resulting from the wet sieving of <0.425-mm composite product samples. The results highlighted the important presence of sphalerite and galena in dimension classes ranging between the 0.250-mm and 0.063-mm classes, as shown in Figure 8a,d. Target Pb and Zn phase concentrations were observed as fluctuating, varying with the reduction in the dimensions of the products. This phenomenon could be framed as a selective comminution behavior of the ore [43], but further data on mineral liberation grades and comminution efficiency should be collected in order to better define the parameters related to this specific possibility.

Sphalerite was the most abundant mineral phase in the composite product samples <0.425 mm obtained by crushing and grinding; its concentration peaked at class 0.180 mm with 70.4 wt %, as plotted in Figure 8c, assuming a decreasing trend toward finer-grained sizes, whereas galena and Pb-related compound concentrations, plotted in Figure 8a, showed an increasing trend towards finer-grained sizes, peaking at class <0.063 mm with a total concentration of 21.6 wt %

The sum of the ZnS and Pb-related compounds', corresponding to the valuable mineral phases present in the ore, peak concentration was reached in the 0.212–0.180-mm class with 82.7 wt %, while their lowest occurrence of 73.1 wt % was in the <0.063-mm particle sizes, as shown in Figure 8d. Their trend is mainly dominated by sphalerite concentration oscillations.

The calcite concentration, plotted in Figure 8b, trend appeared as V-shaped, encountering a descending behavior until the 0.212–0.180-mm class, strictly related to the total valuable phase ZnS + Pb-tot peak.

Moreover, according to the quantitative analyses, important information was obtained for the evaluation of the most abundant mineral phases present in the material and their collocation among the grain size classes of the comminuted samples.

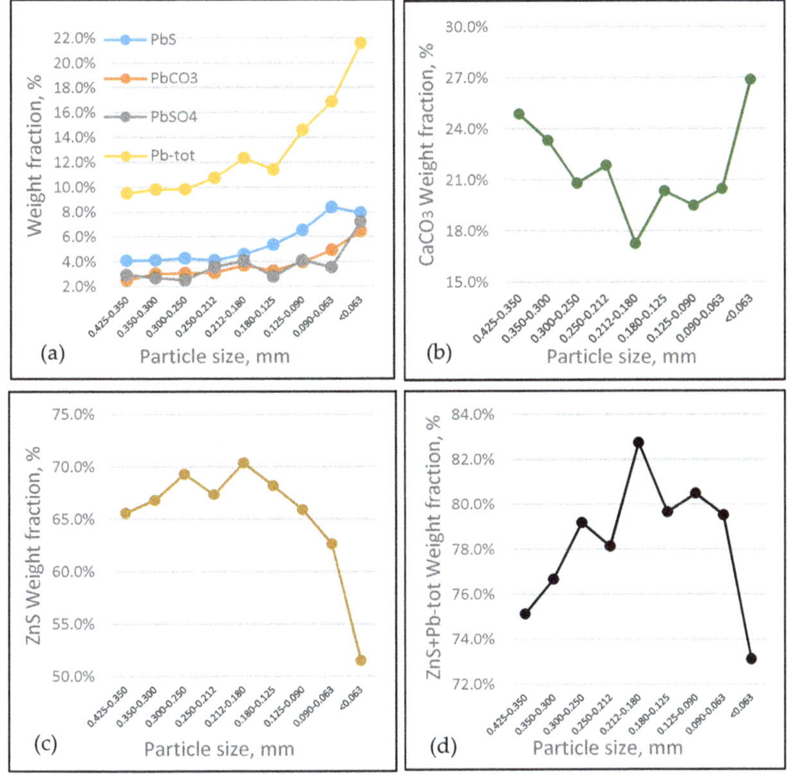

Figure 8. Mineral phase weight fraction variations among the dimension classes in comminution products below 0.425 mm: (**a**) Pb-related compounds (Pb-tot), (**b**) calcite ($CaCO_3$), (**c**) sphalerite (ZnS), and (**d**) valuable phases assumed as a sum of the sphalerite and Pb-related compounds (ZnS+Pb-tot).

5. Conclusions

The purpose of this study was the characterization of ore materials after crushing and grinding force applications in terms of the distribution of target minerals among different-sized classes. Petrographic observations were carried out on polished thin sections by optical microscopy and scanning electron microscopy. Crushing and grinding equipment was used in order to reproduce the comminution forces typically present in a comminution plant, and the output material samples were characterized by means of XRPD quantitative analysis. Important trends in the valuable mineral phase fluctuations among different-sized classes were highlighted from results.

Further detailed studies should be taken into consideration. Comminution configurations should be properly assessed in terms of grindability studies on materials, specific energy consumptions, and industrial-relevant comminution flowsheet set-ups. Mineral liberation studies should also be implemented in order to have a clearer understanding of the comminution behavior of this material under different conditions.

Generally, applicable separation processes for this kind of ore, such as froth flotation or gravity separation methods, should be taken into account [1,44,45]. Concerning the presence of Pb-related compounds in very fine classes, a nonconventional flotation [46,47] should be considered. In general, the amount of valuable minerals present in the sampled area is relevant and worthwhile for industrial purposes.

Author Contributions: Conceptualization, G.B. and P.M.; methodology, G.B. and P.M.; investigation, G.B. and O.B.; resources, P.M.; data curation, G.B.; writing—original draft preparation, G.B.; writing—review and editing, G.B., O.B., and P.M.; and supervision, P.M. All authors have read and agreed to the published version of the manuscript.

Funding: This research received no external funding.

Acknowledgments: The authors are thankful to Alta Zinc Ltd. and Energia Minerals (Italia) s.r.l. for their availability in welcoming access, interest, recovery, and the sharing of relevant information related to their industrial activities.

Conflicts of Interest: The authors declare no conflict of interest.

References

1. Wills, B.A. Comminution in the minerals industry—An overview. *Miner. Eng.* **1990**, *3*, 3–5. [CrossRef]
2. Ozgur, O.; Hakan, B. Comparison of different breakage mechanisms in terms of product particle size distribution and mineral liberation. *Miner. Eng.* **2013**, *49*, 103–108.
3. Hoşten, Ç.; Özbay, C. A comparison of particle bed breakage and rod mill grinding with regard to mineral liberation and particle shape effects. *Miner. Eng.* **1998**, *11*, 871–874. [CrossRef]
4. King, R.P. Comminution and liberation of minerals. *Miner. Eng.* **1994**, *7*, 129–140. [CrossRef]
5. Nadolski, S.; Bern, K.; Amit, Z.; Zorigtkhuu, D. An energy benchmarking model for mineral comminution. *Miner. Eng.* **2014**, *65*, 178–186. [CrossRef]
6. Tromans, D. Mineral comminution: Energy efficiency considerations. *Miner. Eng.* **2008**, *21*, 613–620. [CrossRef]
7. Mwanga, A.; Rosenkranz, J.; Lamberg, P. Testing of ore comminution behavior in the geometallurgical context—A review. *Minerals* **2015**, *5*, 276–297. [CrossRef]
8. King, R.P. *Modeling and Simulation of Mineral Processing System*; Elsevier: Amsterdam, The Netherlands, 2001; pp. 127–212.
9. Bazin, C.; Grant, R.; Cooper, M.; Tessier, R. A method to predict metallurgical performances as a function of fineness of grind. *Miner. Eng.* **1994**, *7*, 1243–1251. [CrossRef]
10. Moradi, S.; Monhemius, A.J. Mixed sulphide–oxide lead and zinc ores: Problems and solutions. *Miner. Eng.* **2011**, *24*, 1062–1076. [CrossRef]
11. Barnov, N.G.; Lavrinenko, A.A.; Lusinyan, O.G.; Chikhladze, V.V. Effect of a Crushing Technique on Lead–Zinc Ore Processing Performance. *J. Min. Sci.* **2018**, *53*, 771–777. [CrossRef]
12. Palm, N.A.; Shackleton, N.J.; Malysiak, V.; O'Connor, C.T. The effect of using different comminution procedures on the flotation of sphalerite. *Miner Eng.* **2010**, *23*, 1053–1057. [CrossRef]
13. AMMI S.p.a. *Impianto di Trattamento per Minerali di Piombo e Zinco (Laveria)*; AMMI S.p.A.: Milan, Italy, 1971. (In Italian)
14. Wei, Y.; Sandenbergh, R.F. Effects of grinding environment on the flotation of Rosh Pinah complex Pb/Zn ore. *Miner. Eng.* **2007**, *20*, 264–272. [CrossRef]
15. Mudenda, C.; Mwanza, B.G.; Kondwani, M. Analysis of the effects of grind size on production of copper concentrate: A case study of mining company in Zambia. In Proceedings of the International Conference on Chemical Processes and Green Energy Engineering (ICCPGEE'15), Harare, Zimbabwe, 14–15 July 2015.
16. Liu, L.; Tan, Q.; Liu, L.; Cao, J. Comparison of different comminution flowsheets in terms of minerals liberation and separation properties. *Miner. Eng.* **2018**, *125*, 26–33. [CrossRef]
17. Leach, D.L.; Bechstädt, T.; Boni, M.; Zeeh, S. Triassic-hosted Mississippi Valley-type zinc-lead ores of Poland, Austria, Slovenia, and Italy. In *Europe's Major Base Metal Deposits*; Ashton, J., Boland, M., Cruise, M., Earls, G., Fusciardi, L., Kelly, J., Stanley, G., Andrew, C., Eds.; Irish Association of Economic Geologists (IAEG): Dublin, Ireland, 2003; pp. 169–213.
18. Brigo, L.; Kostelka, L.; Omenetto, P.; Schneider, H.J.; Schroll, E.; Schulz, O.; Strucl, I. Comparative reflections on four alpine Pb-Zn deposits. In *Time and Strata-Bound Ore Deposits*; Klemm, D.D., Schneider, H.J., Eds.; Springer: Berlin/Heidelberg, Germany, 1977; pp. 273–293.
19. Mondillo, N.; Lupone, F.; Boni, M.; Joachimski, M.; Balassone, G.; De Angelis, M.; Zanin, S.; Granitzio, F. From Alpine-type sulfides to nonsulfides in the Gorno Zn project (Bergamo, Italy). *Miner. Depos.* **2019**, *55*, 953–970. [CrossRef]
20. Alta Zinc. Mining in the North of Italy. Available online: http://www.altazinc.com/ (accessed on 21 July 2020).

21. Rietveld, H. A profile refinement method for nuclear and magnetic structures. *J. Appl. Cryst.* **1969**, *2*, 65–71. [CrossRef]

22. Young, R.A. *The Rietveld Method*; Oxford University Press: Oxford, UK, 1993.

23. McCusker, L.B.; Von Dreele, R.B.; Cox, D.E.; Louer, D.; Scardi, P. Rietveld refinement guidelines. *J. Appl. Cryst.* **1986**, *32*, 36–50. [CrossRef]

24. ICDD. *PDF-4/Minerals*; International Centre for Diffraction Data: Newtown Square, PA, USA, 2020.

25. Sphalerite Mineral Data. Available online: http://webmineral.com/data/Sphalerite.shtml (accessed on 21 July 2020).

26. Calcite Mineral Data. Available online: http://webmineral.com/data/Calcite.shtml (accessed on 21 July 2020).

27. Cerussite Mineral Data. Available online: http://webmineral.com/data/Cerussite.shtml (accessed on 21 July 2020).

28. Anglesite Mineral Data. Available online: http://webmineral.com/data/Anglesite.shtml (accessed on 21 July 2020).

29. Smithsonite Mineral Data. Available online: http://webmineral.com/data/Smithsonite.shtml (accessed on 21 July 2020).

30. Ohkouchi, N.; Kuroda, J.; Taira, A. The origin of Cretaceous black shales: A change in the surface ocean ecosystem and its triggers. *Proc. Jpn. Acad. Ser. B Phys. Biol. Sci.* **2015**, *91*, 273–291. [CrossRef]

31. Pérez-Garibaya, R.; Ramírez-Aguileraa, N.; Bouchardb, J.; Rubio, J. Froth flotation of sphalerite: Collector concentration, gas dispersion and particle size effects. *Miner. Eng.* **2017**, *57*, 72–78. [CrossRef]

32. Duarte, A.C.P.; Grano, S.R. Mechanism for the recovery of silicate gangue minerals in the flotation of ultrafine sphalerite. *Miner. Eng.* **2007**, *20*, 766–775. [CrossRef]

33. Collins, G.L.; Jameson, G.L. Experiments on the flotation of fine particles, the influence of particle size and charge. *Chem. Eng. Sci.* **1976**, *31*, 985–991. [CrossRef]

34. Arbiter, N. *Problems in sulfide ore processing. Beneficiation of Mineral Fines—Problems and Research Needs*; Somasundaran, P., Arbiter, N., Eds.; AIME: Ann Arbor, MI, USA, 1979; pp. 139–152.

35. RRUFF. Sphalerite R040136. Available online: https://rruff.info/R040136 (accessed on 21 July 2020).

36. RRUFF. Calcite R040070. Available online: https://rruff.info/R040070 (accessed on 21 July 2020).

37. RRUFF. Galena R070325. Available online: https://rruff.info/R070325 (accessed on 21 July 2020).

38. RRUFF. Cerussite R040069. Available online: https://rruff.info/R040069 (accessed on 21 July 2020).

39. RRUFF. Anglesite R040004. Available online: https://rruff.info/R040004 (accessed on 21 July 2020).

40. Monecke, T.; Köhler, S.; Kleeberg, R.; Herzig, P.M.; Gemell, J.B. Quantitative phase-analysis by the Rietveld method using X-ray powder-diffraction data: Application to the study of alteration halos associated with volcanic-rock-hosted massive sulfide deposits. *Canad. Mineral.* **2001**, *39*, 1617–1633. [CrossRef]

41. Zhao, P.; Lu, L.; Liu, X.; De la Torre, A.G.; Cheng, X. Error analysis and correction for quantitative phase analysis based on Rietveld-internal standard method: Whether the minor phases can be ignored? *Crystals* **2018**, *8*, 110. [CrossRef]

42. Paradis, S.; Hannigan, P.; Dewing, K. Mississippi Valley-type lead-zinc deposits (MVT). In *Mineral Deposits of Canada*; Geological Association of Canada: St. John's, NL, Canada, 2007; pp. 185–203.

43. Hesse, M.; Popov, O.; Lieberwirth, H. Increasing efficiency by selective comminution. *Miner. Eng.* **2017**, *103–104*, 112–126. [CrossRef]

44. Trahar, W.J.; Warren, L.J. The floatability of very fine particles-a review. *Int. J. Miner. Process.* **1976**, *3*, 103–131. [CrossRef]

45. Gaudin, A.M.; Groh, J.O.; Henderson, H.B. Effect of particle size on flotation. *AIME Tech. Pub.* **1931**, *414*, 3–23.

46. Song, S.; Lopez-Valdivieso, A.; Reyes-Bahena, J.L.; Lara-Valenzuela, C. Floc Flotation of galena and sphalerite fines. *Miner. Eng.* **2001**, *14*, 87–98. [CrossRef]

47. Song, S.; Lopez-Valdivieso, A.; Reyes-Bahena, J.L.; Bermejo-Perez, H.I.; Trass, O. Hydrophobic flocculation of galena fines in aqueous suspensions. *J. Colloid Interface Sci.* **2000**, *227*, 272–281. [CrossRef]

Article

Fracture Analysis of α-Quartz Crystals Subjected to Shear Stress

Giovanni Martinelli [1,2,3,*], **Paolo Plescia** [4], **Emanuela Tempesta** [4], **Enrico Paris** [5] and **Francesco Gallucci** [5]

[1] INGV Istituto Nazionale di Geofisica e Vulcanologia, Via Ugo La Malfa 153, 90146 Palermo, Italy
[2] Northwest Institute of Eco-Environment and Resources, Chinese Academy of Sciences, Lanzhou 730000, China
[3] Key Laboratory of Petroleum Resources, Lanzhou 730000, China
[4] CNR-IGAG, Institute of Environmental Geology and Geoengineering, Research area of Rome-1, 00015 Monterotondo, Italy; ilplescia@gmail.com (P.P.); emanuela.tempesta@igag.cnr.it (E.T.)
[5] CREA-ING, Consiglio per la Ricerca in Agricoltura e l'Analisi dell'Economia Agraria, Unità di Ricerca per l'Ingegneria Agraria, 00015 Monterotondo, Italy; enrico.paris@crea.gov.it (E.P.); francesco.gallucci@crea.gov.it (F.G.)
[*] Correspondence: giovanni.martinelli15@gmail.com

Received: 18 July 2020; Accepted: 29 September 2020; Published: 30 September 2020

Abstract: This study assesses the correlations between the intensity of stress undergone by crystals and the morphological characteristics of particles and fracturing products. The effects of the fractures on the microstructure of quartz are also studied. Alpha quartz, subjected to shear stress, is quickly crushed according to a fracturing sequence, with a total fracture length that is correlated to the stress rate. The shear stress generates a sequence of macro and microstructural events, in particular localized melting phenomena, never highlighted before on quartz and the formation of different polymorphs, such as cristobalite and tridymite together with amorphous silica.

Keywords: quartz; shear stress; tribochemistry; fracturing

---

1. Introduction

The possibility that fractured quartz turns into cristobalite and tridymite poses serious problems in the safe management of industrial milling. In the last twenty years, numerous laboratories have sprung up all over the world conducting experiments on rocks in the conditions of friction typical of technologies utilized in rock grinding. To a large extent, these experiments are carried out trying to correlate the friction coefficient with stress (tangential and normal), the speed and the spaces covered by the simulated fault [1,2]. At the same time, attention has grown towards the "dynamometamorphic" phenomena suffered by rocks and linked to stress: heating, gas dissociation, partial or total melting of the friction layers. In the recent past, several working groups have focused their attention on the role of quartz in the sliding phenomena of rocks, and the idea is that the heating of the contact surface by friction produces a partial melting of the material, the formation of a silica layer amorphous which is quickly hydrated by ambient moisture, which thus becomes a lubricating gel that drastically reduces the friction coefficient of the defects [3,4]. In a recent publication, we formulated another hypothesis to explain the lowering of the coefficient of friction in experiments on quartz-containing rocks. This hypothesis is based on the progressive formation of nanocrystalline cristobalite, which works as a solid lubricant [5,6]. The silica polymorph called cristobalite has the particularity of being an auxetic material, i.e., with a negative Poisson ratio, both in the form of low temperature (alpha) and in that of high temperature (beta). If a layer of cristobalite is found in shear conditions, it can reduce its contact area as a result of the volume contraction and equally reduces the friction coefficient.

The hypothesis is that quartz, subjected to a prolonged reticular distortion over time, tends to take on a structure with ever greater disturbance, but formed by cristobalite nanophases. This phase would be nanocrystalline, not visible by X diffraction but visible by Raman spectroscopy. The accumulation of this phase in the interface between the sliding surfaces would cause the friction coefficient to decay more or less rapidly, depending on the intensity of the dynamometamorphic action [6]. Furthermore, quartz's structural decay is marked by the decay of the electromagnetic signals generated by its fracturing, which diminishes over time as fracturing and damage of the original crystalline structure proceed [5,6]. A key element is missing in this theory: how and when the patina of cristobalite is generated. The purpose of present work is to contribute to a better understanding of the processes related to quartz tribochemistry.

2. Experimental Procedure

The task of this experimental work is to understand the mechanisms that link the progressive demolition of the crystal structure of quartz and the appearance of polymorphic phases during the application of sliding stresses. In two previous works [5,6], the authors of this work reported details on the polymorphic phases of silica that are formed during fracturing and on the correlations between fractures and the amount of very low frequency electromagnetic energy emitted by the crystals during the fracturing.

An innovative "piston cylinder" was used to carry out these studies. It is equipped with two steel pistons which compress the sample and which represent the armatures of a condenser, equipped with an insulating Teflon jacket, which allows the containment of the material. Thanks to Teflon, the pressure distribution exerted is anisotropic, and this is due to the materials that make up the cell itself. The materials that exert the pressure (stainless steel) and the containment material (Teflon) are different in terms of elasticity, coefficient of Poisson and compressive strength. This determines a condition of anisotropy in the exercise of compression: the mineral is not subjected to equal stresses in all directions, but greater according to the direction of the piston and less in the directions perpendicular to it. This anisotropy allows one to create a fracturing system by shear, which generates fractures favoring the development of a network of microcracks. The entire piston cylinder is first evacuated and subsequently subjected to an analytical grade nitrogen flow to ensure an anhydrous atmosphere. The system was created taking into account similar systems published in the past by various authors [7–10]. The pressure is applied through a motorized hydraulic press, which develops a maximum load of 4.9 kN.

The quartz-α samples come from natural quartz crystals, purchased at Ward (Ward, West Henrietta, NY, USA; www.wardsci.com), previously analyzed in optical microscopy to detect any defects, such as inclusions and fractures. The crystals were chosen in the weight range from 0.05 to 1.5 g. Each test generated powders and fragments that were analyzed for shape, size, area and volume, through the new Morphology G3ID image analysis system (Malvern, UK) based on optical scanning microscopy. The automated scanning system allows one to measure and analyze about 1 million particles per hour and to know the values of the perimeter, area and volume of each particle, from 300 μm to 1 μm. This scanning microscope has already been used previously by the authors for investigations on asbestos fibers in soils and recently on quartz crystals [5,6]. To determine the morphologies, the particles are dispersed on the surface of the optical plate. Scanning takes place at constant magnification on all particles following an x-y scanning mechanism; the depth of the particles is reconstructed through the "z-stacking" mechanism that allows one to shoot the same image of the particle on different focal planes.

In this work, we used the data of the optical scan to estimate the quantities of "new" surfaces generated by the fracture, the statistics on morphologies (equivalent diameter, area for every particle). These data were compared to the entity of the stress rate. The calculation of the real volume deserves special mention. The instrument software measures the volume by the diameter equivalent to the sphere (SE Volume), but this value is very far from reality. Therefore, we used an algorithm that is based on the measurement of the attenuation of the intensity of light transmitted by the particle.

For non-scattering media, the Beer–Lambert law (BL, or absorption law) is well recognized to describe the relationship between transmittance and sample thickness as:

$$T = \exp(-\mu_a d) \tag{1}$$

where T is the transmittance, d is the sample thickness, and μ_a is the absorption coefficient (cm^{-1}). This expression is based on the random nature of stochastic light absorption, characterized by the rate constant μ_a. Since the transmitted light intensity is measured for each particle, and since we are always in the presence of quartz, which has an absorption coefficient from 664 cm^{-1} at 450 nm to 312 cm^{-1} at 650 nm [10], the thickness "d" can be derived for all particles. Figure 1A shows the effect of fracturing on quartz grains; the images have been modified after Zhao, B. et al. [11]. On the left, we can see the two-dimensional projection of the particles obtained from the compression fracturing of a quartz grain. The perimeter of the granule, obtained from the two-dimensional projection, is expressed in green. The perimeters of the projections of the individual particles obtained from the fracturing are indicated in red. In Figure 1B, images of the original crystal, a tip weighing 207 mg and some of the particles obtained from the fracturing can be observed.

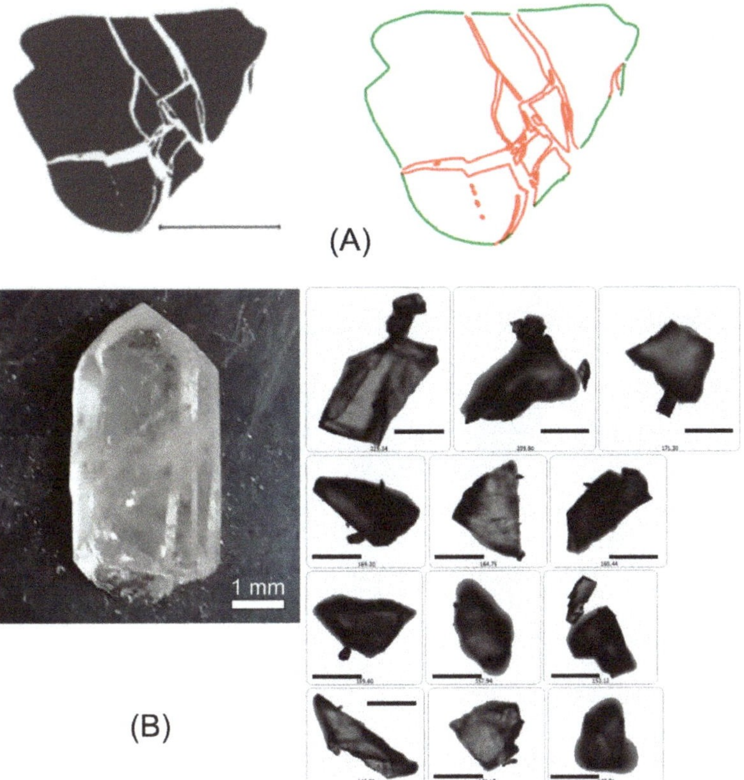

Figure 1. (**A**) Breakage of a grain: on the left, projection images of particles created by the breaking of first grain; on the right, original perimeter (green) and new perimeter (red). (**B**) On the left, the image of the original crystal fragment, 207 mg in weight; on the right, the 12 largest fragments of the left crystal; each fragment of the crystal, from 0.3 mm to 1 µm, was analyzed from the morphological and dimensional point; in total, the left crystal produced over 320,000 fragments; the black bar indicates 150 µm.

The surfaces of the "internal" granules are, in fact, the new fracture surfaces, created when the imposed stress exceeds the critical stress level. The sum of the surfaces of the particles will be equivalent to the sum of the new surfaces created by the fractures, S_{tot}, plus the original surface. In this way, it is possible to calculate the total area created by the fractures. By extrapolating this concept, it is also possible to calculate the total length of the fractures. The total perimeter of all particles is correlated with the total length of the fractures, minus the original perimeter of the crystal and the loss of data due to the two-dimensional transposition of the three-dimensional particles. The latter error is reduced if the particles analyzed by microscopy are in very large numbers. For this reason, the analyses in morphology must be carried out on a high number of particles, usually above 100,000 particles.

In order to determine the nature of silica polymorphs, a Raman microprobe inserted in the Morphology G3ID was used, which uses a 785 nm laser with a spot of 2 microns in diameter. The Raman laser power is 40 mW.

To determine the stress to which the crystal is subjected, a measurement method was developed using the image of the shattered crystal printed on a pressure sensitive film (Pressurex Inc., Madison, NJ, USA; www.sensorprod.com). These films have an ink layer that produces a color, and the intensity of which is proportional to pressure. Figure 2 shows the pressure marks left on the film by different quartz crystals

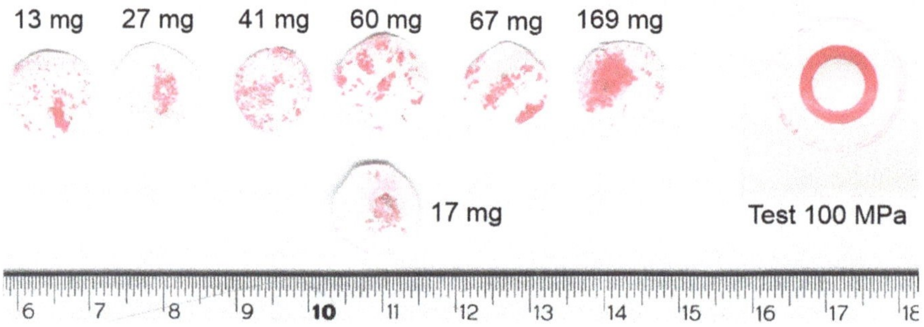

Figure 2. Images of the films impressed by several compressed crystals; on the right, a calibration test. The numbers represent the weight of samples in mg.

Finally, we used scanning electron microscopy to determine the microscopic characteristics of fractures at different stress levels (Zeiss EVO MA10, Zeiss Gmbh, Ulm, Germany). The samples to be analyzed in the SEM were left in their original condition, without cleaning, since debris on the surfaces of the granules can provide important information.

3. Results

3.1. Particle Size Distributions Related to Stress Conditions

The sizes of the quartz particles show complex correlations with pressure undergone and linear correlations between the rate of increase in pressure and the average size.

The particle size curves (Figure 3) obtained from the tests show a distribution of variable shape:

- For particles from samples subjected to a more intense stress rate (MPa·s^{-1}), the distribution is shifted towards the larger particles, and the shape of the curve is monomodal.
- For the particles that have undergone a slower pressure rate, the particles have had time to fracture more, and the distribution is shifted towards smaller particles, with a bimodal curve shape.

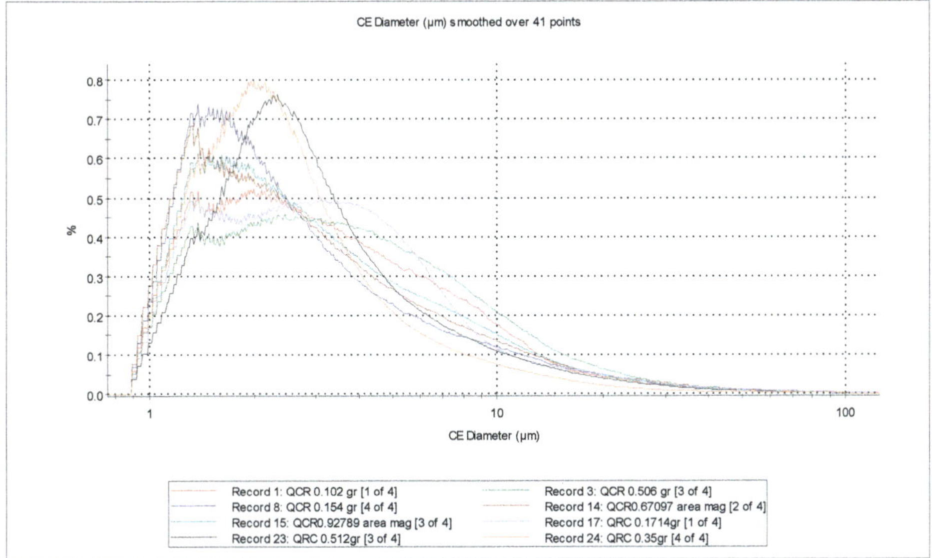

Figure 3. CE diameter distribution curves of quartz particles (the meaning of CE is "circle equivalent"; the curves are obtained from the smoothing by average of 41 readings; the results are expressed in *x-y* graphs and not in histograms, due to the high number of reading channels; and gr is the weight of sample, e.g., 0.102 gr as grams, 102 mg).

Figure 4 shows the correlation between the total length of the fractures per unit of quartz weight (calculated on the basis of the total perimeter of the particles) and stress rate.

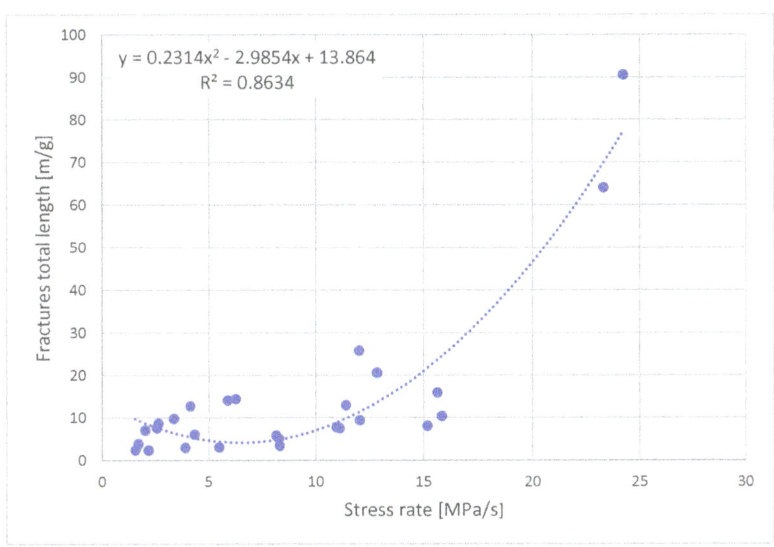

Figure 4. Correlation between the stress rate and total fracture length.

3.2. Determination of the Morphological Characteristics of the Fractures

The morphologies of the fractures in the quartz granules after the cutting effort are extremely varied and interesting, showing that the fracture phenomenon of this mineral is very complex and connected to a large number of parameters. We analyzed the data obtained from the analysis of the images, starting from the samples that underwent the least intense pressures, to end with those that underwent the greatest pressure.

In Figure 5, the fracture forms left on the quartz granules by stress rates of about 0.5 MPa/s are observed. Stress creates a first series of conchoidal-shaped fractures. Around it, there are "sawtooth" shapes of homogeneous dimensions. The average height of the teeth is 2.3 μm. The area filled with these teeth is extremely large and covers several hundred μm^2. We initially thought that these forms were related to a Dauphine twinning; after careful examination we decided not to include this result, as it is still unclear. Dauphine gemination is however compatible with the stress regime to which quartz crystals are subjected in our experiments; in particular, Laughner et al. [12] demonstrated that the stress required to produce twins can be lower than 88 MPa. Since the strength of quartz on compression is 200–300 MPa, Dauphiné twinning generally precedes brittle fracture.

In samples subjected to stress of greater intensity and rate (5–50 Mpa/s), simpler fractures are generated that extend for distances from a few units to a few tens of μm. Figure 6 is a mosaic of eight photos taken on a quartz particles with some ridges and fractures. These fractures (Figure 6) show a profile raised above the surface with a vaguely triangular section and an average length of 20–30 μm from the beginning of the fracture until the first bifurcation, which invariably occurs at angles of 27°. In larger grains, the fractures end with euhedral shaped craters of very variable sizes, from 150 nm to 1–2 μm (Figure 7). The shapes of the craters are regular, with hexagonal, square and trigonal geometries; only the smaller craters are elliptical (Figure 7). Most of these craters are aligned in directions inclined with respect to the direction of the fractures. These alignments reach conspicuous dimensions, often exceeding 100 μm. In some areas, where the density of craters is of the order of one every 2–3 μm, the craters unite and form open fractures, ranging from a few to ten μm wide and proportional lengths of up to 100 μm. In the fragments of the samples subjected to the maximum stress, the presence of "bubbles" is observed on the surface of the material with fractures on the surface (Figure 7C). These bubbles open in some places, forming small craters (50–100 nm), resembling real "hot spots" with a raised edge of apparently melted material (Figure 8). At the apex of some fractures, there are also real "protrusions" (Figure 9), while in rare areas, there are open fractures and craters filled with quartz crystals, that are elongated and small, with diameters of 0.5–0.8 μm and lengths of 1–1.5 μm, and often twinned (Figure 10). The same phenomenon can be observed in some open fractures, completely covered by newly formed quartz crystals, from which the crystals are arranged according to a design that mimics an exit direction of the precursor material. Along the fractures, detachments of filamentous material are observed, starting from the starting point of the fracture (Figure 11). These filaments are composed of silica alone. These fibers are characterized by diameters between 50 and 200 nm and lengths of from a few μm to tens of μm, and they follow the fractures until they detach, and, then, they can take on twisted, curled or meandering shapes. Their section is slightly flattened, almost as if they were ribbons. On the points of the greatest concentration of fractures, hundreds of these filaments are observed, which start from the fractures in a crystallographic direction and then tend to detach (Figure 11).

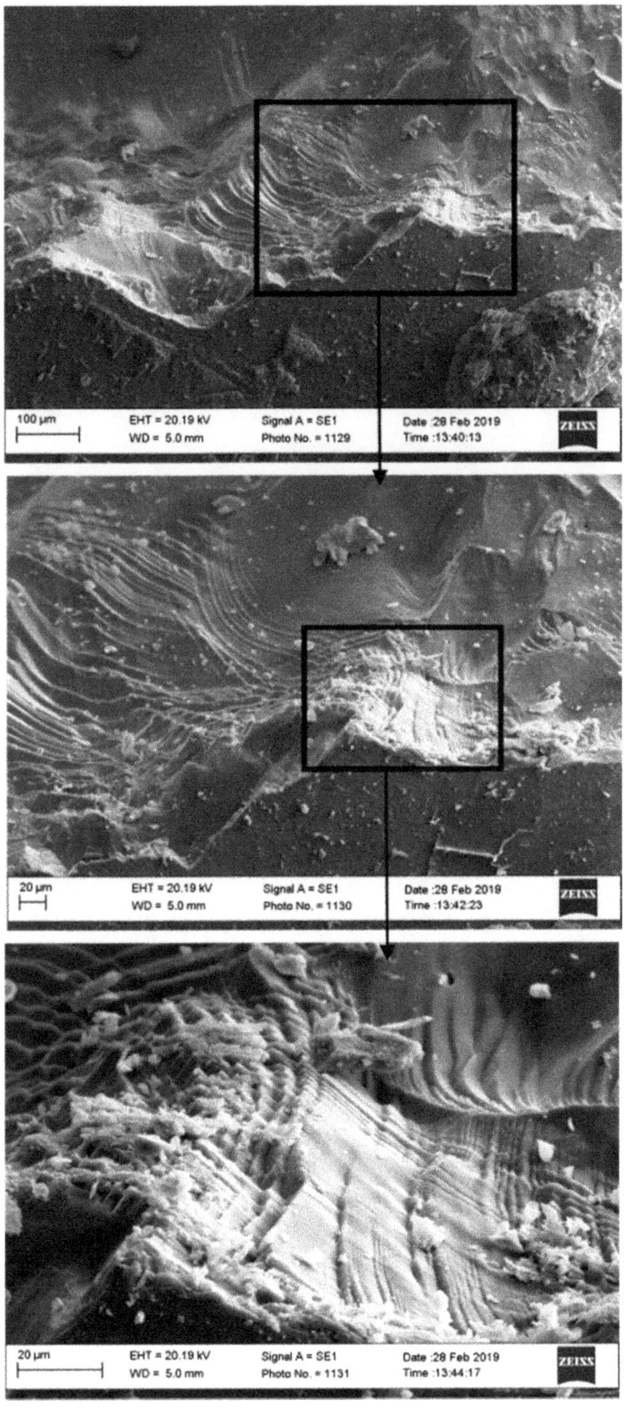

Figure 5. Sawtooth forms generated by the combination of two conchoidal fractures on the surface of the quartz granules; stress rate: 0.5 MPa/s.

Figure 6. Fractures on the surface of a quartz granule subjected to 100 MPa; the bifurcations show angles of about 27° (taken from the work of Martinelli et al. [5]).

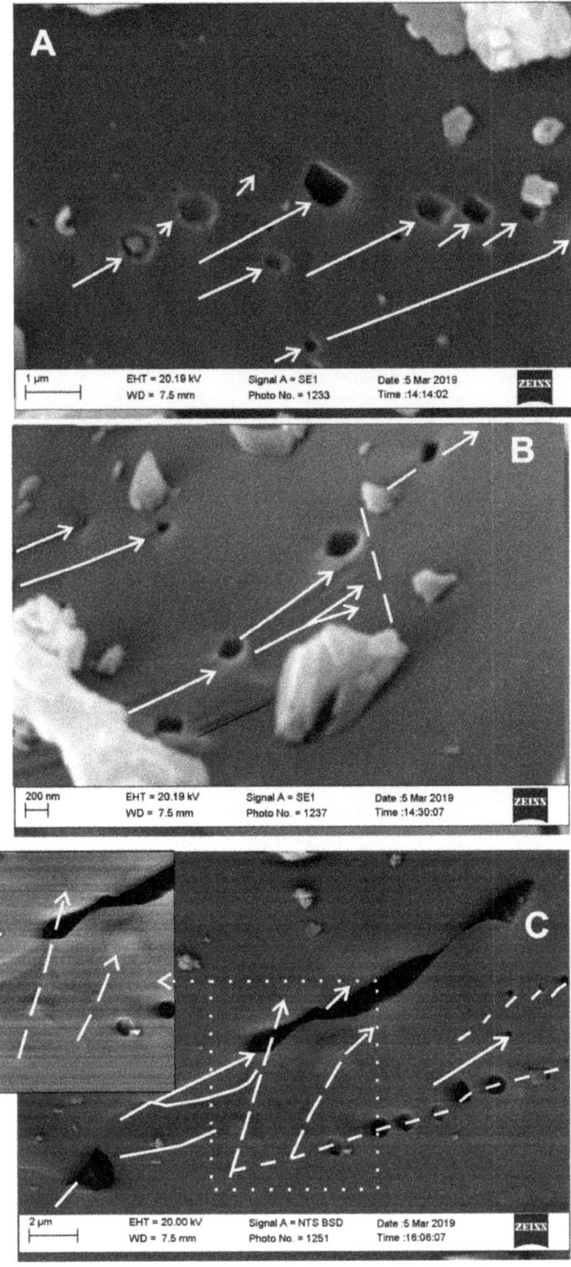

Figure 7. Fractures in the granules at 150–200 MPa; in photo (**A**) the euhedral shapes of the craters at the apex of fractures of a few microns are well observed. Photo (**B**) shows the formation of several generations of fractures starting from the craters. In photo (**C**), there are open fractures formed by the coalescence of the craters that form at the end of the fractures; on the left, aligned bubbles, barely visible on the surface of the granules. In white, the apparent directions taken by the fractures, in black the visible directions.

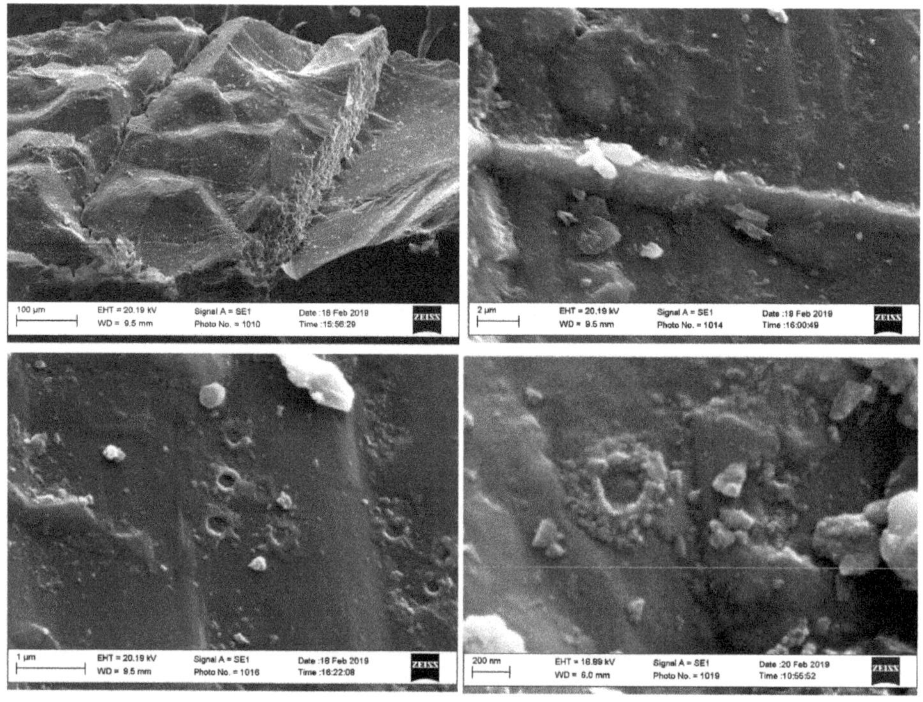

Figure 8. Quartz, 120 MPa: "hot spot" on quartz particle surfaces.

Figure 9. Quartz, 220 MPa: amorphous silica extruded from a fracture on quartz surface.

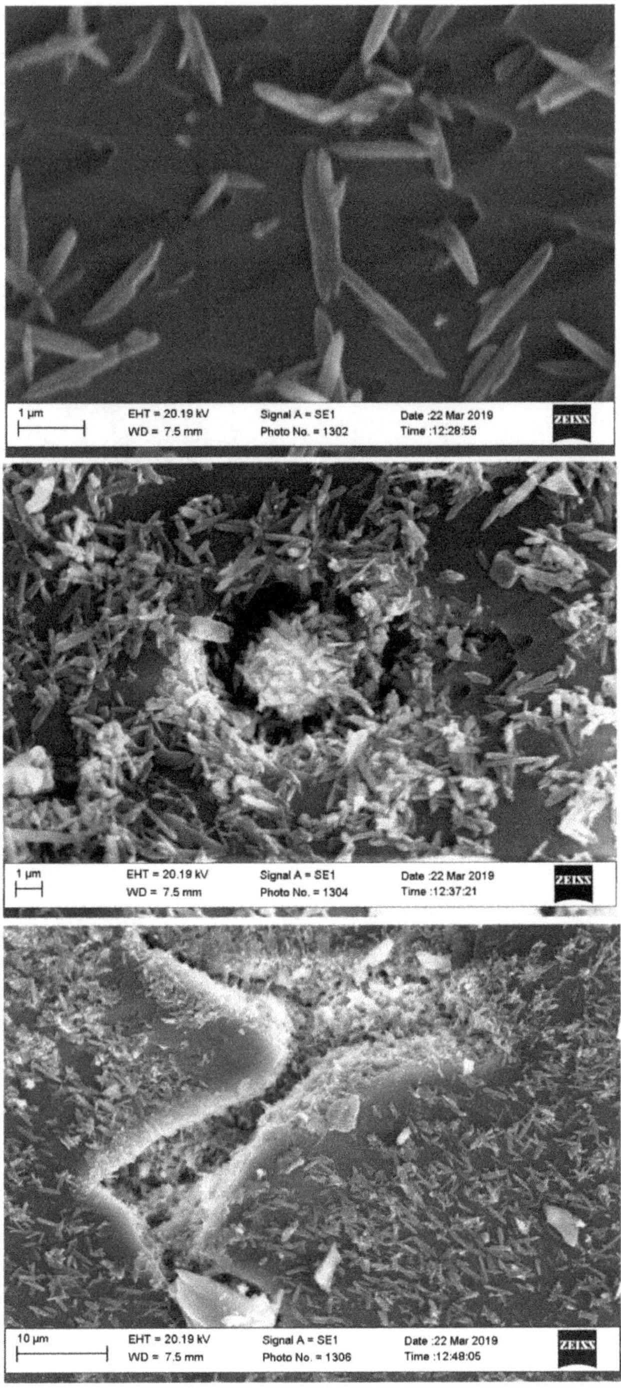

Figure 10. Quartz, 230 MPa: crystallization and growth of quartz microcrystals from silica near fractures and hot spots.

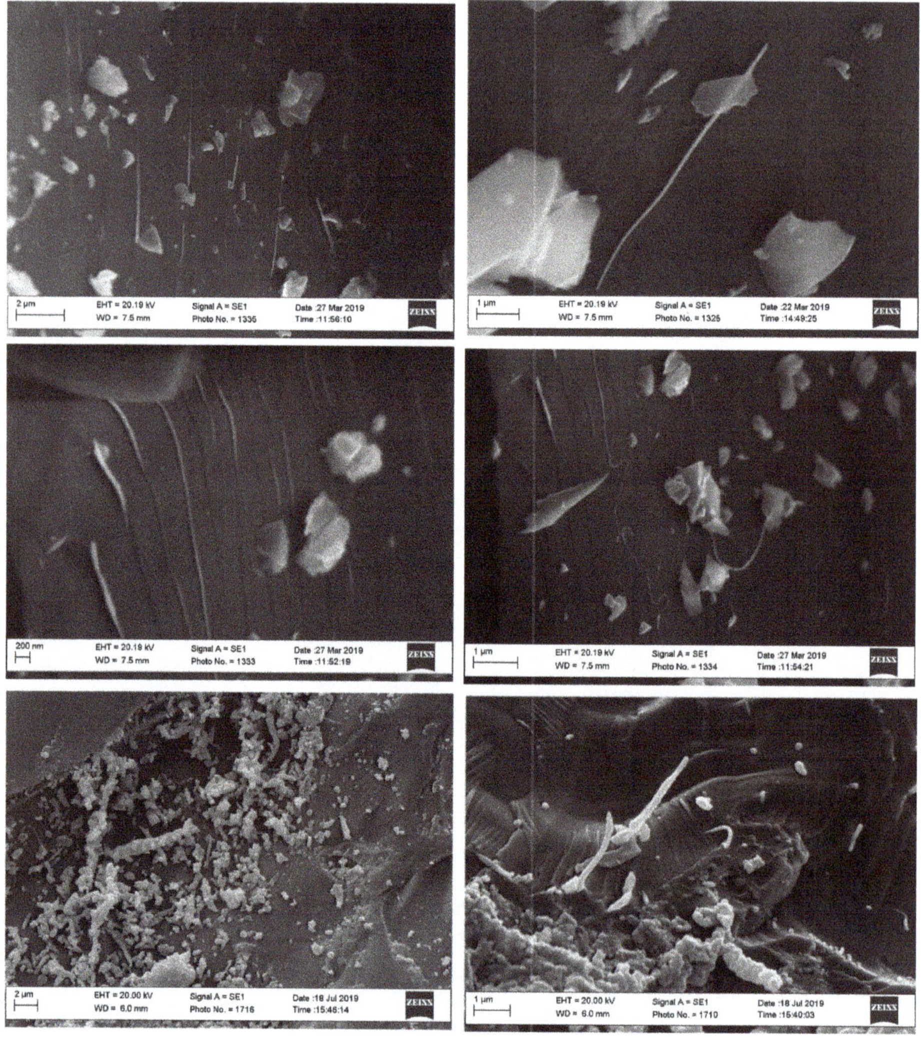

Figure 11. Filamentous forms of silica that are an integral part of fractures of the type shown in Figure 6.

3.3. Raman Analysis of Solid Phases on Fracture Surfaces

Raman analyses were performed on fracture surfaces that appear dark under an optical microscope in polarized light and on areas where fracture streaks with fibrous traces are present. Figure 12 shows some spectra taken respectively in dark isotropic areas. Given the size of the Raman laser spot used ($\approx$2 µm), the analysis cannot select only the filament areas that are smaller than 1/10 of spot, but will also include areas where there is certainly undisturbed quartz. Despite this, the analysis clearly shows the presence of α-cristobalite and tridymite in very variable proportions (Figure 12). In the first spectrum (white) the 465, 404 and 365 cm^{-1} bands of alpha quartz are observed. It belongs to an optically non-isotropic zone. The underlying spectra, indicated as Black1, Black2 and Black3 in Figure 12, are three selected areas within optically isotropic areas of approximately 4 µm^2 each. In the spectra we observe the presence of the most intense bands of quartz, but also of tridymite and cristobalite, albeit of reduced intensity. The identified cristobalite is of the alpha type, while the tridymite is similar to the

PO10 polytype, as per Ruff standard of said mineral. On the right, an enlargement of the spectra is observed in the range of 440 to 320 cm^{-1}. The Black1, 2, 3 spectra have been recorded at 200 Mpa; the quantity of these isotropic patches seems to increase with increasing pressure and with time of stay under stress.

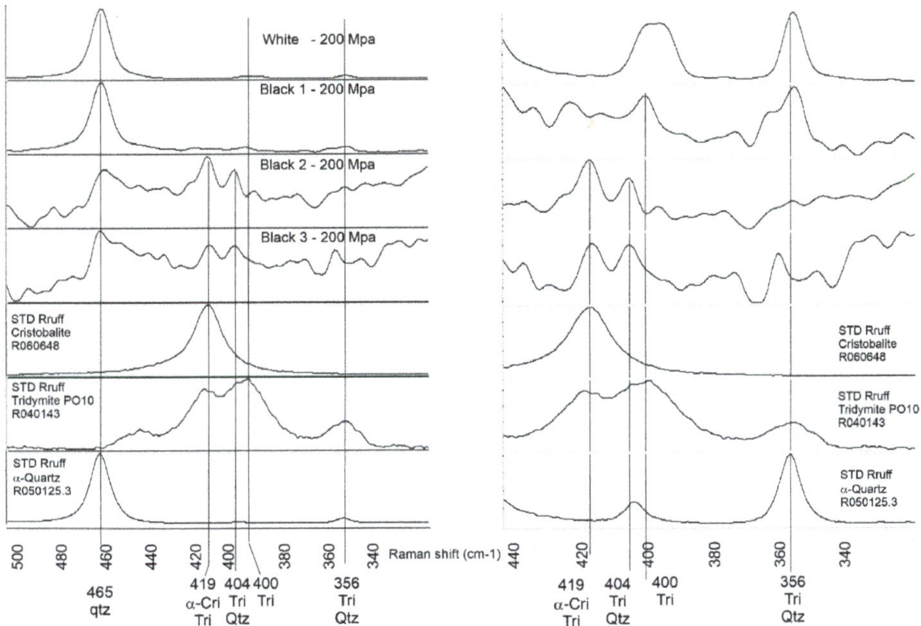

Figure 12. Raman spectra of stressed quartz fragments; the three spectra below derive from the collection of Raman spectra included in the RRUFF collection (RRUFF Project, Department of Geosciences, University of Arizona, 1040 E 4th, Tucson, AZ, USA. 85721-0077): (Cristobalite (low)-R060648, Tridymite PO10-R040143, Quartz (low)-R050125.3); on the right, an enlargement of the area between 440 and 320 cm^{-1}.

4. Discussion of Data

The data obtained in this study should be seen in the light of the most recent literature on the effects of shear stress on alpha quartz, literature to which the authors have also contributed [5,6]. The fractures we have observed show an evolution which, we believe, depends substantially on the stress rate rather than on the pressure itself. In fact, we have verified that even a pressure of 50 MPa, supplied in times exceeding one minute, generates a more intense fragmentation than a pressure of 100 MPa supplied in times of a few tens of seconds. The most relevant phenomena observed on quartz granules that have undergone increasing stress are as follows:

- Formation of craters with euhedral shape at the apex of the fracture;
- Coalescence of the craters to form open fractures;
- Detachment of apparently amorphous siliceous material, fibrous in shape from the ridges and traces of fractures;
- Bubble formation on the surface, some of which show amorphous silica extrusions;
- Formation of craters on the surface from which amorphous silica is expelled;
- Formation of secondary quartz crystallizations, which radiate from "emission" points, such as craters and open fractures.

From Figure 4 it can be observed that the fracture length increases exponentially with the increase in the stress rate. This is reasonable, since the total length is related to the number of particles and this to the number of bifurcations of the cracks that are generated during the breaking of the quartz crystals. Cracks that reach critical speed tend to divide into two cracks (bifurcations) with an acute angle between them, as demonstrated by Tromans and Meech [13–15]. Consequently, if the stress velocity is greater, the cracks' opening speed will also be greater, and this will produce more fractures and a greater number of particles.

In Figure 6, the ridges indicate the presence of fractures under the surface. The crests indicate an increase in volume propagated linearly along the fractures. The fractures separate into two sections with an acute angle of 27°. As is known, the angle of forking varies with the stress state. In particular, the bifurcation angle indicates the relationship between shear stress and normal stress τ/σ_n. An angle of 27° corresponds to a ratio from −0.3 to −0.5, which essentially indicates a bending stress (Richter, 2003) [16]. Both the total length of the fractures and the amplitude of the bifurcation angle are parameters to be framed in a wider context of energy analysis of the fracturing phenomenon, for which the reader is referred to the cited bibliography. In this context, it is sufficient to remember that the fracture length is connected with the fracturing speed [13–15]. The cited authors calculate that in alpha quartz, the fractures split in two when the propagation speed reaches the *climit* (limit speed of 1990 m/s). At that moment, the ratio between the size of the new fracture, a_i and the initial size, a, is equal to 2, and the speeds of the two daughter fractures is reduced to 30–50% of *climit*. The two fractures thus formed increase their speed and must again fork when they reach *climit* and at an a_i/a ratio of 4, according to relation:

$$(a_i/a)_{\text{branch4}} = 2(a_i/a)_{\text{branch3}} = 2(a_i/a)_{\text{branch2}} = 2(a_i/a)_{\text{branch1}} = 2(a_i/a)_{climit}$$

It is important to note that an increase in the propagation speed leads to a higher frequency of bifurcations and therefore to a greater fragmentation. This is confirmed by the observations made with the analysis of the morphological image of the particles, where it is evident that a higher stress rate increases the number of particles and the total fracture length (Figure 4). It is noteworthy that the propagation of fractures in the minerals causes an accumulation of deformation energy, and this energy must dissipate; in mineral mechanics, the ways of disposing of the plastic deformation energy are essentially two: the bifurcation of fractures and, above all, the increase in temperatures in the deformation areas at the fracture tip [17,18].

The formation of the craters visible in Figure 7A–C is consequential to the propagation of fractures. In Figure 7A,B it is observed that the fractures end with a regular-shaped crater, often mimicking a crystalline form with five to six faces. The volume of matter removed from these craters seems independent of the visible (or superficial) length of the fractures. We believe that the loss of fragments of material with crystalline forms is linked to the presence of fracture lines already existing in the crystal, parallel to crystallographic directions. The reader should also note the correspondence between the apex of the fracture and the corresponding crater. We believe this is due to the undermining of the material by the wave at ultrasonic speed connected to the propagation of the fractures. Figure 7C shows the presence of surface bubbles that show cracks of about 40 nm width and lengths of about 1 μm. The bubbles also appear aligned, according to directions that lie at 60° with respect to the fractures. We believe that these bubbles are the superficial manifestation of the melting effect at ultrasonic speed that occurs inside the quartz grains during the crack propagation and which produces most of the phenomena we are describing. The same phenomenon is responsible for the formation of the craters aligned with the amorphous silica extrusions, visible in Figure 8, the silica extrusions in Figure 9, the recrystallized quartz from silica extrusion in Figure 10 and, finally, the filamentous fibers in Figure 11 that are, perhaps, the most interesting part of the discoveries made in this work. The Raman analysis performed on the silica fibers associated with the fractures shows that this silica is organized in the form of α-cristobalite and tridymite (Figure 12). The presence of cristobalite had already been highlighted by the authors of this work [5,6], where the structural changes of quartz in nanocrystalline cristobalite were described when alpha quartz was subjected to shear stresses. In the aforementioned work, we have shown that the action of prolonged shear stress on the quartz crystals determines a reticular distortion so large as to lead the quartz itself to an amorphization; in this "amorphous silica", the radial distribution function analysis (RDF) has highlighted the formation of short-term clusters with a six-tetrahedron organization, similar to cristobalite and tridymite and no longer to four tetrahedrons [6]. It has also been shown that, if these amorphous phases are brought to 1200 °C, they tend to transform nano-cristobalite into a well-crystallized cristobalite, well visible in X-ray diffraction analysis [6]. The same association of minerals had been reported in a paper of Brodie and Rutter (2000) on quartz samples subjected first to tensile stress and then to heating at 1200 °C [17].

To explain the data collected, we hypothesized a mechanism that can materialize thanks to the speed of propagation of the fractures. The heart of the mechanism is tensile stress, which induces the opening of a large number of fractures in the quartz volume. Fracture propagation occurs at an estimated speed between 50% and 60% of the shear wave, therefore between 650 and 1000 m/s, and for this reason we can speak of ultrasonic speed. At these speeds, the fractures tend to fork after just 25–30 μm, and the crack tip transit times are therefore in the order of 30 ÷ 50ns. In such a short fraction of time, the amount of heat produced by the passage of the crack tip cannot have transferred to the volume outside the fracture. Weichert had already measured temperatures above 2000 K on the fractured quartz, and this has been confirmed by other authors [18,19]. These temperature increases generate channels of fused silica, which has a decidedly lower density than quartz (2.2 g/cm^3 against 2.65 g/cm^3). The decrease in density causes an immediate expansion in volume which causes surface fracturing. In actual fact, we can say that during the melting, the molar volume increases from 22.688 cm^3/mol to 27.20 cm^3/mol, with an expansion of 16.5% [19], more than enough to open the surface over the crack and bring out the fused and clotted silica filament that is inside it.

The events we have described are summarized in the diagrams in Figure 13. In the first one, we observe the simulation of the silica filaments growth, starting from the fractures; Figure 13B shows the formation of fractures ending with the euhedral craters and the formation of bullous-shaped growths. Figure 13C shows the pattern of crater formation ("hot spot") and silica protrusion from subsurface melted areas. Our data demonstrate that the friction action applied to quartz in totally anhydrous conditions leads to the formation of amorphous silica, which quickly organizes itself into nanostructures of cristobalite and tridymite [5,6]. Cristobalite has a structure consisting of rings with 6 tetrahedra, with structural arrangements that show an auxetic behavior, that is, with a negative Poisson's ratio [20,21] (Table 1). This means that a tangential compression produces a reduction in length in the normal direction and not an expansion, as expected from solids with a positive Poisson's ratio. The effect is particularly felt by the alpha cristobalite, whose Poisson ratio reaches the considerable negative value of −0.169. In practice, if the tangential contraction is 100 mm, the normal contraction will be 16.9 mm; this reduces the contact surface and correspondingly reduces its friction. It should also be taken into account that the phenomena whose generation we have verified are extremely fast, because they occur in the time interval that elapses during the final phase of propagation of the fractures. The speed of these contractions is linked to the propagation speed of the ultrasonic wave.

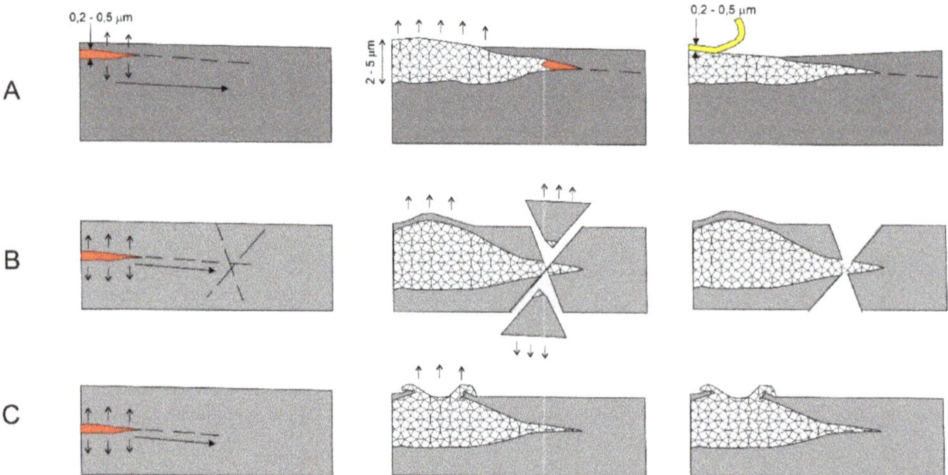

Figure 13. Scheme of propagation of fractures in quartz granules; (**A**) beginning of the phenomenon, as the fractures spread; (**B**) formation of fractures ending with the euhedral craters and superficial "elevations"; (**C**) formation of craters ("hot spots") and the protrusion of silica from the molten areas below.

Table 1. Mechanical parameters for silica polymorphs [13,14,20], G = Shear modulus, E = Young modulus, n = Poisson ratio, ϱ = density, ε = strain.

Silica Polymorphs	G	E	υ	ϱ	ε
	Gpa	Gpa	n	g/cm^3	%
quartz alpha	44.1	95.6	0.084	2.65	8.4
quartz beta	41.5	99.1	0.194	2.53	19.4
cristobalite alpha	37.1	61.3	−0.169	2.32	−16.9
cristobalite beta	32.6	62.3	−0.044	2.20	−4.4
tridymite low	28.8	58.1	0.004	2.27	0.4
tridymite high	26.7	52.8	−0.011	2.26	−1.1
Silica glass	33.4	79.3	0.187	2.20	18.7

If the application of stress is continuous, there is an accumulation of amorphous silica that quickly transforms into cristobalite and tridymite near the fracture and sliding areas. The mechanisms described also explain the so-called "flash melting" which is observed by numerous authors in the shear stress tests on quartz and, above all, the sudden drop in the coefficient of friction that all the authors observe when the quartz is subjected to severe friction conditions [1]. Various authors support the thesis that the friction, exerted on quartz-rich rocks, generates a sort of silica gel lubricating film, an amorphous material rich in water, deriving from the environment where the test takes place. The concept of the lubricating gel is also taken up by [4], who had detected the presence of amorphous silica in the flow tracks of experiments by means of tribometric rotating friction apparatuses, such as the "pin-on-disk" on quartz [3,4]. Authors detected, via Raman spectroscopy, the peaks attributable to moganite (metastable phase of silica, quickly converted into cristobalite and tridymite) and to amorphous silica, as well as reticular distortions of alpha quartz. In addition, the authors showed spectra in FTIR microspectrophotometry to prove the presence of water in the amorphous silica of these traces of flow. The results obtained, albeit preliminary, do appear interesting. In particular, the fact that amorphous silica has hydrated with atmospheric humidity is possible, but it does not prove that it is the means of sliding the faults that insist on quartz-based rocks. In our tests, carried out under anhydrous conditions, there can be no water presence, but the same amorphous silica and cristobalite and nanocrystalline tridymite are formed, which, as we have said, have the characteristic of reacting to efforts by decreasing the contact area, with relative reduction of the friction coefficient. At the end of this long discussion of data, we would like to summarize the events that, in our opinion, characterize the transformation of quartz into a low friction coefficient material:

1. the tensile stress field generates fractures that open quickly, spreading over the entire volume of the area subjected to shear stress;
2. the propagation of the fractures generates the local melting of the quartz, producing the formation of amorphous silica, whose density is lower than that of the quartz; the difference in density and thermal expansion causes the fracture of the surfaces and the dispersion of the silica, which quickly solidifies in the form of nanocrystalline cristobalite (which has a marked auxetic behavior) and sometimes tridymite;
3. the accumulation of these polymorphs in the rock volumes subject to greater stress leads to a progressive reduction of friction and, easily, to the triggering of the movement.

In none of these events is the presence of water necessary.

5. Conclusions

In this work we have definitively shown that anisotropic stress on quartz crystals causes the formation of amorphous silica and, from it, at least two silica polymorphs, cristobalite and tridymite, in nanocrystalline form. These phases accumulate with the advancement of the fractures and can profoundly modify the behavior of the solid in terms of friction coefficient. Amorphous silica is, therefore, not a simple friction melting product of the surface portion of the rock subject to friction, but it is a precursor of far more complex phases that represent the memory of the event. This observation contrasts sharply with what had been hypothesized by several authors, who aim to explain the lowering of the friction coefficient of the quartz subjected to shear stress due to the simple presence of silica gel.

Given the extreme speed with which cristobalites and tridymites are formed, we can argue that these mineral phases are generated while the fracture phase is still active and that their accumulation is such as to heavily modify the rock's mechanical behavior.

Cristobalite can thus be the main cause of the collapse of the friction coefficient occurring in the friction tests on quartz and on rocks rich in quartz.

Author Contributions: Conceptualization: G.M. and P.P.; Methodology: G.M. and P.P.; Analyses and investigation: P.P., E.T., E.P.; Writing—original draft preparation: P.P.; Writing-review and editing: G.M. and P.P.; Resources and project administration: F.G. All authors have read and agreed to the published version of the manuscript.

Funding: This research received no external funding.

Acknowledgments: Thanks are due to two anonymous reviewers who contributed, with their suggestions, to improve the quality of the present paper. The Chinese Academy of Sciences Visiting Professorship partially supported Giovanni Martinelli for Senior International Scientists (2018VMA0007).

References

1. Di Toro, G.; Goldsby, D.L.; Tullis, T.E. Friction falls towards zero in quartz rock as slip velocity approaches seismic rates. *Nature* **2004**, *427*, 436–439. [CrossRef]
2. Hirose, T.; Shimamoto, T. Growth of molten zone as a mechanism of slip weakening of simulated faults in gabbro during frictional melting. *J. Geophys. Res. Space Phys.* **2005**, *110*. [CrossRef]
3. Hayashi, N.; Tsutsumi, A. Deformation textures and mechanical behavior of a hydrated amorphous silica formed along an experimentally produced fault in chert. *Geophys. Res. Lett.* **2010**, *37*, 12305. [CrossRef]
4. Nakamura, Y.; Muto, J.; Nagahama, H.; Shimizu, I.; Miura, T.; Arakawa, I. Amorphization of quartz by friction: Implication to silica-gel lubrication of fault surfaces. *Geophys. Res. Lett.* **2012**, *39*, 21303. [CrossRef]
5. Martinelli, G.; Plescia, P.; Tempesta, E. Electromagnetic emissions from quartz subjected to shear stress: Spectral signatures and geophysical implications. *Geosciences* **2020**, *10*, 140. [CrossRef]

6. Martinelli, G.; Plescia, P.; Tempesta, E. "Pre-Earthquake" Micro-Structural Effects Induced by Shear Stress on α-Quartz in Laboratory Experiments. *Geosciences* **2020**, *10*, 155. [CrossRef]

7. Tavares, L.M.; King, R.P. Single particle fracture under impact loading. *Int. J. Miner. Process.* **1998**, *54*, 1–28. [CrossRef]

8. Tavares, L.M. Optimum routes for particles breakage by impact. *Powder Technol.* **2004**, *142*, 81–91. [CrossRef]

9. Tugcam Tuzcu, E.; Rajamani, R.K. Modeling breakage rates in mills with impact energy spectra and ultra fast load cell data. *Miner. Eng.* **2011**, *24*, 252–260. [CrossRef]

10. Marcos, L.V.R.; Larruquert, J.I.; Méndez, J.A.; Aznárez, J.A. Self-consistent optical constants of SiO_2 and Ta_2O_5 films. *Opt. Mater. Express* **2016**, *6*, 3622–3637. [CrossRef]

11. Zhao, B.; Wang, J.; Coop, M.R.; Viggiani, G.; Jiang, M. An investigation of single sand particle fracture using X-ray micro-tomography. *Géotechnique* **2015**, *65*, 625–641. [CrossRef]

12. Laughner, J.W.; Newnham, R.E.; Cross, L.E. Mechanical Twinning in small quartz crystals. *Phys. Chem. Min.* **1982**, *8*, 20–24. [CrossRef]

13. Tromans, D.; Meech, J. Enhanced dissolution of minerals: Stored energy, amorphism and mechanical activation. *Miner. Eng.* **2001**, *14*, 1359–1377. [CrossRef]

14. Tromans, D.; Meech, J.A. Fracture toughness and surface energies of minerals: Theoretical estimates for oxides, sulphides, silicates and halides. *Miner. Eng.* **2002**, *15*, 1027–1041. [CrossRef]

15. Tromans, D.; Meech, J.A. Fracture toughness and surface energies of covalent minerals: Theoretical estimates. *Miner. Eng.* **2002**, *17*, 1–15. [CrossRef]

16. Richter, H. Fractography of Bioceramics. In *Key Engineering Materials*; Dusza, J., Ed.; Trans Tech Publications: Baech, Switzerland, 2003; Volume 223, pp. 157–180.

17. Weichert, R.; Shonert, K. Heat generation at the tip of a moving crack. *J. Mech. Phys. Solids* **1978**, *26*, 151–161. [CrossRef]

18. Rittel, D. On the conversion of plastic work to heat during high strain deformation of glassy polymers. *Mech. Mater.* **1999**, *31*, 131–139. [CrossRef]

19. Brodie, K.H.; Rutter, E.H. Rapid stress release caused by polymorphic transformation during the experimental deformation of quartz. *Geophys. Res. Lett.* **2000**, *27*, 3089–3092. [CrossRef]

20. Pabst, W.; Gregorova, E. Elastic Properties of silica polymorphs—A review. *Ceram. Silik.* **2013**, *57*, 167–184.

21. Yeganeh-Haeri, A.; Weidner, D.J.; Parise, J.B. Elasticity of a-Cristobalite: A Silicon Dioxide with a Negative Poisson's Ratio. *Science* **1992**, *257*, 650–652. [CrossRef] [PubMed]

Article

Development of a More Descriptive Particle Breakage Probability Model

Murray M. Bwalya [1] and Ngonidzashe Chimwani [2,*]

1   School of Chemical and Metallurgical Engineering, University of the Witwatersrand, Johannesburg 2050, South Africa; mulenga.bwalya@wits.ac.za
2   Institute of the Development of Energy for African Sustainability (IDEAS), a Research Centre of the University of South Africa (UNISA), Florida Campus, Private Bag X6, Johannesburg 1710, South Africa
*   Correspondence: ngodzazw@gmail.com; Tel.: +27-731838174

Received: 8 June 2020; Accepted: 6 August 2020; Published: 12 August 2020

Abstract: Single-particle breakage test is becoming increasingly popular, as researchers seek to understand fracture response that is purely a function of the material being tested, instead of that which is based on the performance of the comminution device being used. To that end, an empirical breakage probability model that builds on previous work was proposed. The experimental results demonstrate the significance of both energy input and the number of repeated breakage attempts. Four different materials were compared, to gain a better insight into the breakage response. This modelling work goes further from previous research of the authors, by showing that not only does size related threshold energy and repeated impacts characterize particle breakage properties, but each material exhibits unique trends in terms of how its threshold energy and its rate of deterioration varies with particle size and each impact, respectively. This behaviour can be attributed to the different mechanical characteristics of the material and their flaw distribution. The importance of these aspects was highlighted.

Keywords: single particle breakage; energy input; drop-weight tester; breakage modelling; grinding prediction

---

1. Introduction

For quite some time, researchers have been pursuing an understanding of particle fracture at a fundamental level. This approach is necessary, as it would help in the design of equipment that achieves the most desirable outcomes. There has been appreciable progress to some extent, however, in most cases, the understanding is more focused on the equipment comminuting the particles. The JKMRC introduced a single breakage test equipment to establish a way of separating material breakage properties from the influence of equipment. This, however, introduces the challenge of relating material breakage response data obtained independent of the target comminution device to the performance to be expected. Thus, as Delboni and Morell [1] point out, the identification of the mechanisms involved in comminution equipment is imperative for the successful modelling of its comminution process.

In the 19th century, Rittinger [2] and Kick [3] proposed some comminution energy laws; Kick proposed comminution energy to be a function of the volumetric reduction ratio, while Rittinger related it to the new surface area produced. It was however not possible based on these laws to predict how much energy would be required to achieve a particular level of size reduction. Reasonable success has somewhat been achieved by Bond's energy law [4], which was suggested to be based on Griffith's law of crack propagation [5].

The theory proposed by Griffith's highlighted the importance of pre-existing flaws within particles. Using this theory, energy balance around the crack could be calculated and energy required to extend the crack and consequently produce a new surface could be thus determined. It will just be mentioned in passing that, while grinding is directed towards forming new surface area, most of the energy invested in the process is absorbed by other aspects unrelated to comminution, as Rumpf [6] observed. This somewhat complicates the problem of relating input energy to the resulting product. In addition, no two particles will have an identical distribution of inherent cracks, and thus some statistical modelling of these tests is inevitable. It should also be pointed out that the tests must be representative of the stressing mechanism, that is to be expected in the comminution device being targeted by the test; otherwise, there would be little correlation between the test and the actual equipment performance.

The JKMRC initially introduced the pendulum test and refined the technique, by developing the drop-weight tester [7,8]. Further enhancement of the test procedure has been made by the introduction of the JK rotary breakage tester [9], which has reduced the time required for testing significantly, though this test now relies more on a single impact than double impact.

Bwalya et al. [10] explored a modified equation of Weichert [11], and applied it to drop-weight test breakage data. This model went further by recognizing repeated breakage attempts as being equally important as energy input in determining the probability of particle breakage. This was the first time this factor was included, and this has since then appeared in later models [12–14]. The tests were done on some specific material, which led to a successful demonstration of how this could be combined with the discrete element method (DEM) energy spectral output to predict milling rates of grinding processes. The model was described by this equation:

$$P_b = 1 - e^{-(0.006m+5)n^{0.525}\left(\frac{E-E_{x0}}{E_{x0}}\right)}$$

(1)

where P_b is the probability of breaking a particle of mass m in grams, and every particle which lost 25% or more by volume to daughter fragments due to breakage was recorded as broken n and E (J) are number of impact attempts and energy input level, respectively. E_{x0} (J) (particle threshold energy) is the minimum energy level required to initiate any form of breakage in a particle. This is derived by plotting energy levels against the probability of breakage; the point where the extrapolated curve crosses the x-axis is assumed to be the E_{x0} value.

To illustrate this method, a graph after Bwalya [15] is shown in Figure 1, and an example of the application of this same technique based on present work is presented in Figure A1. Surprisingly, the E_{x0} value for any particle size for the gold waste-rock was described accurately by the equation:

$$E_{x0} = 0.19m^{0.76}$$

(2)

where m is the mass in grams.

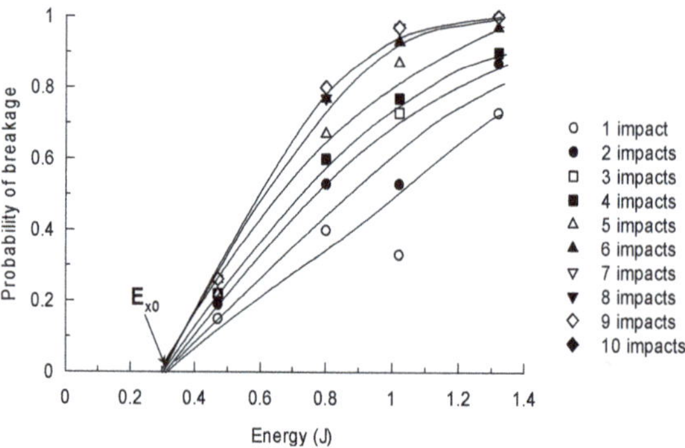

Figure 1. Determination of E_{x0} (minimum energy required for fracture) after [15].

The main feature that emerged from that work is that both input energy level and number of impact attempts significantly affect the probability of breakage. The higher the energy input level, the faster the breakage probability approaches the asymptotical maximum value of 1 with each repeated impact. With a lower energy input level, the breakage probability slope increase is much slower, thus, requires several more impact attempts to reach the value of 1. These characteristics are illustrated in Figure 2, after Bwalya and Moys [16]. While it is impressive that the probability of breakage can be well correlated with drop-weight test data, there is uncertainty with the determination of the energy that is actually used up in breakage.

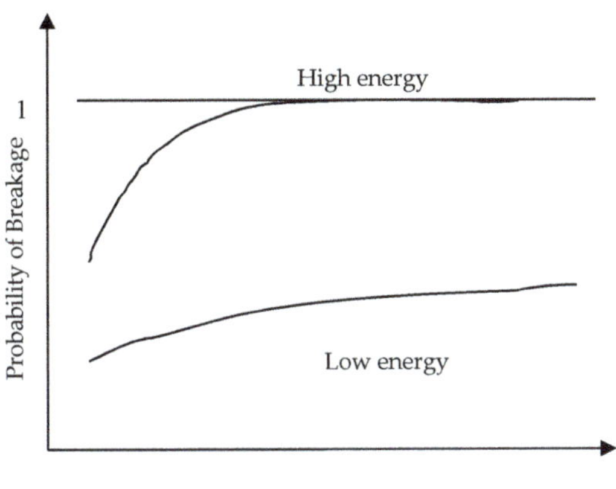

Figure 2. The influence of energy input and breakage attempts required to break a particle.

The ultra-fast load cell (UFLC) that has been developed at The University of Utah [17], has the advantage of capturing the fracture events at high resolution. A force-time profile is recorded during the contact event between impactor and particle, which is used to accurately calculate the energy input to the point of fracture. The force-time profile can also be used to study other events, such as secondary fracture, and for this reason, it has been a leading tool in characterizing single particle breakage with great detail [18,19].

Though the major goal of this research is to develop a scheme for using DEM to predict grinding rate in ball mills, the present work is limited to adequately modelling particle breakage response to varying levels of input energy and repeated stress application in drop weight tests. As observed by Genc et al. [20], the drop weight tester has the advantages of being more flexible and generally yielding consistent results. A good overview of the current status of drop weight testing was given by Shi [21]; Chandramohan et al. [22]; Hosseinzadeh and Ergun [23], whilst quantitative descriptions of breakage probability by repeated impacts were proposed by Bonfils et al. [24]; Tavares and King [25]. Shi's [21] review demonstrates how the adaptation of Equation (1) to the prediction of the breakage function has been successfully applied by the JKMRC to model numerous industrial cases.

The researchers cited above generally recognize the significance of both size related threshold energy, as well as cumulative impact effect. In this work, a modelling approach is developed that allows independent variation of both size related threshold energy and damage deterioration phenomenon, to cater for a wider range of materials. It remains to be seen if this approach will be able to correlate with material structure and flaws distribution.

To model particle breakage response to successive impacts, three key factors are considered:

- Threshold energy
- Rate of deterioration with each impact
- Energy relationship with size

The concept of rate of deterioration is based on the assumption that, for particles of the same material and same size, the resistance to breakage is supposed to be the same, and the only explanation for variation is that some will have larger existing flaws, causing them to break at earliest attempts. For some materials, to extend these flaws only requires a few impact attempts, while for other materials, crack growth with successive impacts will be slower. Plastic deformation may also, to some extent, contribute to the overall process that leads to eventual particle failure.

Figure 2 shows a typical increase in the probability of breakage as a function of impact energy, and Equation (2) shows how particle size relates with threshold energy. An explicit equation that builds on Equations (1) and (2) has been developed, which is written as follows:

$$P_b = 1 - exp\left(-an^b\left(\frac{E - cm^d}{cm^d}\right)\right) \tag{3}$$

where a and b are parameters modelling deterioration with each impact, c and d model how threshold energy varies with particle size. Moreover, m as defined before is the particle mass in grams. An inquiry is thus made in this research to determine if the developed model can successfully establish the relationship between impact energy and the particle breakage and show how the threshold energy varies with particle size. A programme of experimental work on different materials is presented in a subsequent section to test this model.

2. Materials and Methods

This section contains a general procedure that was used for preparing the three different materials (silica sandstone, dolomite and low-grade coal sample) that were tested. A fourth material (gold waste rock) is considered from Bwalya's thesis [15] where a similar procedure was used. The crushed material was reduced to test size, using either rotary sampler or Riffler, and graded to different sizes using screens with a fourth root screen ratio limitation between the upper and lower screen. Each mass group comprised 150 particles, which were then subdivided into three groups of 50 particles. Figure 3 shows the schematic of the preparation steps. Table 1 gives the size classes that were prepared.

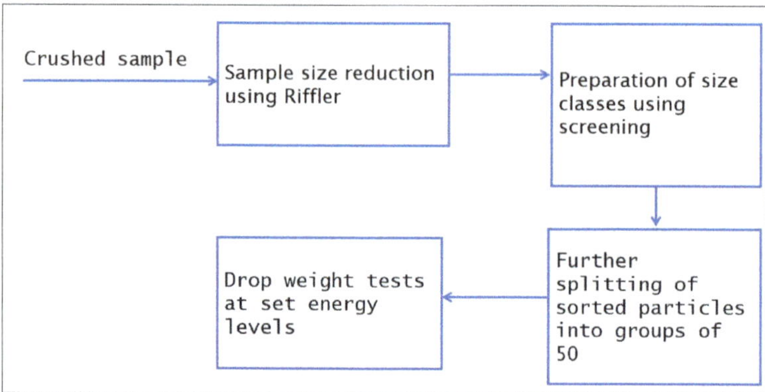

Figure 3. Sample preparation steps.

Table 1. Size classes used of the different material used for drop weight tests.

Material	Size Class in mm	Approximate Density (g/cm^3)
Silica Sandstone	−26.5 + 22.4, −19 +16, −11.2 + 9.5	2.65
Dolomite	−31.5 + 26.5, −26.5 + 22.4, −19 + 16, −11.2 + 9.5	2.84
Low-Grade Coal	−31.5 + 26.5, −26.5 + 22.4, −19 + 16, −11.2 + 9.5	1.7
Gold Waste Rock	−75 +63, −45 + 37.5, −37.5 + 31.5, −19 + 16, −16 + 13.2, −13.2 + 11.2, −11.2 + 9.5	2.75

The particle groups were then subjected to breakage tests, particle by particle, in an in house- built drop weight tester manufactured in the Chemical and Metallurgical Engineering Department at the University of Witwatersrand (Johannesburg, South Africa). This is illustrated in Figure 4. Since apart from density, particle shape is another important criterion that affects average mass of material retained on a screen, Figure 5 is included to show the samples of material used, the size ranges of which are specified in Table 1.

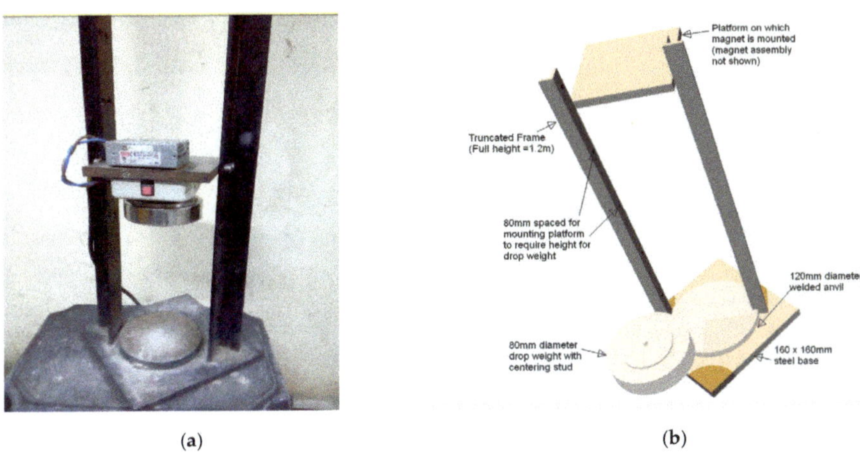

(a) **(b)**

Figure 4. (**a**) Picture of the drop weight tester, (**b**) with illustration to highlight its main features.

Figure 5. Display of the samples of material used in the drop weight tests.

For each particular sub-group, the energy input was varied between 0.63 and 4.4 J by altering the mass of the weight and its drop height. The two steel disc weights used weighed 1.6 kg and 3.07 kg. Energy input was controlled by varying the drop height between 0.1 and 1.2 m, in steps of 0.08 m.

These steps correspond to the position of holes in the vertical steel frame that allow the platform holding the magnet to be positioned flexibly, and tightened by pins.

To each of the particle groups prepared for the test, one breakage attempt was made on each particle, and every particle which lost 25% or more by volume to daughter fragments due to breakage was recorded as broken, as mentioned in the previous section. Those that survived were subjected to another impact and the number of broken particles at the second attempt was recorded once again. The maximum number of repeated attempts was pegged at 5. Upon completion of the test for a particular group, the resulting fragments from all the particles were combined and sieve analysed to determine the size distribution of the fragments, which will be the topic of a future publication by the authors.

The energy level would then be changed by adjusting the drop height or the drop weight and the procedure repeated for the next group of 50 particles. Thus, each mass group had a low, medium and high energy input level from which a breakage probability table was compiled (see Table 2 as an example). Input energy for any particle was calculated using Equation (4):

$$E = Mg(h_1 - h_2) \tag{4}$$

where M is the steel weight mass (kg), g is the gravitational acceleration, and h_1 is the height above the platform where the drop weight is mounted. Moreover, h_2 is the final rest height after breakage, and this is dependent on the residue fragment size distribution.

Table 2. Parameters obtained for the different materials.

Parameters	Silica	Dolomite	Low-Grade Coal	Gold Waste Rock
c	0.03	0.33	0.07	0.19
d	0.99	0.37	0.42	0.72
a	0.12	0.34	0.23	0.41
b	0.83	0.70	0.50	0.52

3. Results

The probability of breakage was varied with energy input level, as well as impact attempts for one particular particle size group. It was observed that with a low energy input level of 0.65 J, only 18% of the particles were broken at first attempt, and with the second attempt, a further 13% was broken, bringing the cumulative total to 31%. It was also observed that at the end of the 5th attempt, only 53% of the particles are broken. These results are represented by the lowest experimental data points in Figure 6.

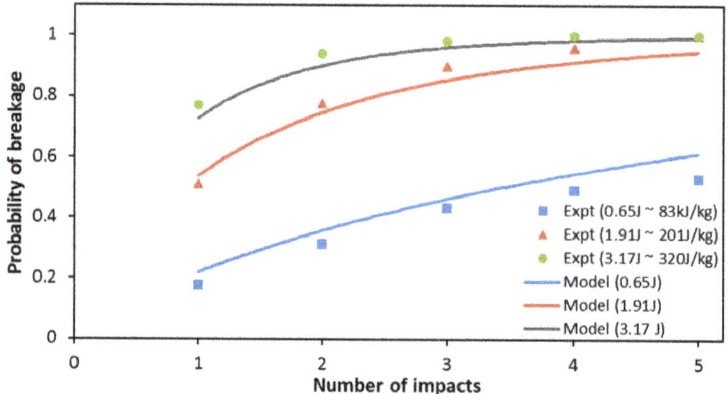

Figure 6. Comparison of experiment breakage probabilities with the model prediction for the −19 + 16.0 mm silica particles.

It was seen that, when higher energy levels are used, the probability of breakage increases accordingly. With the energy level of 3.17 J, all particles were broken within 3 attempts, as can be seen for the highest experimental data points in Figure 6. This trend was also observed in Figure 7 and is typical of all the other size groups, albeit with correspondingly higher energy input levels used for bigger particles.

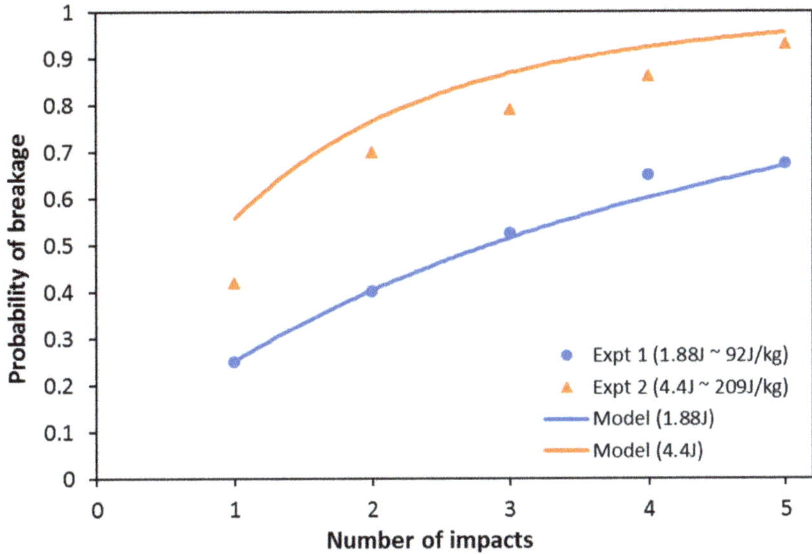

Figure 7. Comparison of experiment breakage probabilities with the model prediction for the −26.5 + 22.4 mm silica particles.

The trends of breakage probability increasing with both magnitude of energy input and number of successive breakage attempts described in the foregoing paragraphs are similar for the different size groups. Similar trends were also observed for the different materials, as can be seen in Figures 8–10 plotted in the next section. The only difference being the magnitudes of energy input responsible for those effects.

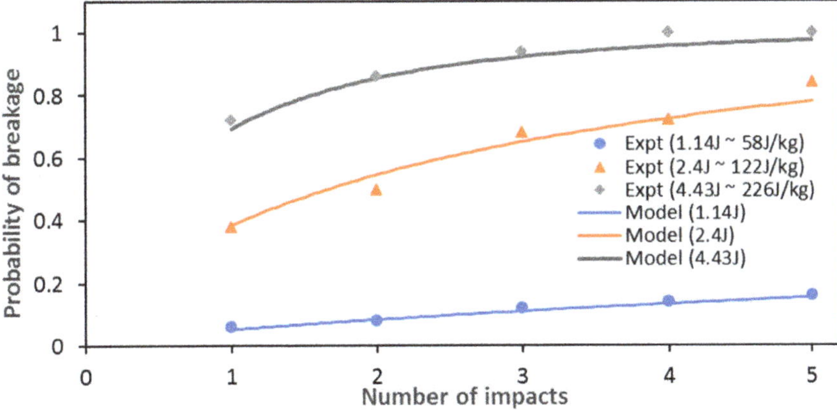

Figure 8. Experimental and Model Breakage probability curves for a −26.5 + 22.4 mm Dolomite sample.

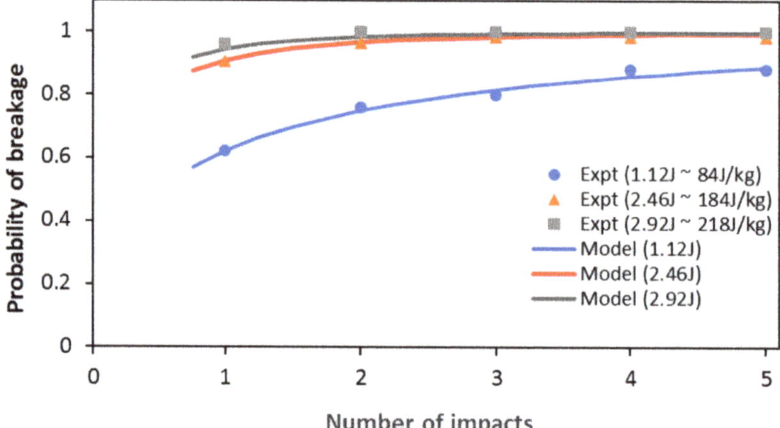

Figure 9. Experimental and Model Breakage probability curves for a −26.5 + 22.4 mm Coal sample.

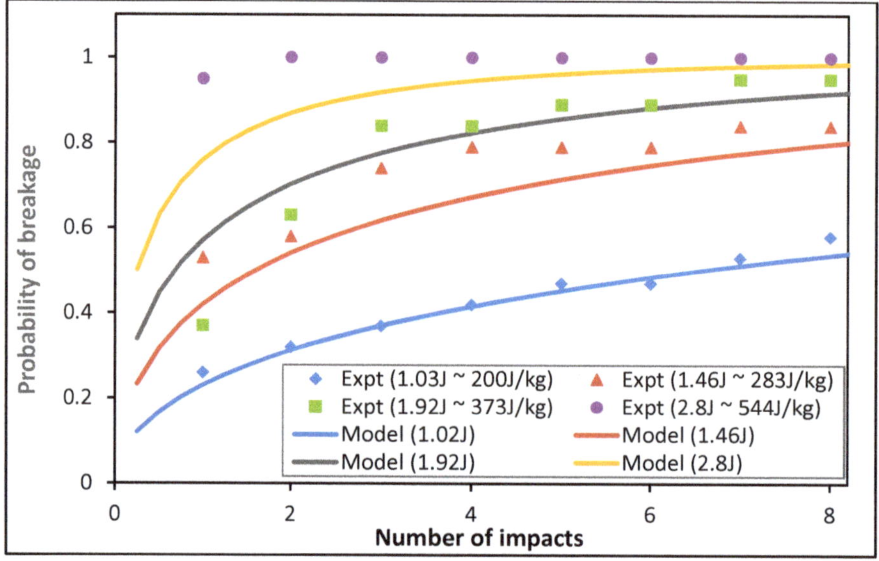

Figure 10. Experimental and Model Breakage probability curves for a −16 mm + 13.2 mm Gold waste rock sample.

3.1. Modelling Results

The model parameters presented in Table 2 were obtained for the four different materials, using the least sum of squares of the differences between model and the experimental values. The Excel Solver add-in was activated to iterate using the general reduced gradient non-linear routine, to find a solution that minimised the differences between the experimental and predicted data. The results for all the materials that were tested are presented in Figures 6–10 and the parameter values, compiled for these different materials, are presented in Table 2.

Figures 6 and 7 show how the model, to a great extent, has successfully described the Silica breakage data for the two sizes shown. It is clearly seen that though the graphs are similar, it is not easy to tell if for similar specific energy input per kilogram; one size class breaks more easily than the other. It will be shown later that for some materials, the specific energy (J/kg) required for breakage is size indifferent, while for most materials, smaller particles require more energy relatively, to achieve similar breakage.

Using appropriate material parameters, it was seen that the model also described these other data very well. The same size range for both the dolomite and coal samples was selected to facilitate comparison among the three materials.

Though the energy inputs do not correspond exactly, it is evident that coal is the weakest material. The gold waste rock sample is definitely the toughest of the lot, as its smaller particles exhibits similar resistance to breakage as other samples under similar impact loading. As can be seen in Figure 10, its specific energy input is higher than the other materials. A more in-depth analysis is carried out in the next section.

3.2. Discussion

3.2.1. Effect of Size on Particle Strength

To appreciate the significance of threshold energy, size normalised E_{x0} was plotted in Figure 11 and it was observed that generally the materials get weaker with increasing particle size. Most researchers attribute this to an increase in flaw density with increase in particle size. Apparently, it is not the case with the silica, which has the same threshold energy for the entire size range.

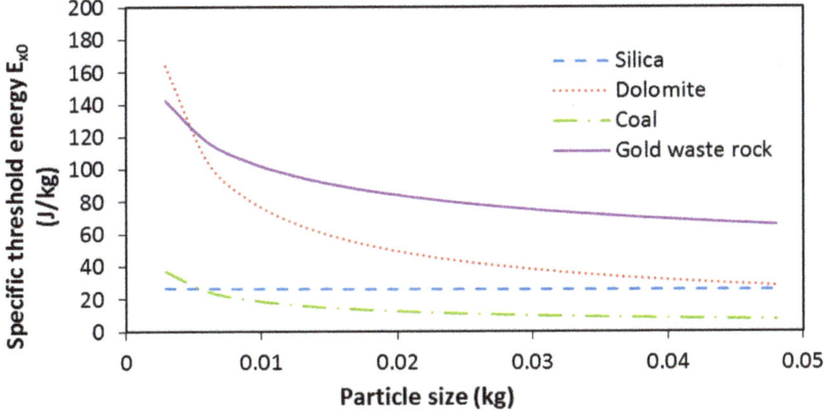

Figure 11. Comparison of specific threshold energy (J/kg).

It was observed for the coal that the increase in relative strength with size was gradual. Extrapolation of the model suggests that this starts to increase exponentially as particles become smaller than 10 g. As for the gold waste rock and the dolomite, the relative increase in toughness with decreasing particle size is at a much higher rate. To show the implication of this phenomenon, specific probability breakage results for both gold waste rock and silica were compared.

In Figure 12, breakage probabilities at similar specific energy input are compared for the gold waste rock for −13.2 + 11.2 mm and −19 + 16 mm particles; it is seen clearly that the breakage probabilities for similar specific energy input are higher for the coarser particles. In other words, the bigger particles are relatively weaker for this material [26]. In Figure 13, a similar comparison is made for silica material for the −26.5 + 22.4 mm and −19 + 16 mm particles, and it is seen that the breakage probability differences for similar specific energy input are not as drastic as those seen for the gold waste rock in Figure 12. Apparently, the smaller particles offered stronger resistance to breakage at low energy input while at higher energy input, they were broken relatively more easily.

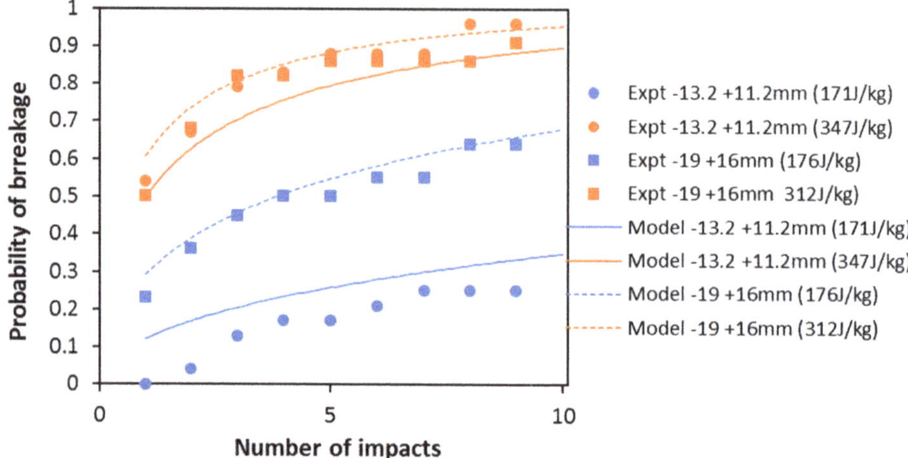

Figure 12. Comparison of gold waste rock breakage probabilities for two particle sizes at specific impact energy input.

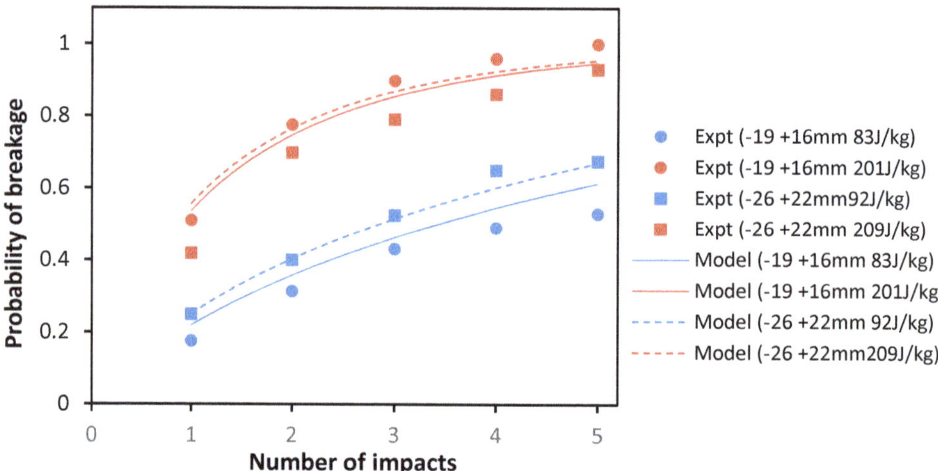

Figure 13. Comparison of silica breakage probabilities at similar specific impact energy input.

3.2.2. Difference in Damage Accumulation in Materials

The rate of deterioration with each impact can also be compared by normalising the part comprising energy terms in Equation (3) to get the following equation:

$$P_b = 1 - e^{(-an^b)} \tag{5}$$

when parameters a and b are applied to this equation for the different materials, the result is what is seen in Figure 14. This reveals some interesting aspects about the materials; with the exception of silica, the other materials appear to become more difficult to break. Here, the silica stands out due to its brittle nature; it is seen that as long as the threshold energy is exceeded, each impact causes significant cumulative damage. However, it was noticed that, for both the coal and gold waste rock, the relative damage is minimal with each impact. This difference may probably be due to non-brittle deformation behaviour or the tendency of some materials to plastically deform, thus requiring further breakage attempts before yielding.

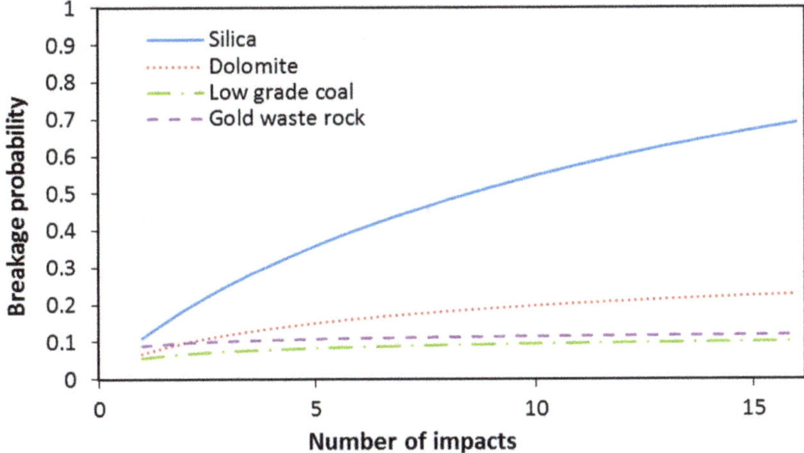

Figure 14. Comparison of the rate of deterioration with each impact for different materials.

It is natural to assume that a high E_{x0} value would usually indicate that a particular material is difficult to break, but as demonstrated above, E_{x0} varies differently for some materials, and one material which is stronger at a smaller size can end up being stronger at a different size. It was also observed that a material with a low E_{x0} can still be difficult to break if it has a low rate of deterioration with successive impacts. As such, it is recommended to test particles contained in various size ranges, to capture the observed size effect.

The model described by Equation (3) was also tested against published data from Morrison et al. [14]; using the Excel solver to estimate model parameters, the model successfully described the data. Thus, there is a good indication that this model could be of wide applicability, and future tests on more different types of material are planned.

In Appendix B is included Figure A2, which presents an algorithm that has been proposed by Bwalya [15] that applies breakage probability data to DEM simulation to predict the comminution rate in size reduction equipment.

4. Conclusions

The fracture response of particles to energy input is dependent on the material types and flaw size and density, which varies with particle size. This has been the basis of most of the probability fracture models that have been developed recently. The new model has demonstrated that it is possible to characterize particle breakage properties from the size related threshold energy point of view. The model has revealed that materials exhibit unique trends in terms of how their threshold energy and rate of deterioration vary with particle size and each impact, respectively. Its ability to predict particle fracture probability has also been successfully demonstrated on four different materials. Among the materials tested, gold waste rock proved to be the toughest, and its relative increase in toughness with decreasing particle size was also higher than the other three materials. The difference in cumulative damage beyond the threshold energy may be attributed to non-brittle deformation behaviour or plastic deformation of particles. The purely brittle material will require a few impacts to disintegrate, while those with plastic deformation tendencies will endure several more impacts before complete failure. As for the differences in threshold energies, the existence of flaws which are relatively bigger and more frequent in larger particles was considered to be a key factor. The damage accumulation as a result of repeated impacts is likely a function of properties such as the shape, composition, flaw distribution, and the presence of mineral grains, whose effects will be the object of future research by the authors.

Author Contributions: Conceptualization, M.M.B. and N.C.; methodology, M.M.B. and N.C.; software, M.M.B.; validation, M.M.B. and N.C.; formal analysis, M.M.B. and N.C.; investigation, M.M.B. and N.C.; resources, M.M.B. and N.C.; data curation, M.M.B. and N.C.; writing—original draft preparation, M.M.B. and N.C.; writing—review and editing, M.M.B. and N.C.; visualization, M.M.B. and N.C; supervision, M.M.B. and N.C.; project administration, M.M.B. and N.C.; funding acquisition, Y.Y. All authors have read and agreed to the published version of the manuscript.

Funding: This research received no external funding.

Acknowledgments: MINTEK is greatly acknowledged for allowing the authors to use the drop weight tester and their laboratory facility. The authors particularly mention Portai Mudau for ensuring that all their needs were met during the time of experiments.

Conflicts of Interest: The authors declare no conflict of interest.

Nomenclature

a	dimensionless material parameter that is related to material resistance to damage
b	dimensionless material parameter that is related to material brittleness
c	parameter that models particle size effect on breakage
d	second parameter that models particle size effect on breakage
E	energy (J)
E_{x0}	minimum energy (J) required to initiate any form of particle damage
g	gravitational acceleration (m/s^2)
h	height (m) applies to all subscripts
M	mass (kg)
n	number of breakage attempts made on a particle
P_b	probability of breaking a particle (in our case by at least 20%)
X	particle mass (g)
X_m	particle mass (kg)

Appendix A

Determination of E_{x0} applied to the dolomite sample.

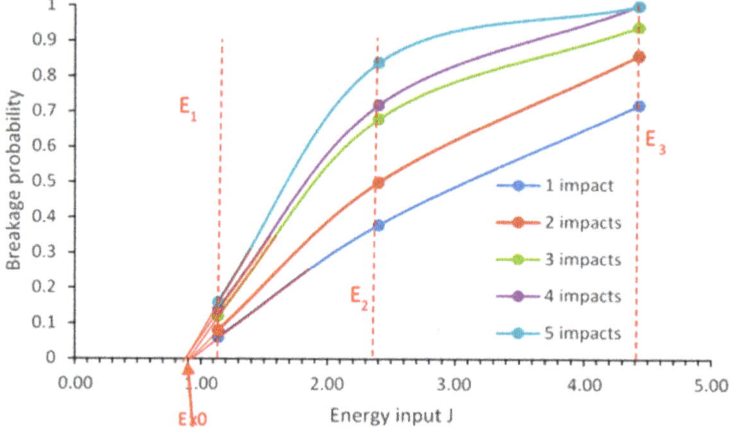

Figure A1. Determination of E_{x0} applied to the dolomite sample.

It is important to note that the lowest energy level (E_1) must be as low as possible, to make the determination of E_{x0} more accurate. This graph differs from the probability graphs that are in the main report, as this highlights the importance of magnitude of impact energy. At Energy level 1 (E_1), even after five impacts the probability of breakage is less than 0.2. This suggests that one may probably need hundred attempts to get breakage. At E_2 and E_1 each impact contributes significantly to raising the breakage probability, and only a few impacts are required to reach the probability of 1. One can extrapolate to the right to find the lowest energy that ensures 100% breakage probability with one impact. However, the main purpose of this graph is to establish E_{x0} and from the extrapolations of the lines convergence at point on the X axis, giving a value of E_{x0} to be 0.9 J.

Appendix B

Suggestion of how drop weight test can be combined with DEM to predict comminution.

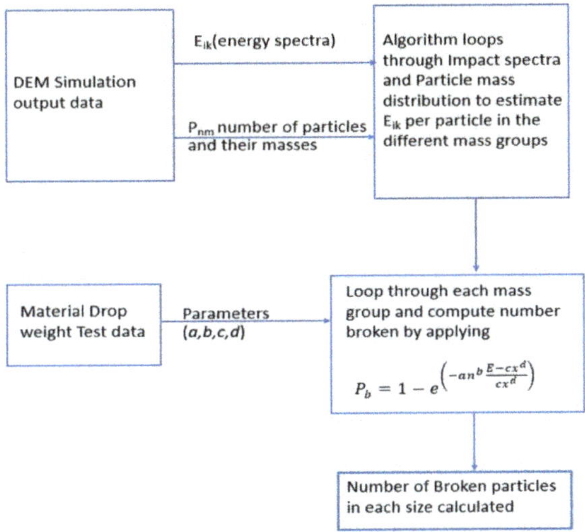

$$P_b = 1 - e^{\left(-an^b \frac{E - cx^d}{cx^d}\right)}$$

Figure A2. Algorithm that applies drop weight test and DEM simulation data to calculate milling rate.

References

1. Delboni, H.; Morell, S. A Load-Interactive Model for predicting the performance of autogenous and semi-autogenous mills. *Kona Powder Technol.* **2002**, *20*, 208–220. [CrossRef]
2. Von Rittinger, P.R. *Lehrbuch der Aufbereitungs Kunde*; Ernst and Korn: Berlin, Germany, 1867.
3. Kick, F. *Des Gesetz der Proportionalem Widerstnad und Seine Anwendung*; Felix: Leipzig, Germany, 1885.
4. Bond, F.C. The third theory of comminution. *Trans. AIME* **1952**, *193*, 484–494.
5. Griffith, A. The phenomena of rupture and flow in solids. *Phil. Trans. R. Soc.* **1921**, *221*, 163.
6. Rumpf, H. Physical aspect of comminution and a new formulation of a law of comminution. *Powder Technol.* **1973**, *7*, 145–159. [CrossRef]
7. Narayanan, S.S. Single particle breakage tests: A review of principles and applications to comminution modelling. *Bull Proc. Austr. Inst. Min. Metall.* **1986**, *291*, 49–58.
8. Leung, K. An Energy Based Ore Specific Model for Autogeneous and Semiautogenous Grinding. Ph.D. Thesis, University of Queensland, St Lucia, Australia, 1987.
9. Shi, F.; Kojovic, T.; Larbi-Bram, S.; Manlapig, E. Development of a rapid particle breakage characterisation device—The JKRBT. *Miner. Eng.* **2009**, *22*, 602–612. [CrossRef]
10. Bwalya, M.M.; Moys, M.H.; Hinde, A. The use of the DEM and fracture mechanics to improve grinding rate prediction. *Miner. Eng.* **2001**, *14*, 565–573. [CrossRef]
11. Weichert, R. Fracture physics in comminution. In Proceedings of the 7th European Symposium Comminution, Ljubljana, Slovenia, 12–14 June 1990; pp. 3–20.
12. Peukert, W.; Vogel, L. Breakage behaviour of different materials—Construction of a master curve for the breakage probability. *Powder Technol.* **2003**, *129*, 101–110.
13. Vogel, L.; Peukert, W. Determination of material properties relevant to grinding by practicable labscale milling tests. *Int. J. Miner. Process.* **2004**, *74S*, S329–S338. [CrossRef]

14. Morrison, R.D.; Shi, F.N.; Whyte, R. Modelling of incremental rock breakage by impact: For use in DEM models. *Miner. Eng.* **2007**, *20*, 303–309. [CrossRef]

15. Bwalya, M.M. Using the Discrete Element Method to Guide the Modelling of Semi and Fully Autogenous Milling. Ph.D. Thesis, University of Witwatersrand, Johannesburg, South Africa, 2005.

16. Bwalya, M.M.; Moys, M.H. The Use of DEM in Predicting Grinding Rate. In Proceedings of the XXII International Mineral Processing Congress, Cape Town, South Africa, 29 September–3 October 2003; pp. 1612–1617.

17. King, R.P.; Bourgeois, F. Measurement of fracture energy during single-particle fracture. *Miner. Eng.* **1993**, *6*, 353–367. [CrossRef]

18. Tavares, L.M.; King, R.P. Modeling of particle fracture by repeated impacts using continuum damage mechanics. *Powder Technol.* **2002**, *123*, 138–146. [CrossRef]

19. Tavares, L.M.; Carvalho, R. Modelling breakage rates of coarse materials in ball mills. *Miner. Eng.* **2009**, *22*, 650–659. [CrossRef]

20. Genc, O.; Ergun, L.; Benzer, H. Single particle impact breakage characterization of materials by drop-weight testing. *Physicochem. Probl. Miner. Process.* **2004**, *38*, 241–255.

21. Shi, F. A review of the applications of the JK size-dependent breakage model Part 1: Ore and coal breakage characterization. *Int. J. Miner. Process.* **2016**, *155*, 118–129. [CrossRef]

22. Chandramohan, R.; Lane, G.S.; Foggiatto, B.; Bueno, M.P. Reliability of Some Ore Characterization Tests—Ausenco. In Proceedings of the SAG mill conference, Vancouver, BC, Canada, 20–24 September 2015.

23. Hosseinzadeh, H.; Ergun, L. Determination of Breakage Distribution Function of fine chromite ores with Bed Breakage Method. In Proceedings of the XIII International Mineral Processing Symposium, Bodrum, Turkey, 10–12 October 2012.

24. Bonfils, B.; Ballantyne, G.R.; Powell, M.S. Developments in incremental rock breakage testing methodologies and modeling. *Int. J. Miner. Process.* **2016**, *152*, 16–25. [CrossRef]

25. Tavares, L.M.; King, R.P. Single-particle fracture during impact loading. *Int. J. Miner. Process.* **1998**, *54*, 1–28. [CrossRef]

26. Baumgardt, S.; Buss, B.; May, P.; Schubert, H. *On the Comparison of Results in Single Grain Crushing under Different Kinds of Load*; XI. IMPC: Cagliari, Italy, 1975.

Article

Confined Bed Breakage of Fine Iron Ore Concentrates

Túlio M. Campos [1], Gilvandro Bueno [1,2] and Luís Marcelo Tavares [1,*]

[1] Department of Metallurgical and Materials Engineering, Universidade Federal do Rio de Janeiro–COPPE/UFRJ, Cx. Postal 68505, Rio de Janeiro CEP 21941-972, RJ, Brazil; tulio_uca2013@poli.ufrj.br (T.M.C.); gilvandro.bueno@vale.com (G.B.)

[2] Vale S.A., Complexo de Tubarão, Vitória CEP 29090-911, ES, Brazil

* Correspondence: tavares@metalmat.ufrj.br; Tel.: +55-2290-1544

Received: 26 June 2020; Accepted: 24 July 2020; Published: 27 July 2020

Abstract: High-pressure grinding rolls (HPGR) have gained great popularity in the mining industry in the last 25 years or so. One of the first successful applications of the technology has been in iron ore pressing prior to pelletization. Piston-and-die tests can provide good insights on the material response in an HPGR. This work analyzed confined bed breakage of four iron ore concentrates under different conditions. Saturation in breakage of particles contained in the top size in the tests was observed to occur at specific energies of about 2 kWh/t, whereas full saturation in breakage, with no additional increase in specific surface area of the material, occurred at energies above about 6 kWh/t. An expression was proposed to characterize the propensity of a material to break under confined bed conditions. The phenomenology involved in confined bed breakage of such materials was then analyzed in light of the results.

Keywords: bed breakage; iron ore; comminution; saturation; piston-and-die; compaction; compression; breakage

---

1. Introduction

Comminution machines that rely on confined bed breakage, namely high-pressure grinding rolls (HPGR) and the vertical roller mill (VRM), have gained great popularity in the cement industry [1–7] and, in the case of the HPGR, also in the minerals industry [8–13]. Their main advantages include good energy efficiency, high throughput, and ability to operate in size reduction operations from secondary crushing down to fine grinding. One particularly important characteristic of these technologies is that, as the applied specific energies are raised, the efficiency of energy usage drops, partially owing to the phenomenon of breakage saturation.

One important approach to gain insights into the operation of comminution machines that rely on confined breakage is to conduct tests using the piston-and-die apparatus [14,15]. Through these tests, it is possible to analyze how the applied energy is dissipated under controlled conditions and is used to create new surfaces. Schubert [16] highlighted that the total energy input in a particle bed can be divided into several microprocesses, namely friction losses, plastic deformation work, elastic recovery, energy dissipated during breakage, and compaction [16]. As such, in order to assess the particle breakage behavior resulting from the application of different applied vertical stresses, several authors [14,15,17–24] used the force–displacement curve and the size reduction results from piston-and-die tests for different materials. Kalala et al. [15] presented results indicating an elastic recovery of around 40% of the total input energy for pressing iron ores in a piston-and-die apparatus with compressive forces of 1700 kN and with particle sizes of about 12 mm.

Great attention has been dedicated to the study of the compaction behavior of particulate materials under confined conditions. Such compaction occurs during processes that include metal powder smelting, as well as tableting and pelletizing in the pharmaceutical, ceramic, and food industries.

Indeed, compaction has been studied and modelled for inorganic powders [25], metal powders [26], ceramic powders [27], fine grinding processes [17,28,29], and organic powders in the manufacture of pharmaceutical tablets [30–33]. The particle bed compaction behavior can be influenced by different parameters including the deformation or stress–strain behavior [34,35] as well as particle size and shape [33,36]. Assessing limestone breakage response under confined conditions, Cabiscol et al. [29] observed that the packing density reached a maximum of around 90% of the specific gravity of the material with particles with mean initial size of 236 μm, being equivalent to the value found by Wünsch et al. [28] for different materials used in the production of pharmaceutics. On the other hand, recent works have attempted to assess particle bed behavior under confined conditions from computational simulations using the discrete element method (DEM) [37–39]. Besides DEM simulations, Garner et al. [37] also showed several experimental results highlighting the ability to reach relative densities around 0.95 for a vertical stress of 150 MPa. Indeed, the complexity of this process is the main reason for the poor predictability and the limited understanding yet available.

One particularly successful application of confined bed breakage is the HPGR in association with ball milling in either pregrinding or regrinding of iron ore concentrates for pellet feed production [8,13]. The present work investigated breakage of iron ore concentrates in confined conditions in single and multiple pressings in a piston-and-die apparatus with the aim of identifying the onset of saturation during confined breakage.

2. Experimental Section

2.1. Materials

Samples were collected of four iron ore concentrates that are fed into Vale's pelletizing plants at Complexo de Tubarão (Vitória, Brazil). Three of these concentrates (Itabira, Brucutu, and Timbopeba) are produced by flotation of ores from the Iron Quadrangle from the state of Minas Gerais (Brazil), having a top particle size of 1 mm. The fourth consists of the result of the preparation of ore from the Carajás mineral province (Pará, Brazil).

Specific gravity was measured by Helium pycnometry. Bulk (apparent) density was measured by placing each weighed sample in a beaker and measuring the apparent volume after vibration. Prior to testing, all samples were dried in air.

2.2. Piston-and-Die Tests

Particles contained in three size ranges were tested, namely 150–125 μm, 106–75 μm, and 53–45 μm, which were prepared by careful wet sieving. Each previously dried sample, containing 30 g, was placed in the die for testing. The piston had a 40 mm diameter and the resulting initial bed height was about 13 mm. As such, these tests were conducted under ideal confined particle bed conditions, as precluded by Schönert [14,40], that is, the height of the bed is larger than six times the top size of the samples and the bed diameter is larger than three times the bed height, so that wall effects were minimized.

Figure 1 illustrates the servo-hydraulic press (Shimadzu, Kyoto, Japan) used, showing the linear variable differential transformer (LVDT) employed to measure the particle bed displacement (a) and the piston-and-die apparatus in detail (b). The elastic deformation of the system was subtracted from the value measured using the LVDT.

The compressive forces were applied in the range of 50–1000 kN (40–800 MPa), and the deformation rate was 5 mm/min. The output of the test was the force–displacement curve. The specific input energy in each test was calculated from the numerical integration up to the maximum load. Upon completion of the test, the bed of particles was removed from the die, dispersed in water, and the size distribution measured. Size analyses were conducted directly of the material from the test by laser scattering in a Malvern Mastersizer 2000 (Malvern Instruments Inc., Malvern, UK) while the Blaine specific surface area (BSA) was measured using a PCBlaine-Star (Zünderwerke Ernst Brün GmbH, Marl, Germany).

In addition to that, each pressed sample was also wet sieved using the bottom size of the interval in order to assess the proportion of material broken.

An additional set of tests consisted of conducting multiple pressings of material contained in the 106–75 µm size range. The initial bed was pressed up to 200 kN (159 MPa), the load removed, the material dispersed and analyzed, then it was reloaded to the die and the same procedure repeated. This specific energy was selected since it is in the range of values used in the pelletizing industry [12].

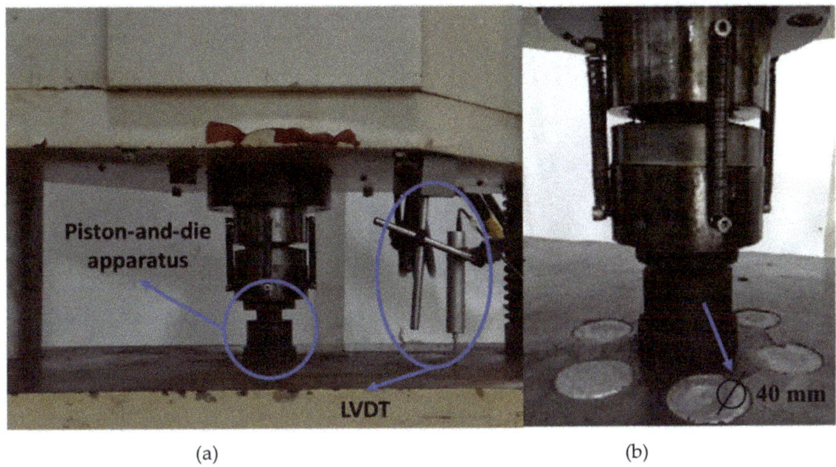

Figure 1. Press used in the piston-and-die tests highlighting the linear variable differential transformer (LVDT) used to measure the (**a**) particle bed displacement and (**b**) piston and die apparatus.

3. Results

3.1. Material Characteristics

A summary of the characteristics of the samples is presented in Table 1. The Carajás sample has the lowest values of density, whereas Brucutu and Itabira have the highest. Details about the composition of the samples may be found elsewhere [41].

Table 1. Summary of physical characteristics of the samples.

Sample	Specific Gravity (g/cm^3)	Apparent Density (g/cm^3)
Itabira	5.07	3.01
Brucutu	5.03	3.02
Timbopeba	4.80	2.85
Carajás	4.55	2.70

3.2. Force–Displacement Profiles

In order to analyze bed response in compressive bed breakage in greater detail, force–deformation profiles are analyzed as follows. Figure 2 presents force–displacement curves for different compressive forces, highlighting the load and unload (relief) curves. It is evident that, when the bed is compressed up to relatively small loads, namely below 100 kN (80 MPa), no elastic recovery appears, with no recoil of the bed. Mütze [17] observed that, in this case, all energy applied to the bed is either dissipated in rearranging the particles or in producing particle breakage. Under such conditions, the thickness of the bed after unloading progressively reduces as loads increase. Beyond this point, as loads increase, the force–displacement curves become steeper. In this case, unloading exhibits progressively more elastic response, with the bed presenting ever more elastic recovery, which becomes evident from the lower slopes of the unloading curves.

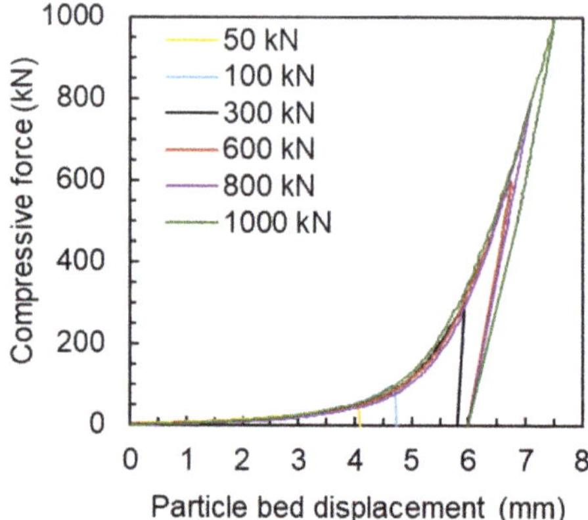

Figure 2. Load and unload force-displacement curves in tests up to different applied compressive forces for the Itabira sample contained in the size range of 150–125 μm.

Force-deformation curves such as those in Figure 2 may be presented more suitably on the basis of the relationship between the packing density (ratio between apparent density of the bed and specific gravity) and the vertical stress applied. Results are presented in Figure 3 for the Itabira and Carajás samples contained in the narrow particle size range of 150–125 μm. For a vertical stress of 800 MPa (1000 kN), the Itabira sample presented a maximum packing density of around 0.88, whereas a maximum of 0.95 was reached for the Carajás sample. The reasonable difference in the curves indicates a softer response for the Carajás sample under compressive loads. In addition, both results are able to show that, even though different vertical stresses were applied, the final packing density after total relief was nearly constant for each material, being equal to around 0.70 for the Itabira and 0.75 for the Carajás samples.

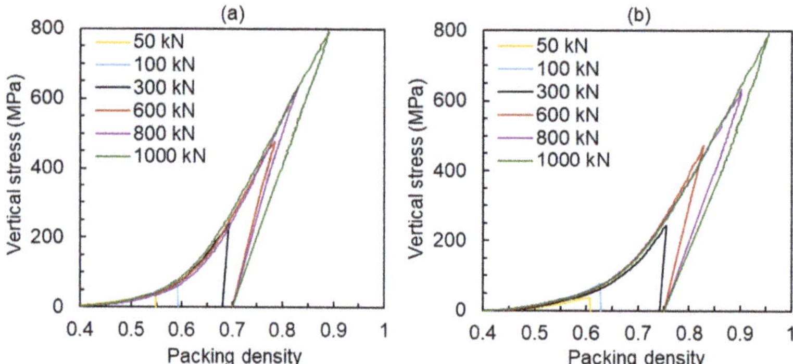

Figure 3. Relationship between the vertical stress and the packing density for different compressive forces applied for the (a) Itabira and (b) Carajás samples in the narrow size range of 150–125 μm.

A more detailed examination of the results is possible by calculating the areas below the curves. The input energy is given by numerical integration of the curves up to the maximum vertical stress, whereas the elastic energy is given by the area corresponding to the unloading of the piston. The inelastic or dissipated energy is simply given by the difference between the two. Results are presented in Figure 4, which shows the rapid increase in elastic energy with the increment in vertical stress. The results also show that for low compressive forces the elastic energy is almost negligible, whereas it corresponds to around 60% of the input energy for the highest compressive forces analyzed.

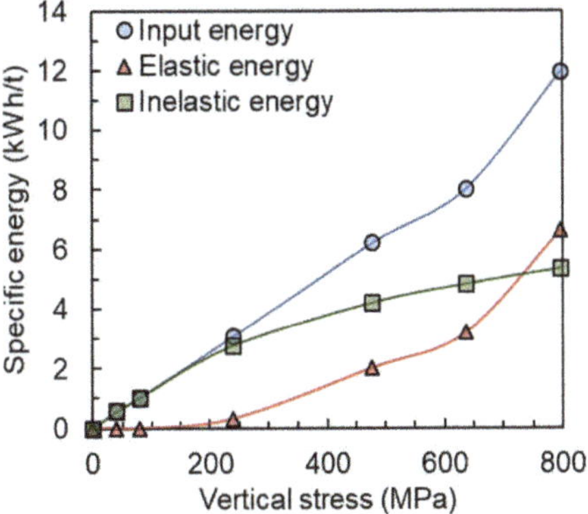

Figure 4. Variation of the input, elastic, and dissipated energy in the particle bed as a function of vertical stress for the Itabira sample for the narrow size range of 150–125 µm.

In order to assess the particle bed behavior under multiple pressing cycles, Figure 5 shows the relationship between the packing density with the vertical stress applied in seven repeated pressing stages. For a vertical stress of up to 160 MPa (200 kN), the different pressing stages showed a great distinction in the initial bed configuration with the packing density reaching a maximum value of around 0.7 for the last stage. This result is consistent with the maximum packing density found after the single-stage pressing process presented in Figure 3a for the Itabira sample. Indeed, there is a marked relationship between the initial feed size distribution and the progressive change in packing density of the material. The results from Figure 5 indicate that, for the fine feed size distributions used, there is an increment in packing density caused by the reduction in the voids fraction within the particle bed. As the multiple stages of pressing were applied, the force–displacement profile started to superimpose, with this effect being potentially associated with the high particle bed packing. The dispersion of the material following each pressing cycle, coupled with the application of a relatively low vertical stress in each cycle, allowed to prevent particle bed saturation.

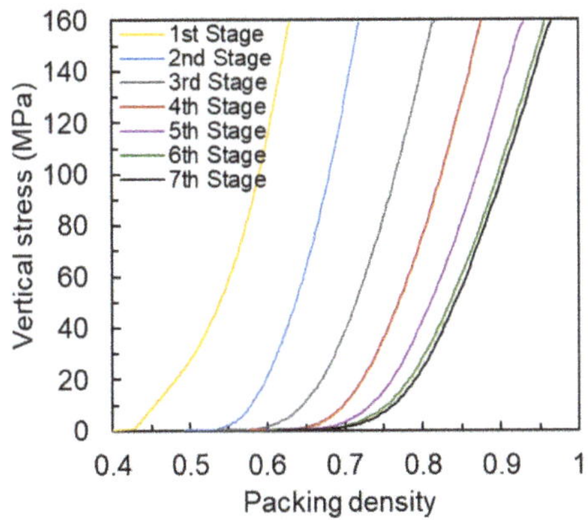

Figure 5. Relationship between vertical stress and packing density for different stages of pressing for the Itabira sample in a narrow size range of 106–75 µm.

3.3. Size Analyses

Figure 6 presents the product size distributions for the Brucutu and Carajás samples at different stressing conditions. They demonstrate the increase in fineness as compressive forces increase, as well as the onset of the breakage saturation that is associated with the application of compressive forces higher than about 300 kN (240 MPa). The higher propensity of the Carajás sample to breakage in the piston-and-die test is evident from the larger proportion of fines produced.

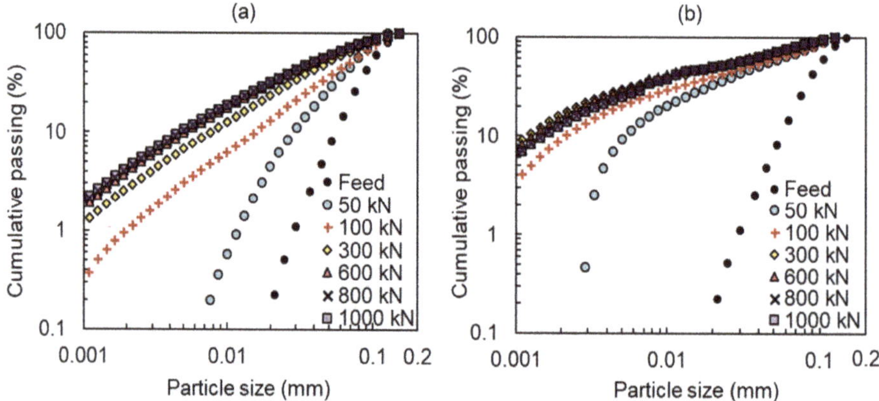

Figure 6. Product size distributions for different maximum compressive forces for the (**a**) Brucutu and (**b**) Carajás samples contained in the size range of 150–125 µm in piston-and-die tests.

3.4. Breakage of Top Size Particles

After each piston-and-die test, the proportion passing the original narrow size range was recorded by sieving. The proportion broken was then plotted as a function of specific energy in Figure 7, following the approach used by Liu and Schönert [18] and Dundar et al. [21].

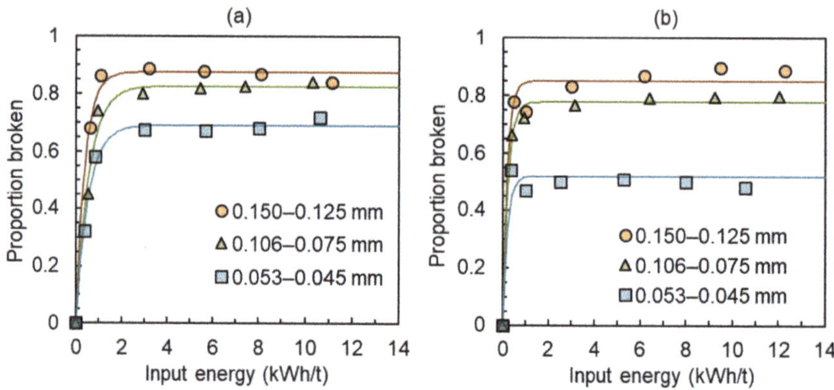

Figure 7. Proportion broken out of the original size for (**a**) Brucutu and (**b**) Carajás samples for different specific energies and feed particle sizes.

The figure shows that the proportion of particles broken increases significantly at low specific input energies, but reaches a maximum value, which becomes nearly constant with increasing input energies. Beyond this point, increasing the input energy does not lead to more breakage of particles contained in the original size range, as they become stabilized by neighboring particles. Tavares [42] observed that, whereas little or no additional breakage occurred at these higher energy inputs in HPGR experiments, particles may become progressively weaker. Furthermore, as observed by Liu and Schönert [18], it is evident that this maximum proportion broken varies with size, reducing significantly for the finer size range studied. Indeed, in the finest size range studied (53–45 µm), less than half of the particles broke for the Carajás sample (Figure 7b), in spite of the energy applied, showing the significant size effect on particle stabilization in confined bed breakage. The figure also shows that, at specific energies in the order of 2 kWh/t, saturation is reached on the maximum proportion of particles broken. This specific energy corresponds to a maximum load in the order of 200 kN (159 MPa), coinciding with the conditions in which the bed starts to recoil partially elastically (Figure 3) and also the condition under which the multiple stage pressings (Figure 5) were carried out.

Figure 8 compares results on the proportion broken for the Itabira sample as a function of input energy in individual pressings at progressively higher vertical stresses and results from multistage pressings. It shows that the sequential pressing and dispersion of the material prior to another pressing stage allowed the additional breakage of particles contained in the top size fraction, preventing saturation.

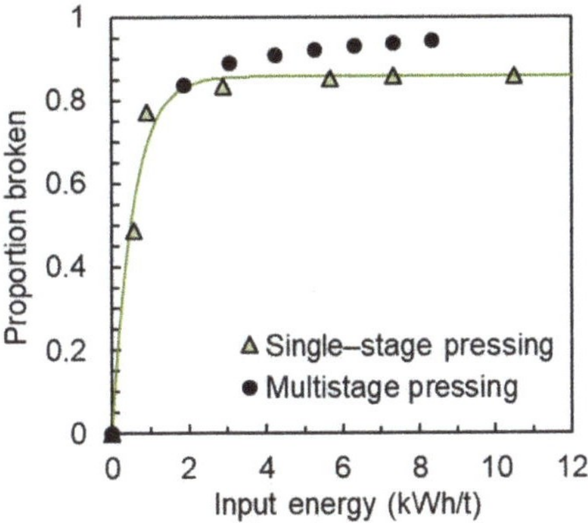

Figure 8. Proportion broken out of the original size for the Itabira sample in the narrow particle size range of 106–75 µm for single-stage pressing and multistage pressing. Line fits data from single-stage pressing.

3.5. Blaine Specific Surface Area

The Blaine specific surface area (BSA), an important parameter used to characterize the fineness of powdered materials, is commonly used to control pellet feed quality in iron ore pelletizing plants [43]. Figure 9 shows the relationship between the BSA increase from the feed BSA and the specific energy applied for all samples in piston-and-die tests. It is evident that the BSA increase is nearly proportional to the input specific energy up to a point, beyond which the slope of the line reduces. As such, in analogy to Figure 7, the data approach a maximum value, but in this case only at specific energy inputs above about 6 kWh/t. Such a value corresponds to maximum vertical loads in the order of 600 kN (480 MPa).

As already reported by Schönert [14] and recently observed by Zhou et al. [38] using DEM simulations, the main cause for the drop in energy efficiency in particle bed breakage is the reduction in the voids that are caused by fine debris relocating themselves as a result of the application of high normal applied stresses. Such a drop in energy efficiency in compressed bed breakage is also evident in Figure 9 for specific energies higher than about 6 kWh/t.

With the aim of analyzing in greater detail the issue of saturation in compressed bed breakage, a comparison between single and multiple pressing results for the Itabira sample is presented in Figure 9b. It compares the results from piston-and-die tests carried out at different maximum pressures to results from multiple compressive cycles at a constant maximum loading force of 200 kN (159 MPa). Dispersion of the material after each pressing cycle allowed reaching significantly higher BSA values than those obtained in a single pass at higher pressures, since the dispersion prevented particles from stabilizing in the bed. This effect becomes noticeable at specific energies above about 3 kWh/t. These results demonstrate the value, for iron ore concentrates, of using multiple passes as a way of increasing the specific surface area of the product, in contrast to a single pass.

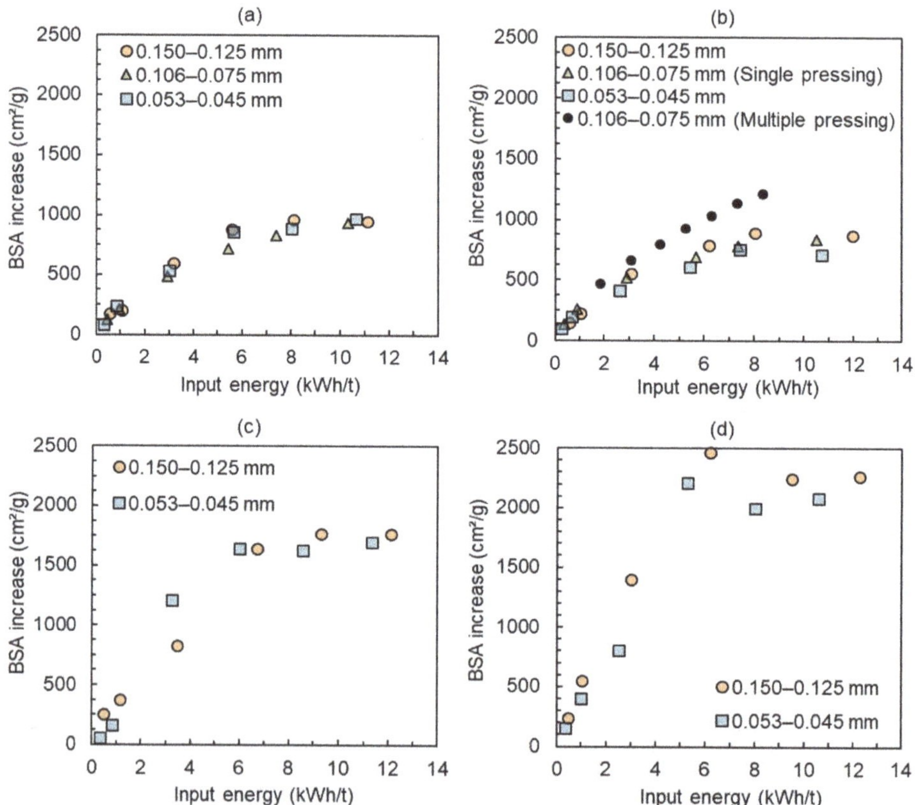

Figure 9. Relationship between Blaine specific surface area (BSA) increase and specific input energy for (**a**) Brucutu, (**b**) Itabira, (**c**) Timbopeba, and (**d**) Carajás samples.

Given the various effects observed, including that of the mean particle size, the size distribution of the original material, the degree of its dispersion, and the observed trends towards saturation at higher pressures, a parameter, called pressing factor κ, is proposed to capture the bed propensity to pressing in a test. It is given by the following:

$$\kappa = \left(\phi_p - \phi_o\right)(1 - \phi_o) \tag{1}$$

where ϕ_o is the packing density of the originally pressed material and ϕ_p is the final packing density of the bed after unloading the piston, that is, of the permanently deformed bed. ϕ_o varies as a function of particle size and size distribution, degree of dispersion, if previously in loose form or preloaded. In the experiments in the present work, it was estimated from the value corresponding to a preload of 20 MPa, in order to remove any bias due to uneven bed surface in the beginning of the test (Figure 3).

The higher the value of κ, the greater the expected size reduction for a given material. As such, it increases with the increase in pressure and stressing energy applied to the bed, the maximum achievable packing density, whereas it decreases with an increase in initial bed packing density.

As such, it acknowledges that pressing results change if the bed was previously loose or preloaded in the beginning of a test.

Figure 10 presents the results from Figure 9, now as a function of the parameter κ. It shows that the increase in Blaine specific surface area varies as a function of parameter κ, relatively independently of initial particle size and final pressure. It demonstrates that it is possible to represent data not only from single but also from multiple pressings (Figure 10b) in the same curve. As such, it suggests that κ is a parameter that may be used to characterize the potential of a material to undergo breakage under confined conditions.

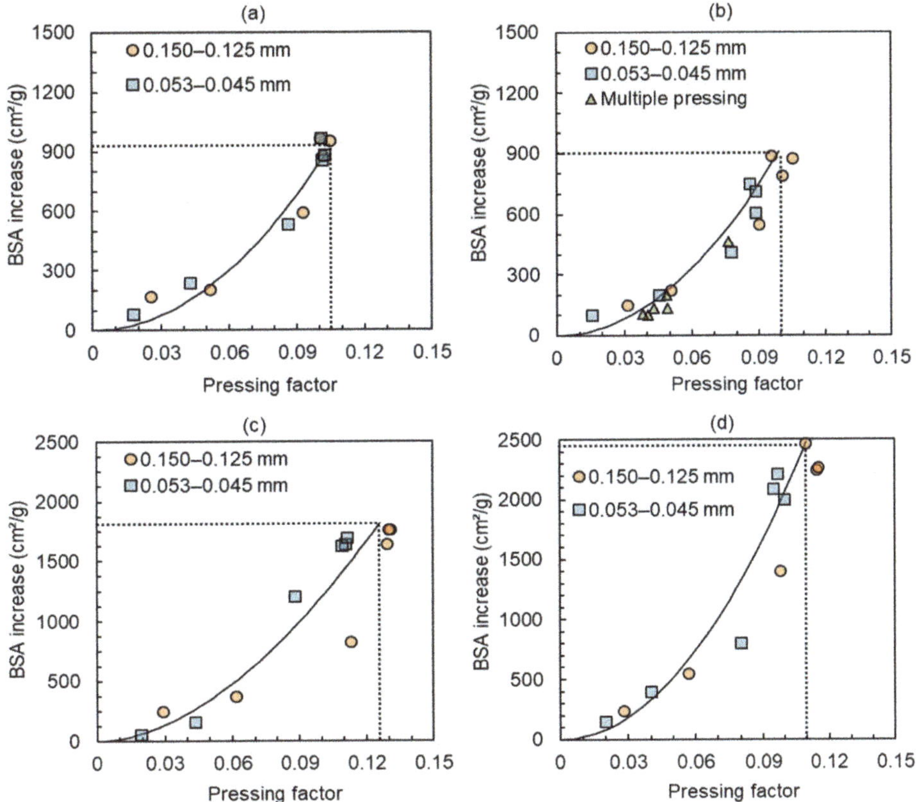

Figure 10. Relationship between BSA increase and pressing factor κ for the (a) Brucutu, (b) Itabira, (c) Timbopeba, and (d) Carajás samples.

Since data included experiments with narrow size beds that presented the lowest initial packing density ϕ_0 and that were subjected to pressures beyond those responsible for breakage saturation (Figure 9), a maximum achievable value of Blaine specific surface area increase in a single loading stage could also be identified, as shown in Figure 10. Such a value corresponded to the maximum achievable magnitude of pressing factor. It varied according to material, being about 0.10–0.11 for Brucutu and Itabira and about 0.11–0.13 for Timbopeba and Carajás, identifying the greater pressing propensity of the latter in comparison to the former (Figure 10).

Figure 11 then shows the variation of the pressing factor as a function of the input energy in pressing for one of the samples. It demonstrates that the data, including those from multiple pressings, follow approximately the same general trend, with a maximum pressing factor reached with specific energies above about 6 kWh/t, varying only marginally with initial particle size. Alternatively, the relationship between the pressing factor κ and the input energy could be estimated by a model describing the force–displacement profile for both loading and unloading of the bed [17], which would make the method suitable for predicting results in the piston-and-die apparatus.

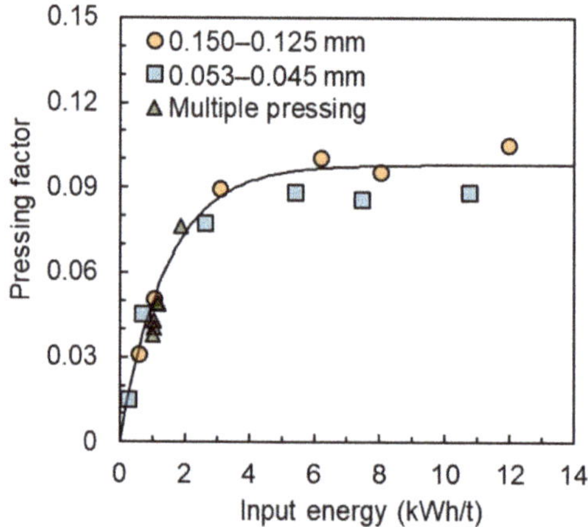

Figure 11. Relationship between the factor κ and specific input energy for the Itabira sample, including data for single and multiple pressings (106–75 μm).

3.6. Energy Utilization

Assuming the validity of Rittinger's law, Rumpf [44] proposed a definition of energy utilization as the ratio between the increment in surface area from the feed and the energy spent in comminution. This definition has been used by Campos et al. [12] with a minor modification presented in Equation (2), from the ratio between the BSA increase and the specific energy spent on the process:

$$\text{Energy utilization} = \frac{BSA_{product} - BSA_{feed}}{\text{Specific energy consumption}} \tag{2}$$

For iron ore pelletizing operations, this relationship is commonly used as a metric for characterizing the comminution process efficiency [12,41].

In order to analyze the energy utilization in greater detail, Equation (2) is used considering both the total specific energy consumption and also only the inelastic energy, that is, the result of subtracting the elastic recovery from the input specific energy, as shown in Figure 4. These results are presented in Figure 12, which shows that energy utilization was maximum when the bed was subjected to the lowest pressures. Such high energy utilization at the lowest pressures may be explained by the fact that stressing energy is used to break the most brittle particles contained in the original size range. Figure 12, however, shows that beyond the minimum energy applied, energy utilization decreased, in particular for pressures above 150 MPa, even when the elastic restitution was subtracted from the input stressing energy. This is evidence of the progressively lower energy efficiency in size reduction as beds approach saturation.

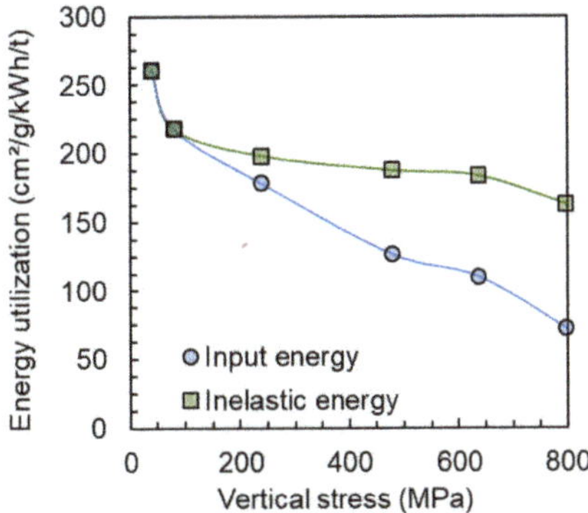

Figure 12. Comparison of energy utilization for different applied vertical stresses for the Itabira sample for the narrow size range of 150–125 μm, considering both the input and the inelastic energy.

4. Discussion

From the various analyses, it becomes feasible to describe the main features observed during the pressing of confined beds containing fine iron ore concentrates. This is illustrated in Figure 13, which shows that the first step in compression of the particle bed corresponds to a rearrangement of the particles, which occurs up to about point A. This key feature can be observed in Figure 3 for the different materials tested and, as already reported by Mütze [23], is associated predominantly with low stress levels (below about 40 MPa), when particles rearrange themselves within the bed by rotating and sliding in respect to each other. This is associated with very low energy inputs but is responsible for reasonable increases in the packing density of the bed.

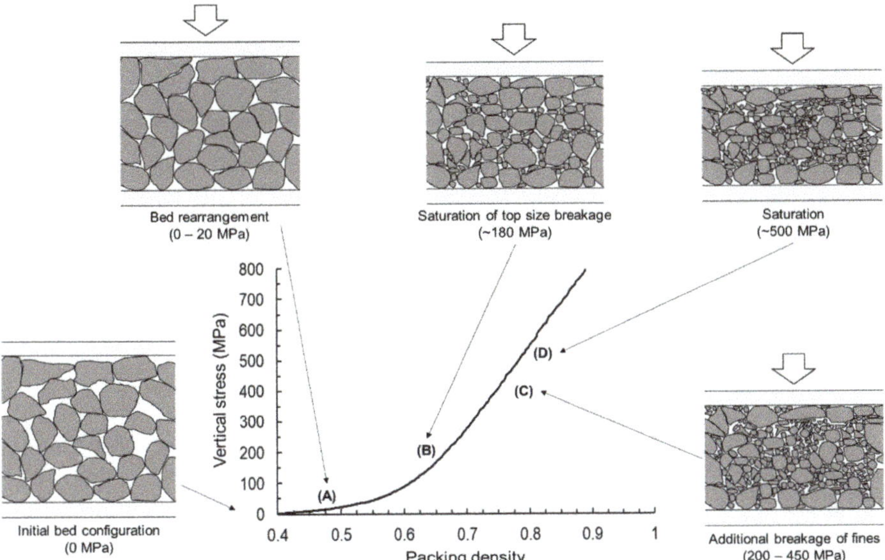

Figure 13. Schematic summarizing the main features of the particle bed pressing behavior of fine iron ore particles.

After initial rearrangement of the bed and within the interval from points A to B in Figure 13, the stress chains connecting the particles are formed, with stresses increasing significantly with deformation. As the critical force or stressing energy required to fracture each of the particles is reached, the weakest particles are progressively broken, leading to a rearrangement of the stress field, which is transmitted by the skeleton formed by the remaining unbroken particles. Point B in the figure corresponds to the saturation in breakage of particles contained in the top size. This occurs at pressures of around 180 MPa (2 kWh/t), as already shown in Figure 7. Such stresses and specific energies would have been sufficient to break all particles contained in the original size, if stressed individually in a micro compression tester [41]. Beyond this point, a proportion of particles contained in the top size, which varied from about 10% for the coarsest sizes tested to about 50% for the finest, remained unbroken, even as stressing energies increased. This is associated with the well-known phenomenon in confined particle bed breakage where the finer debris from breakage of the coarser particles are able to prevent the remaining coarser particles from further breaking [18,23]. The larger proportion of finer particles remaining in comparison to the coarser ones demonstrates the greater ability of the finer particles to dissipate the stresses. Until this point, the bed behaves perfectly inelastically, with all energy applied dissipated in either bed rearrangement or breakage.

The interval from point B to C in Figure 13 corresponds to further breakage of the debris from the initial breakage of the top size material in the case of narrow size beds, with additional increase in fines generation and specific surface area, as is evident in Figures 6 and 9, without additional breakage of particles contained in the top size class. The same effect of saturation in breakage of the top size then appears to happen with progressively finer sizes, as the increase in applied pressures results in only the progressive generation of finer particles.

In the interval from point B to C, deformations start to show an elastic component, with the bed recoiling partially after reaching the maximum stress during loading. As stresses are progressively increased, bed breakage reaches complete saturation at about point D, with no measurable increase in fines generation or specific surface area as a result of applying higher loads. At pressure beyond point D, a substantial increase in the elastic energy stored in the bed, which can correspond to over half of the energy input (Figure 4), is observed. Unlike the interval from A to B, in this interval the increasing inelastic energy is progressively used less in breakage, but more in dissipative phenomena, such as friction, plastic deformation, and interlocking and cold-welding of the particles, forming a tough final agglomerate, quite familiar in tableting studies [30–33] and consistent with final compaction.

5. Summary and Conclusions

The breakage of beds made up of narrow sizes of iron ore concentrates in confined conditions was studied in great detail. Stress versus relative density curves exhibited deformations that were purely inelastic for pressures of up to about 180 MPa. Up to this value, breakage of the top size fraction increased proportionally with input energy, reaching saturation at about 2 kWh/t. The saturation corresponded to proportions remaining in the top size that were as low as 10% for the 150–125 µm size range and as low as half for the 53–45 µm size range, depending on material.

At higher pressures, breakage of the top size fraction reached a maximum, whereas generation of additional fines from breakage of the progeny from the initially broken particles occurred. In this interval, progressively larger elastic deformations were observed, besides a more modest increase in specific surface area, reaching a point of full bed saturation (compaction) at specific energies beyond about 6 kWh/t, beyond which no increase in BSA occurred. This was evident from the achievement of the maximum increase in BSA under such conditions. Results also show the reasonable difference between the energy utilization calculated on the basis of the input energy and the inelastic energy (elastic energy subtracted from the input energy), even showing that a part of the inelastic energy may be used in dissipative processes such as friction, plastic deformation, and particle packing and not contributing to new surface area generation.

A pressing parameter κ was then proposed to quantify the propensity of a material contained in a bed to reduce in size when loaded under confined conditions. It correlated well with the increase in the Blaine specific surface area for beds with different initial size distributions, including narrow sizes and wide size ranges (multiple pressings) as well as different final pressures. It also captured the bed saturation condition in the maximum achievable value of κ observed in the experiments, which varied from about 0.10 to 013 for the materials studied.

A comparison of stressing at progressively higher pressures and multiple pressings at an intermediate pressure (160 MPa) showed that it is possible to enhance breakage of the top size material, as well as reach higher values of energy utilization, when multiple pressing stages are used. These results provide the technical justification and motivation for the application of multiple stages of pressing in HPGRs.

Author Contributions: Conceptualization, T.M.C., G.B., and L.M.T.; methodology, T.M.C. and G.B.; formal analysis, T.M.C. and L.M.T.; investigation, T.M.C., G.B., and L.M.T.; data curation, T.M.C. and L.M.T.; writing—original draft preparation, T.M.C. and L.M.T.; writing—review and editing, T.M.C., G.B., and L.M.T.; supervision, L.M.T. All authors have read and agreed to the published version of the manuscript.

Funding: This research was funded by Vale S.A. The authors would also like to thank the Brazilian Agencies CNPq (grant number 310293/2017-0) and FAPERJ (grant number E-26/202.574/2019).

Acknowledgments: The authors would like to thank Vale S.A. for financial and technical support for the research.

Conflicts of Interest: The authors declare no conflict of interest.

References

1. Kellerwessel, H. High pressure material bed comminution in practice. *ZKG Int.* **1990**, *43*, 71–75.
2. Tamashige, T.; Obana, H.; Hamaguchi, M. Operational results of OK series roller mill. *IEEE Trans. Ind. Appl.* **1991**, *27*, 416–424. [CrossRef]
3. McIvor, R.E. High Pressure Grinding Rolls—A review. In *Comminution Practices*; Kawatra, S.K., Ed.; SME: Littleton, CO, USA, 1997; pp. 95–98.
4. Ito, M.; Sato, K.; Naoi, Y. Productivity increase of the vertical roller mill for cement grinding. In Proceedings of the IEEE/PCA Cement Industry Technical Conference. XXXIX Conference Record (Cat. No.97CH36076), Hershey, PA, USA, 20–24 April 1997.
5. Benzer, H.; Ergun, L.; Lynch, A.J.; Oner, M.; Gunlu, A.; Celik, I.B. Modelling cement grinding circuits. *Miner. Eng.* **2001**, *14*, 1469–1482. [CrossRef]
6. Aydogan, N.A.; Ergun, L.; Benzer, H. High pressure grinding rolls (HPGR) applications in the cement industry. *Miner. Eng.* **2006**, *19*, 130–139. [CrossRef]
7. Altun, D.; Benzer, H.; Aydogan, N.; Gerold, C. Operational parameters affecting the vertical roller mill performance. *Miner. Eng.* **2017**, *103*, 67–71. [CrossRef]
8. Van der Meer, F.P. Roller press grinding of pellet feed: Experiences of KHD in the iron ore industry. In Proceedings of the AusIMM Conference on Iron Ore Resources and Reserves Estimation, Perth, Australia, 25–26 September 1997; pp. 1–15.
9. Morley, C. High-Pressure Grinding Rolls—A technology review. In *Advanced in Comminution*, 1st ed.; Kawatra, S.K., Ed.; SME: Littleton, CO, USA, 2006; pp. 15–40.
10. Michaelis, H.V.O.N. How energy efficient is HPGR? In Proceedings of the World Gold Conference, Gauteng, South Africa, 26–30 October 2009; pp. 7–15.
11. Powell, M.S.; Hilden, M.M.; Evertsson, C.M.; Asbjörnsson, G.; Benzer, A.H.; Mainza, A.N.; Tavares, L.M.; Davis, B.; Plint, N.; Rule, C. Optimisation opportunities for high pressure grinding rolls circuits. In *We Are Metallurgists, Not Magicians!*, 1st ed.; Australian Institute of Mining and Metallurgy (Organisation): Carlton, Australia, 2017; Volume 1, pp. 483–498.
12. Campos, T.M.; Bueno, G.; Barrios, G.K.; Tavares, L.M. Pressing iron ore concentrate in a pilot-scale HPGR. Part 1: Experimental results. *Miner. Eng.* **2019**, *140*, 105875. [CrossRef]
13. Thomazini, A.D.; Trés, E.P.; Macedo, F.A.D.; Athayde, M.; Bueno, G.; Fernandes, R.B.; Nunes, R.A.P. Development of a novel grinding process to iron ore pelletizing through HPGR milling in closed circuit. *Min. Metall. Explor.* **2020**, *37*, 933–941. [CrossRef]

14. Schönert, K. The influence of particle bed configurations and confinements on particle breakage. *Int. J. Miner. Process.* **1996**, *44*, 1–16. [CrossRef]

15. Kalala, J.T.; Dong, H.; Hinde, A.L. Using piston die tests to predict the breakage behavior of HPGR. In Proceedings of the 5th International Conference Autogenous and Semi-Autogenous Grinding, Vancouver, BC, Canada, 25–28 September 2011; pp. 1–12.

16. Schubert, H. Zu Einigen Fragen der Kollektivzerkleinerung. *Chem. Technol.* **1967**, *19*, 595–598.

17. Mütze, T. Energy dissipation in particle bed comminution. *Int. J. Miner. Process.* **2015**, *136*, 15–19. [CrossRef]

18. Liu, J.; Schönert, K. Modelling of interparticle breakage. *Int. J. Miner. Process.* **1996**, *44*, 101–115. [CrossRef]

19. Fuerstenau, D.W.; Gutsche, O.; Kapur, P.C. Confined particle bed comminution under compressive loads. *Int. J. Miner. Process.* **1996**, *44*, 521–537. [CrossRef]

20. Mütze, T.; Husemann, K. Compressive stress: Effect of stress velocity on confined particle bed comminution. *Chem. Eng. Res. Design* **2008**, *86*, 379–383. [CrossRef]

21. Dundar, H.; Benzer, H.; Aydogan, N. Application of population balance model to HPGR crushing. *Miner. Eng.* **2013**, *50*, 114–120. [CrossRef]

22. Davaanyam, Z. Piston Press Test Procedures for Predicting Energy-Size Reduction of High Pressure Grinding Rolls. Ph.D. Thesis, University of British Columbia, Vancouver, BC, Canada, 2015.

23. Mütze, T. Modelling the stress behaviour in particle bed comminution. *Int. J. Miner. Process.* **2016**, *156*, 14–23. [CrossRef]

24. Benzer, H.; Dundar, H.; Altun, O.; Tavares, L.M.; Mazzinghy, D.B.; Russo, J.C. HPGR simulation from piston-die tests with an itabirite ore. *REM Int. Eng. J.* **2017**, *70*, 99–107. [CrossRef]

25. Train, D. An investigation into the compaction of powders. *J. Pharm. Pharmacol.* **1956**, *8*, 745–761. [CrossRef]

26. Heckel, R.W. Density–pressure relationships in powder compaction. *Trans. Metall. Soc. AIME* **1961**, *221*, 671–675.

27. Cooper, A.R.; Eaton, L.E. Compaction behavior of several ceramic powders. *J. Am. Ceram. Soc.* **1962**, *45*, 97–101. [CrossRef]

28. Wünsch, I.; Finke, J.H.; John, E.; Juhnke, M.; Kwade, A. A mathematical approach to consider solid compressibility in the compression of pharmaceutical powders. *Pharmaceutics* **2019**, *11*, 121. [CrossRef]

29. Cabiscol, R.; Shi, H.; Wünsch, I.; Magnanimo, V.; Finke, J.H.; Luding, S.; Kwade, A. Effect of particle size on powder compaction and tablet strength using limestone. *Adv. Powder Technol.* **2020**, *31*, 1280–1289. [CrossRef]

30. Kawakita, K.; Lüdde, K.H. Some considerations on powder compression equations. *Powder Technol.* **1971**, *4*, 61–68. [CrossRef]

31. Armstrong, N.A.; Haines-Nutt, R.F. Elastic recovery and surface area changes in compacted powder systems. *Powder Technol.* **1974**, *9*, 287–290. [CrossRef]

32. Vachon, M.G.; Chulia, D. The use of energy indices in estimating powder compaction functionality of mixtures in pharmaceutical tableting. *Int. J. Pharm.* **1999**, *177*, 183–200. [CrossRef]

33. Patel, S.; Kaushal, A.M.; Bansal, A.K. Effect of particle size and compression force on compaction behavior and derived mathematical parameters of compressibility. *Pharm. Res.* **2007**, *24*, 111–124. [CrossRef]

34. Tye, C.K.; Sun, C.C.; Amidon, G.E. Evaluation of the effects of tableting speed on the relationships between compaction pressure, tablet tensile strength, and tablet solid fraction. *J. Pharm. Sci.* **2005**, *94*, 465–472. [CrossRef]

35. David, S.T.; Augsburger, L.L. Plastic flow during compression of directly compressible fillers and its effect on tablet strength. *J. Pharm. Sci.* **1977**, *66*, 155–159. [CrossRef]

36. Podczeck, F.; Sharma, M. The influence of particle size and shape of components of binary powder mixtures on the maximum volume reduction due to packing. *Int. J. Pharm.* **1996**, *137*, 41–47. [CrossRef]

37. Garner, S.; Strong, J.; Zavaliangos, A. Study of the die compaction of powders to high relative densities using the discrete element method. *Powder Technol.* **2018**, *330*, 357–370. [CrossRef]

38. Zhou, W.; Wang, D.; Ma, G.; Cao, X.; Hu, C.; Wu, W. Discrete element modeling of particle breakage considering different fragment replacement modes. *Powder Technol.* **2020**, *360*, 312–323. [CrossRef]

39. Barrios, G.K.; Jiménez-Herrera, N.; Tavares, L.M. Simulation of particle bed breakage by slow compression and impact using a DEM particle replacement model. *Adv. Powder Technol.* **2020**, *31*, 2749–2758. [CrossRef]

40. Schönert, K. Physical and technical aspects of very and micro fine grinding. In Proceedings of the 2nd World Congress Particle Technology, Society of Technology, Kyoto, Japan, 19–22 September 1990; pp. 557–571.

41. Campos, T.M.; Bueno, G.; Alfonso, V.R.; Mayerhofer, F.; Kwade, A.; Tavares, L.M. Relationships between particle breakage characteristics and comminution response of fine iron ore concentrates. *Miner. Eng.* **2020**, submitted for publication.
42. Tavares, L.M. Particle weakening in high-pressure roll grinding. *Miner. Eng.* **2005**, *18*, 651–657. [CrossRef]
43. Meyer, K. *Pelletizing of Iron Ores*, 1st ed.; Springer: New York, NY, USA, 1980.
44. Rumpf, H. Physical aspects of comminution and new formulation of a law of comminution. *Powder Technol.* **1973**, *7*, 145–159. [CrossRef]

Article

Fit-for-Purpose VSI Modelling Framework for Process Simulation

Simon Grunditz *, Gauti Asbjörnsson, Erik Hulthén and Magnus Evertsson

Department of Industrial and Materials Science, Chalmers University of Technology, SE-41296 Göteborg, Sweden; gauti@chalmers.se (G.A.); erik.hulthen@chalmers.se (E.H.); magnus.evertsson@chalmers.se (M.E.)
* Correspondence: simon.grunditz@chalmers.se

Abstract: The worldwide shortage of natural sand has created a need for improved methods to create a replacement product. The use of vertical shaft impact (VSI) crushers is one possible solution, since VSI crushers can create particles with a good aspect ratio and smooth surfaces for use in different applications such as in construction. To evaluate the impact a VSI crusher has on the process performance, a more fit-for-purpose model is needed for process simulations. This paper aims to present a modelling framework to improve particle breakage prediction in VSI crushers. The model is based on the theory of energy-based breakage behavior. Particle collision energy data are extracted from discrete element method (DEM) simulations with particle velocities, i.e., rotor speed, as the input. A selection–breakage approach is then used to create the particle size distribution (PSD). For each site, the model is trained with two datasets for the PSDs at different VSI rotor tip speeds. This allows the model to predict the product output for different rotor tip speeds beyond the experimental configurations. A dataset from 24 different sites in Sweden is used for training and validating the model to showcase the robustness of the model. The model presented in this paper has a low barrier for implementation suitable for trying different speeds at existing sites and can be used as a replacement to a manual testing approach.

Keywords: VSI; DEM; sand; breakage; modelling

Citation: Grunditz, S.; Asbjörnsson, G.; Hulthén, E.; Evertsson, M. Fit-for-Purpose VSI Modelling Framework for Process Simulation. *Minerals* **2021**, *11*, 40. https://doi.org/10.3390/min11010040

Received: 31 October 2020
Accepted: 28 December 2020
Published: 31 December 2020

1. Introduction

The increased infrastructure development of society has been accompanied by an increase in the use of natural materials for construction [1]. One resource that has seen massive extraction is natural sand deposits. The usefulness of sand in creating quality concrete products has led to the erosion of numerous coastlines caused by aggressive sand dredging [2]. Some governments such as Sweden are trying to incentivize new approaches with increased taxes to minimize the use of natural sand along with limiting permits for existing deposits.

Given the finite nature of natural sand and the increased demand on the market, replacement resources are becoming more common. One approach is replacing natural sand with manufactured sand. The process of creating manufactured sand for concrete includes vertical shaft impactors (VSIs), which are common comminution machines for crushing particles in the tertiary or quaternary stage of aggregate crushing circuits. Autogenous impact crushing results in particles that, compared to compressive crushing, have a smoother surface and a more spherical shape but with an increased amount of fine particulate matter in the process. Manufactured sand has successfully been used to replace natural aggregates in several applications, such as concrete production [3,4]. The main attributes sought after in natural sand particles are sphericity and high volume-to-surface ratio. These attributes minimize the use of cement in the concrete mixes and lead to desirable rheology for pouring [5].

Natural sand requires fewer processing steps than the manufacturing of artificial sand. To allow for a transition to more sustainable products, the process needs to become more

115

efficient. By establishing a framework to adopt crusher models to existing production plants, simulations can be used to optimize settings at real plants, and therefore, a more efficient process can be achieved.

The dynamics of the particles has been shown to determine the product size distribution. A detailed description of the dynamics of the particles inside and exiting a rotor was provided by Rychel [6]. However, Kojovic [7] formulated a predictive model for the VSI crusher that correlated product size with power draw based on breakage of the ore in a drop-weight test. Attempts to model the VSI output by adapting the Whiten model as a basis for modeling have also been proven to be a viable approach [8].

According to Nikolov [9], the essential part of the performance of the VSI crusher is the relationships between the radius of the rotor, angular velocity and the rate of material being fed into the VSI crusher when given a specific particle size distribution for the feed material. Bengtson and Evertsson [10] then proposed a model that also considered the variations in rock properties and predicted capacity and power draw as well as the particle shape distribution. Lindqvist [11] also suggests that impact crushing can significantly reduce the energy required in grinding.

The attempts to model the performance of VSI crushers have often been limited by the lack of information about the interaction between particles inside the crusher. An approach to overcome this issue has been through the use of the discrete element method (DEM). With the DEM, every particle and its dynamics are simulated, making it possible to estimate the particle trajectories, residence times and collision energies when particles travel through the crusher. DEMs have been the main method of choice for other researchers such as Cunha, Cleary and Sinnott [12–14].

In Grunditz et al. [14], the data from DEM simulations of a VSI crusher were analyzed to gain an understanding of how the energy in a VSI crusher is distributed in relation to the different sizes of the particles fed into it and the number of impacts that occur. One of these energy collision spectra can be seen in Figure 1. In addition, an exploration of how the particles travel in the crushing chamber after leaving the rotor was conducted in conjunction with evaluating the residence time of each particle [15,16].

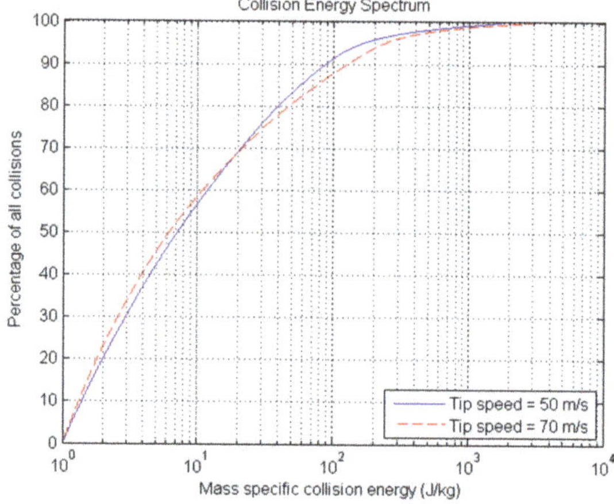

Figure 1. The distribution of particle collision energies inside a discrete element method (DEM) simulation of a vertical shaft impact (VSI). Two different rotor tip speeds were simulated.

To create a robust model, material from numerous sites was collected and used for the calibration and validation of the model. The model accurately predicts the product size distribution and is useful to give operators the opportunity to simulate changes to their feed size, material attributes and machine parameters before committing to them.

The use of data from several different sites of varying geological types enables the mapping of material properties to certain model parameters, achieved by deploying the same crushing process and equipment in the same configuration at multiple different crushing plants. These physical material properties were attributed to f_{mat} and $E^0{}_{min}$ established in previous modelling efforts by Nikolov [9,17].

A selection–breakage approach is used to simulate the effects of impact energy present in VSI crushers. Vogel and Peukert established a master curve for breakage that considers the accumulation of impacts when observing particle breakage [18,19].

The purpose of this paper is to use previously established models and appropriate optimization routines for calibration of a fit-for-purpose VSI crusher model. A framework was proposed to enable crushing operators, plant owners and product managers to explore and test differing machine configurations to reach alternate product goals more systematically.

2. Methodology

2.1. Modeling

Breakage modeling relies on descriptive functions of particle behavior when subjected to high energy impacts. The goal of these studies was to assess how much energy a particle requires to break relative to its initial size. In addition, the size of the daughter fragments following such an event was of interest.

Several previous studies and experiments [17] provide the basis to understand the role of particle size and impact energy on the breakage of the particular materials studied. Each material tested in this way provides properties that apply to only one specific quarry site. Having to run these exhaustive tests for each modelling attempt for a site is costly.

The different sizes into which a particle will statistically split are referred to as an appearance function. An appearance function was included in the optimization routine and assumed that the behavior of the breakage follows a polynomial function, as seen in Equation (1), where $a_{i,j}$ is the fraction of particles of one size being reduced into a smaller subset; β_i values are coefficients, and x_i is the size of the originating particle. The index i represents the number of particle sizes available in the PSD dataset starting from 1, the largest size, to N being the smallest size. This creates an appearance matrix for the selected particles sizes. An example of such an appearance matrix can be seen in Table 1 where each column for progeny Equation (1) is shown.

$$a_{i,j} = \beta_1 + \beta_2 x_i^1 + \beta_3 x_i^2 + \beta_4 x_i^3 \tag{1}$$

Table 1. The appearance matrix for a PSD with 5 different size bins. Each column for progeny Equation (1).

$a_{1,1}$	0	0	0	0
$a_{1,2}$	$a_{2,2}$	0	0	0
$a_{1,3}$	$a_{2,3}$	$a_{3,3}$	0	0
$a_{1,4}$	$a_{2,4}$	$a_{3,4}$	$a_{4,4}$	0
$a_{1,5}$	$a_{2,5}$	$a_{3,5}$	$a_{4,5}$	$a_{5,5}$

The most common approach to estimate the appearance function is with a drop weight test or a Julius Kruttschnitt rotary breakage tester (JKRBT) to generate a family of *t10* curves [20]. In [21], three different ore types were tested with the JKRBT [22], and their response was calculated with a modified equation based on the work of Vogel and Peukert [19] with satisfying results. A similar approach was implemented in this work by integrating the theories established by Vogel and Peukert and further expanded on by Bonfils into the calibration of an appearance function.

The selection–breakage method was implemented to determine whether a particle sustained enough impact energy to suffer a breakage event [10]. By using the kinetic energy and size of particles, a probability distribution of particle breakage can be predicted (see Equations (2) and (3) where B, the breakage, is the appearance matrix. The index *i* represents the number of particle sizes available in the PSD dataset starting from 1, the largest size, to N being the smallest size. The model essentially treats particles that enter the crusher as a binary option: either they are crushed to breakage or the particle withstands the impact and retains its size. The particles that are chosen for breakage are then reduced in size according to the appearance function. The product is then added to the unbroken particles that pass through. This process can be seen in the flowchart in Figure 2. The number of impacts a particle, following its exit from the rotor, is assumed to be 1 based on the results of DEM simulations analyzing the collision energy spectrum. The majority of the collisions are under 10 J/kg and are assumed not to influence the impact breakage. These lower energy collisions are assumed to be after the initial collision with the rockbed in the crushing chamber.

$$P = [BSf + (I - S)f] \tag{2}$$

$$S = \left[1 - exp\left(f_{mat}x_i^\alpha k\left(E_{CS} - \frac{E_{min}^0}{x_i^\alpha}\right)\right)\right] \tag{3}$$

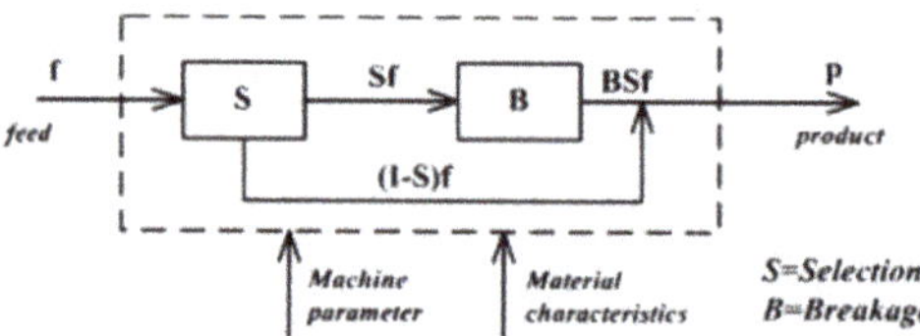

Figure 2. The selection–breakage process was used to determine the product generated from the feed [10].

After the model had been calibrated for different sites, an effort to validate it was performed. The feed data and machine parameters for the VSI crushers were then put into the model, and the product was predicted. The speed of the particles at the rotor tip is transformed to kinetic energy, E_{CS}, as seen in Equation (4).

$$E_{CS} = \frac{v_{rotor}^2}{2} \tag{4}$$

The predicted particle size distributions (PSDs) for the product and the experimental product were then compared to determine how much the model varied from the experiments. The model was based on crushers that employ a rockbox configuration. The input parameters and the overall process of running the model assumed that no prior material characterization had been performed. The approach can be seen in Figure 3.

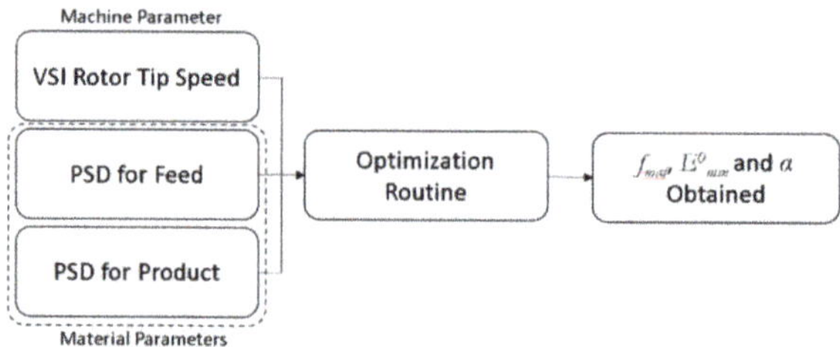

Figure 3. The overall modelling process for a new site where the material has not been calibrated and then adequately mapped to the material database.

Once the material parameters have been obtained, the model can be run with alternative machine parameters to achieve different rotor tip speeds, as seen in Figure 4.

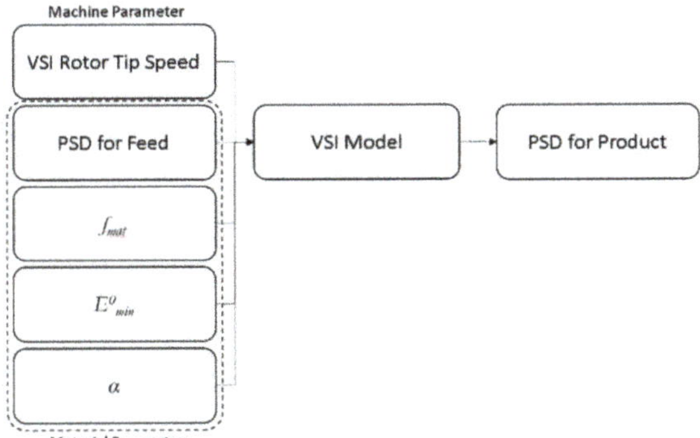

Figure 4. A model overview showing the parameters required to run the VSI crusher model at alternate speeds to predict the product PSD. The model requires a material database with f_{mat}, E^0_{min} and α values obtained from previous calibration efforts.

2.2. Sampling Procedures

Due to the enclosed design and destructive behavior of comminution machines, it is difficult to directly observe what occurs inside with changes in operational parameters. Experimental data that are measurable for a VSI crusher include the incoming material feed rate and particle size distribution, outgoing mass flow and particle size distribution, rotor tip speed, rotor radius and material properties. These parameters are often quantified on a site-by-site basis to establish the range of products that can be expected with the current setup of machines and the specific ore characteristics. For this model, datasets were collected from different sites throughout Sweden to obtain a comprehensive overview of the performance of the VSI crusher. The material from these sites was crushed with a Metso BARMAC 5100 SE VSI crusher (Metso, Helsinki, Finland), which has a diameter of 510 mm and speeds between 69 and 79 m/s. The conveyer before and after the VSI crusher was stopped to allow material sampling. A conveyor length of 1 m was isolated, and all the material was taken according to the SS-EN 933-1 standard. The intricacies of variation between the sites in terms of materials and machine parameters can create uncertainties in the analysis of its effects. Each source of variation, originating from either equipment, operator, material or site production level, contributes to a potential confounding of the results. The locations of the different sites can be seen in Figure 5 and the overall plant layout used in Figure 6.

Previous studies [13,15,17] have investigated the phenomena, conditions and effects of crushing particles at single-energy levels. Several of these studies have been conducted to analyze the inputs of specific energy levels on particles of different sizes. The results have often been reported in useful units such as J/kg to make comparisons between different particle sizes easier. More recent studies have shown simulations and observations of the different energies that the particles are exposed to while being expelled from the rotor and into the crushing chamber [12].

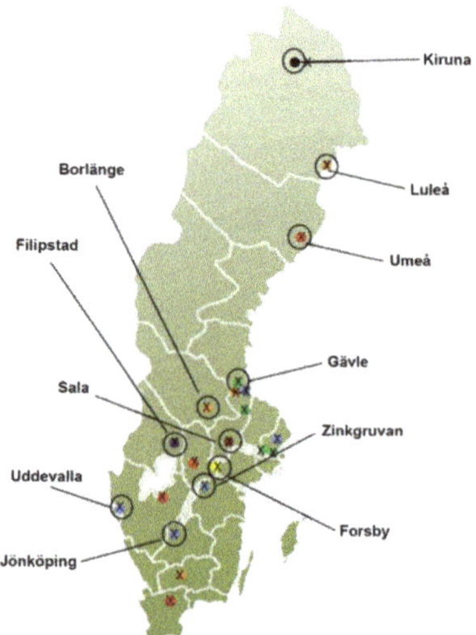

Figure 5. Location of the quarries tested in Sweden. Circles denote where the mobile VSI plant was set up, and the colored dots show where the material originated from.

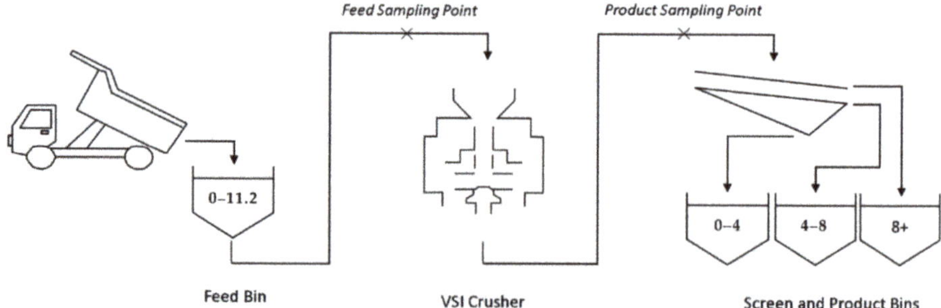

Figure 6. The mobile VSI crusher plant layout used at the different quarries tested.

2.3. Optimization Framework

The purpose of this framework is to enable the rapid deployment of an accurate model without the need for exhaustive levels of sampling and ore characterization. The initial design space for the framework was intentionally kept small to reduce the computational load to run the optimization. Each round of sampling takes considerable time and effort on site for collection. With the data at hand and prior models to simulate certain aspects of impact crushing, a surrogate model was established for the appearance function and the selection breakage. Both of these sets of values were assessed for accuracy, and an error function was formulated (see Equation (5)), where $P_{exp,High}$ and $P_{exp,Low}$ are PSD data from experiments, $P_{sim,High}$ and $P_{sim,Low}$ are PSD data from simulations using the model and the i index is for the different particle sizes. The two datasets were run at two different rotor tip speeds, 69 and 79 m/s, which is indicated with the $_{Low}$ and $_{High}$ indexes, respectively. The framework for running the optimization utilizes the genetic algorithm for minimizing the error function.

$$\min_{\{\beta_i, f_{mat}, E^0_{min}, \alpha\}} E = \sum_{i=1}^{n} \left(\left(P_{i_{Exp,High}} - P_{i_{Sim,High}} \right)^2 - \left(P_{i_{Exp,Low}} - P_{i_{Sim,Low}} \right)^2 \right) \tag{5}$$

The different bounds used in conjunction with the Genetic Algorithm can be seen in Table 2. A population size of 8000 was initiated with a set of 800 generations. The limits for the coefficients for the appearance generation polynomial were reached by initial testing to end up with a feasible yet large solution space. The values for f_{mat}, E^0_{min} and α were derived through the implications of their properties and then expanded on the spectrum at which these types of materials have been documented.

Table 2. Listing the lower and upper bounds of the 6 different variables used during the optimization routine.

Variables	$\beta_1, \beta_2, \beta_3, \beta_4$	f_{mat}	E^0_{min}	α
Lower Bound	−1000	0.001	0	0
Upper Bound	1000	3	30	1

The error function takes the result of the product estimation function using the same optimizer for two different speeds and compares each of them to their experimental data according to Equation (5). The genetic algorithm will then continue to the next generation until it reaches the tolerance limit and finds an optimum.

3. Results

Running calibrations on multiple datasets from multiple sites enabled the quantification of the material properties for the different geological rock types. The property values obtained from a wide array of sites all converge on values within the range that is to be expected of rock materials. In Table 3, the summary of the calibrated material properties per site is listed, where "Default Feed" indicates that only one sample was taken for the feed into the crusher, and the variation was assumed to be negligible, while the term "Individual Feeds" denotes that each rotor tip speed configuration run has matching feed sample data.

Table 3. Material data from all the sites used in the study, along with their locations and calibrated values.

Geological Type	Site Location	Top Size	f_{mat}	Source	E^0_{min}	α
Diabase	Ubbarp	19 mm	0.25	Default Feed	21.04	0.94
Diorite	Borlänge	11.2 mm	0.75	Default Feed	16.29	0.94
Dolomite	Glanshamar	11.2 mm	1.3	Individual Feeds	14.83	0.91
Gneiss	Umeå	19 mm	0.8	Individual Feeds	18.61	0.92
Gneiss	Atle	19 mm	0.68	Default Feed	22.8	0.88
Gneiss	Össjö	4 mm	1.73	Default Feed	29.22	0.99
Gneissic Granite	Skyttorp	11.2 mm	1.7	Default Feed	15.79	0.93
Gneissic Granite	Önnestad	11.2 mm	1.32	Default Feed	26.54	0.86
Granite	Tierp	11.2 mm	2.18	Unique Feeds	16.74	0.96
Granite	Gävle	4 mm	1.45	Unique Feeds	20.48	0.92
Granite	Gävle	11.2 mm	2.08	Unique Feeds	19.39	0.9
Granite	Gävle	11.2 mm	1.63	Unique Feeds	17.91	0.82
Granite	Kållered	16 mm	1.36	Individual Feeds	15.45	0.85
Granitic Gneiss	Enhörna	19 mm	2.02	Default Feed	15.48	0.91
Granitoid	Bro	10 mm	0.06	Individual Feeds	28.02	0.81
Granodiorite	Sunderbyn	11.2 mm	1.92	Individual Feeds	23.48	0.95
Limestone	Forsby	19 mm	2.35	Default Feed	25.22	0.82
Porphyry	Kiruna	10 mm	0.39	Individual Feeds	16.83	0.88
Quartzite	Gåsgruvan	19 mm	0.25	Default Feed	21.04	0.94

The results for each dataset originating from one site can be visualized, as in Figure 7. The dashed lines represent the estimation of the model for the product, and the solid lines are the tested product curves for the corresponding speeds. Lastly, the black solid line indicates the feed or feeds in the dataset. Of the three rotor tip speeds, only two were used to train the model, the highest and lowest speeds, while the middle speed was used to verify the response of the model.

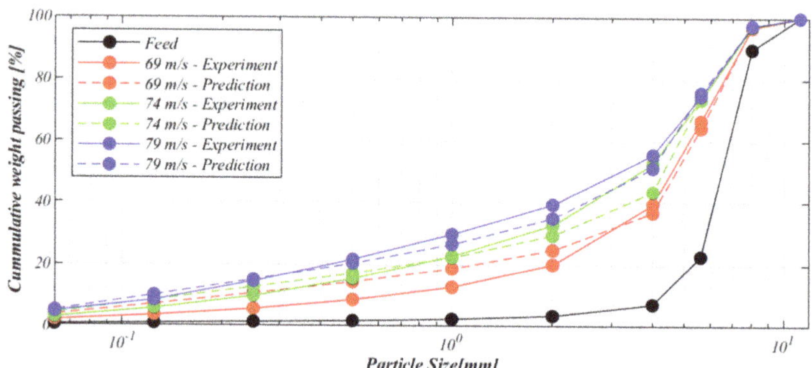

Figure 7. Particle size distributions from one of the sites and the corresponding simulation results. Case data are from Tierp, as seen in Table 3.

By running all the datasets through the same optimization routine, similar results were obtained for each site, like the results achieved for Figure 7 but with different levels of magnitude relating to the feed used and material properties. While these properties include a multitude of factors that affect particle breakage and size progeny, some parameters are more prevalent than others and will vary less between sites. For instance, the feed size PSD has a large impact on the product, but the majority of sites surveyed used an 11.2-mm top size feed. Three factors were calibrated from each site with the use of experimental data.

E^0_{min} is a characterization of the energy level required to achieve particle breakage, while the f_{mat} property encompasses not only the material properties but also the properties arising from different particle shapes and α is a size dependency parameter. These parameters are dependent on many variables within the material but generate a perception of the strength of the different material types.

The models created and calibrated with these methods have created significantly accurate models within normal operative speeds. In each dataset, a speed, intermediate, was excluded from the training data to be used as verification of the model. Values of P20, P50 and P80 were extracted from the simulated product data and were compared to similar points interpolated from the experimental data. The values can be seen in Figure 8. These sets were compared to each other to assess the accuracy and range of the model. In large part, due to the optimization method, the values for P50 were the least differing, while the tail ends were often more spread away from the experimental data. Some variation in the experimental data is also assumed to be linked to the belt cut method of samples taken from the sites but is generally considered to be negligible.

To test the robustness of the model, PSD predictions of speeds exceeding the rotor tip speeds the model was trained with were made. In Figure 9, predictions were made for a tip speed of 85 m/s (15% higher kinetic energy than training data) and 65 m/s (12% less kinetic energy than training data), respectively.

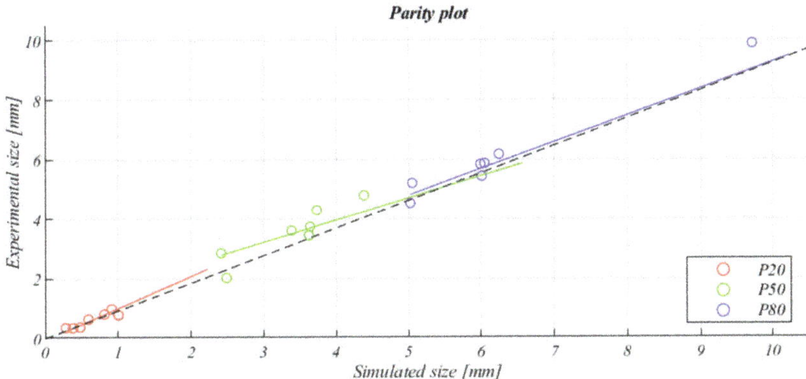

Figure 8. A parity plot of the P20, P50 and P80 values from the experimental data compared to the simulated data using the developed models.

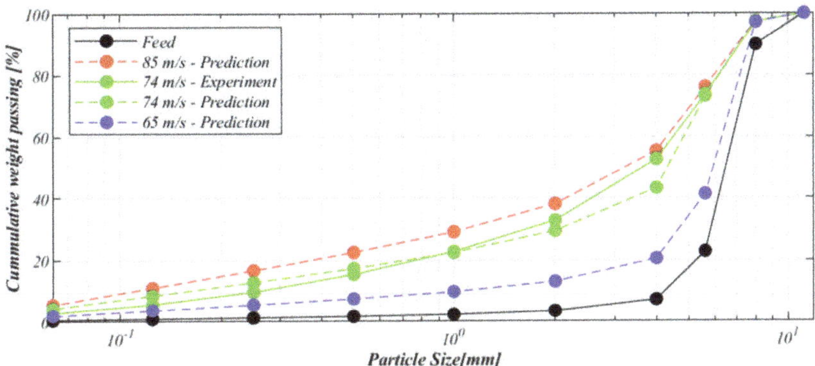

Figure 9. Particle size distributions from one of the sites and the corresponding simulation results for speeds exceeding the training data of the model. Note that the training speeds were still 69 and 79 m/s, not shown in the figure.

4. Discussion

The distribution of the datasets for the different sites and materials in Figure 10 displays an interesting pattern. Materials on the large scale tend to either end up with higher values for E^0_{min} and lower values for f_{mat} or, on the opposite side, with high values for f_{mat} and low values for E^0_{min}. Since f_{mat} can be regarded as the resistance of the material against fracture, materials with lower values will be more pliable, and a higher level of particle size reduction can be expected when breakage does occur. E^0_{min} is regarded as the energy threshold of a material, the level it can withstand before succumbing to comminution breakage. Materials with higher values tend to require higher levels of energy to break, ultimately requiring higher rotor tip speeds relative to materials with lower values.

The model that has been presented can be used to optimize existing plants. The machine parameters, the particle size distribution for the feed and product and the rock type are entered into the model, and the site operator can then change the flow rates, rotor speed or sizes to reach their desired production goals. The practical application of the model is to enable plant owners to estimate the feasibility of operating a VSI crusher on site and to see what different types of products they can expect to achieve and in what tonnages.

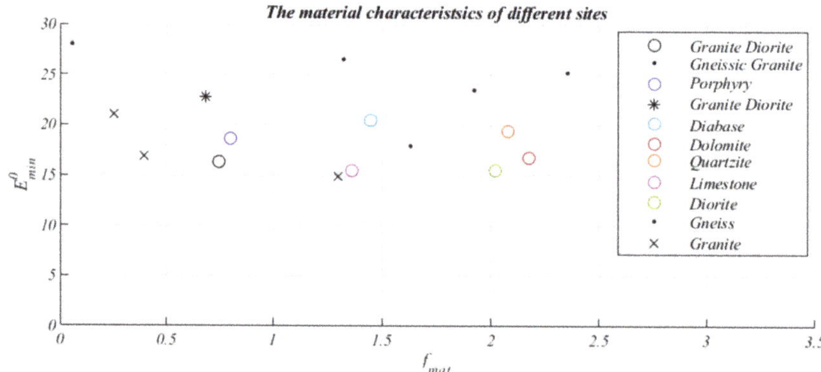

Figure 10. A scatter plot of the f_{mat} and E^0_{min} values for different rock types and sites analyzed.

One possible shortcoming of the model and the theory it uses as a basis is that the properties of materials are summarized into two parameters and the size dependency into one parameter. While useful to make the optimization and helpful to characterize a material, it is possible that certain details or complexities are overlooked or disregarded due to this approach. For instance, the sphericity and flakiness of material going into the crusher compared to what goes out are often very different due to the abrasive nature of the crusher. This factor is included in the material parameters but not isolated as a separate measurement.

Author Contributions: Conceptualization, S.G. and E.H.; methodology, S.G.; software, S.G. and G.A.; writing—original draft preparation, S.G.; writing—review and editing, S.G., E.H. and M.E.; supervision, G.A., E.H. and M.E. All authors have read and agreed to the published version of the manuscript.

Funding: This research was funded by Swerock, Chalmers University of Technology, VINNOVA, SBUF and MinFo/Hesselman.

Institutional Review Board Statement: Not applicable.

Informed Consent Statement: Not applicable.

Acknowledgments: The authors would also like to acknowledge the financial support from Swerock through the 3B project, Vinnova, SBUF and MinFo/Hesselman. The personnel at Swerock are also gratefully acknowledged for their support and insights. This work was carried out within the Sustainable Production Initiative and the Production Area of Advance at Chalmers University of Technology.

Conflicts of Interest: The authors declare no conflict of interest. The funders had no role in the design of the study; in the collection, analyses, or interpretation of data; in the writing of the manuscript, or in the decision to publish the results.

Abbreviations

$a(-)$	Fraction of initial particle volume as daughter fragment size
$\alpha(-)$	Size-dependency parameter
$\beta(-)$	Appearance function coefficients
$x(m)$	Initial particle size
$f_{mat}(kg/Jm)$	Material parameter
$v_{rotor}(m/s)$	Rotor tip speed
$k(-)$	Number of impacts
$S(-)$	Fraction of broken particles
$E(-)$	Error objective of optimization function

$E^0_{min}(Jm/kg)$ Mass-specific threshold energy for breakage
$E_{CS}(J/kg)$ Mass-specific kinetic impact energy
$\mathbf{P}_{exp}(-)$ The particle size from the cumulative particle size distribution from experimental data
$\mathbf{P}_{sim}(-)$ The particle size from the cumulative particle size distribution from simulation data

References

1. Schoning, K.; Lundqvist, L. *Hållbar Ballastförsörjning—Förutsättningar i Stockholms och Uppsala län*; Technical Report; The Geological Survey of Sweden: Uppsala, Sweden, 2018.
2. Masalu, D.C.P. Coastal Erosion and Its Social and Environmental Aspects in Tanzania: A Case Study in Illegal Sand Mining. *Coast. Manag.* **2002**, *30*, 347–359. [CrossRef]
3. Gonçalves, J.; Tavares, L.; Filho, R.T.; Fairbairn, E.; Cunha, E. Comparison of natural and manufactured fine aggregates in cement mortars. *Cem. Concr. Res.* **2007**, *37*, 924–932. [CrossRef]
4. Cepuritis, R.; Jacobsen, S.; Onnela, T. Sand production with VSI crushing and air classification: Optimising fines grading for concrete production with micro-proportioning. *Miner. Eng.* **2015**, *78*, 1–14. [CrossRef]
5. Lima, P.R.L.; Filho, R.D.T.; Gomes, O.D.F.M. Influence of Recycled Aggregate on the Rheological Behavior of Cement Mortar. *Key Eng. Mater.* **2014**, *600*, 297–307. [CrossRef]
6. Rychel, R. Modellierung des Betriebsverhaltens von Rotorschleuderbrechern. Ph.D Thesis, Technischen Universität Freiberg, Freiberg, Germany, 2001.
7. Kojovic, T.; Napier-Munn, T.J.; Andersen, J.S. Modelling cone and impact crushers using laboratory determined energy-breakage functions. In Proceedings of the Comminution Practices Symposium: SME Annual Meeting, Denver, CO, USA, 24–27 February 1997.
8. Segura-Salazar, J.; Barrios, G.K.; Rodriguez, V.; Tavares, L.M. Mathematical modeling of a vertical shaft impact crusher using the Whiten model. *Miner. Eng.* **2017**, *111*, 222–228. [CrossRef]
9. Nikolov, S. Modelling and simulation of particle breakage in impact crushers. *Int. J. Miner. Process.* **2004**, *74*, S219–S225. [CrossRef]
10. Bengtsson, M.; Evertsson, C. Modelling of output and power consumption in vertical shaft impact crushers. *Int. J. Miner. Process.* **2008**, *88*, 18–23. [CrossRef]
11. Lindqvist, M. Energy considerations in compressive and impact crushing of rock. *Miner. Eng.* **2008**, *21*, 631–641. [CrossRef]
12. Da Cunha, E.R.; De Carvalho, R.M.; Tavares, L.M. Simulation of solids flow and energy transfer in a vertical shaft impact crusher using DEM. *Miner. Eng.* **2013**, *43*, 85–90. [CrossRef]
13. Cleary, P.W.; Sinnott, M.D.; Morrison, R.D. DEM prediction of particle flows and breakage in comminution processes. In Proceedings of the AIChE Annual Meeting, San Francisco, CA, USA, 12–17 November 2006.
14. Sinnott, M.; Cleary, P.W. Simulation of particle flows and breakage in crushers using DEM: Part 2—Impact crushers. *Miner. Eng.* **2015**, *74*, 163–177. [CrossRef]
15. Grunditz, S.; Evertsson, M.; Hulthén, E.; Bengtsson, M. Prediction of Collision Energy in the VSI Crusher. In Proceedings of the European Symposium on Comminution and Classification, Gothenburg, Sweden, 7–10 September 2015.
16. Grunditz, S.; Evertsson, M.; Hulthén, E.; Bengtsson, M. The Effect of Rotor Tip Speed of a Vertical Shaft Impactor on the Collision Energy Spectrum. In Proceedings of the Minerals Engineering Conference Computational Modelling, Falmouth, UK, 10–11 June 2015.
17. Nikolov, S. A performance model for impact crushers. *Miner. Eng.* **2002**, *15*, 715–721. [CrossRef]
18. Vogel, L.; Peukert, W. Breakage behaviour of different materials—Construction of a mastercurve for the breakage probability. *Powder Technol.* **2003**, *129*, 101–110. [CrossRef]
19. Vogel, L.; Peukert, W. From single particle impact behaviour to modelling of impact mills. *Chem. Eng. Sci.* **2005**, *60*, 5164–5176. [CrossRef]
20. Genç, Ö.; Ergun, L.; Benzer, H. Single particle impact breakage characterization of materials by drop weight testing. *Physicochem. Probl. Miner. Process.* **2004**, *38*, 241–255.
21. Bonfils, B.; Ballantyne, G.R.; Powell, M.S. Developments in incremental rock breakage testing methodologies and modelling. *Int. J. Miner. Process.* **2016**, *152*, 16–25. [CrossRef]
22. Shi, F.; Kojovic, T.; Larbi-Bram, S.; Manlapig, E. Development of a new particle breakage characterisation device—The JKRBT. In Proceedings of the Comminution 2008 Conference, Falmouth, UK, 17–20 June 2008.

Article

DEM Simulation of Laboratory-Scale Jaw Crushing of a Gold-Bearing Ore Using a Particle Replacement Model

Gabriel Kamilo Barrios [1,*], **Narcés Jiménez-Herrera** [1,*], **Silvia Natalia Fuentes-Torres** [1] and **Luís Marcelo Tavares** [2]

[1] Colombian Geological Survey (CGS), Cali 760001, Colombia; sfuentes@sgc.gov.co
[2] Department of Metallurgical and Material Engineering, Laboratory of Mineral Technology, Universidade Federal do Rio de Janeiro, Rio de Janeiro CEP 21941-972, Brazil; tavares@metalmat.ufrj.br
* Correspondence: gkpbarrios@metalmat.ufrj.br (G.K.B.); njimenez@metalmat.ufrj.br (N.J.-H.); Tel.: +(57)-312-622-5883 (G.K.B.)

Received: 10 July 2020; Accepted: 12 August 2020; Published: 14 August 2020

Abstract: The Discrete Element Method (DEM) is a numerical method that is able to simulate the mechanical behavior of bulk solids flow using spheres or polyhedral elements, offering a powerful tool for equipment design and optimization through modeling and simulation. The present work uses a Particle Replacement Model (PRM) embedded in the software EDEM® to model and simulate operation of a laboratory-scale jaw crusher. The PRM was calibrated using data from single particle slow compression tests, whereas simulations of the jaw crusher were validated on the basis of experiments, with very good agreement. DEM simulations described the performance of the crusher in terms of throughput, product size distribution, compressive force on the jaws surface, reduction ratio, and energy consumption as a function of closed side setting and frequency.

Keywords: crushing; jaw crusher; Discrete Element Method; Particle Replacement Model; comminution; simulation; modeling; primary crushing; particle breakage

---

1. Introduction

Jaw crushers are widely used in the primary crushing stage and, sometimes, even in the secondary for many applications, including the processing of metallic, non-metallic, energetic, and industrial minerals, as well as in the processing of construction and demolition waste.

Despite the technology being over a century old [1], the original designs of the jaw crushers have been maintained nearly unchanged, taking advantage of the simplicity of their structure and mechanical operation. These features result in ease in manufacturing, repairing, dissembling and low capital cost in comparison to other types of crushers [2].

This machine is composed of two metallic plates forming a V-shape. One of them is fixed while the other swings, moving due to action of an eccentric shaft connected to a motor. When in operation, the ore is fed to the top opening and travels down along the chamber. On its path, the ore is crushed in successive cycles of application of stress, primarily compression, which are applied when the moving plate approaches the fixed plate. In the return movement of the moving plate, the ore particles slip down the chamber until they are stressed once again in the following cycle. The process continues until particles reach a size that is finer than the bottom opening so, in that moment, the ore particles drop out of the crushing chamber [3].

Jaw crushers are robust, being an attractive option in operations where the feed has a coarse top size and a moderate reduction ratio without fines is required [2]. Their performance, in terms of capacity or throughput, power and energy consumption, depends on material properties, equipment design and operating parameters. On one hand, the material characteristics are given by density, hardness, bulk density, particle size distribution, particle top size and particle strength (toughness) and crushability of the feed material [4]. The design parameters of the crusher include the size of the top opening, the set, the volume of the crushing chamber, and the type of jaw surface, which may be smooth or corrugated [5]. The equipment operational parameters include the frequency and amplitude of the movable jaw stroke, the feed rate, the closed side setting (CSS) of the discharge opening, among others [6,7]. It is worth mentioning that the material that is used to line the jaws must be hard and tough in order to endure impact and wear during operation. Some aspects of the crusher materials and a failure analysis of a jaw crusher have been investigated by Olawale and Ibitoye [8].

Different mathematical models have been developed to describe the performance of the jaw crusher. The first generation models were based on empirical expressions to predict capacity [9,10] and energy consumption [10–12]. Later, more robust mathematical models were proposed, based on the population balance model, to represent more details of the machine performance, including the full product size distribution [13]. More recently, a model that describes the kinematics of the equipment to predict flow, capacity, power, among others, has been proposed [14].

Over the last few decades mechanistic approaches that rely on the Discrete Element Method have shown great value in the description of the performance of different types of crushers. Fusheng et al. [15] and Legendre and Zevenhoven [16] performed DEM simulations of size reduction of a single particle in a jaw crusher, in which the bonded particle model (BPM) was used to describe particle breakage. Particle breakage models in DEM usually require significant computational effort and their application in systems with multiple particles, such as those found in crushers operating in industry, may be complex [17]. Among the particle breakage model approaches compared in a recent review by Jimenez-Herrera et al. [16], PRM using spheres was identified as the one with the lowest computational cost, making it attractive to simulate the machine operation in which the feed is constituted by a stream of particles. Indeed, a successful application of PRM has been demonstrated to selected compression crushers, including a jaw crusher [18].

The present work describes the modeling and simulation of a laboratory-scale jaw crusher using DEM with the particle replacement model embedded to describe product size distribution, throughput and crusher power. The PRM has been implemented as a modification of the Hertz-Mindlin contact model, through which each spherical mother particle is replaced by a distribution of daughter spherical particles every time the mother particle is subjected to a force that surpasses a maximum set value. A comprehensive description of the model used is presented elsewhere [19]. DEM simulations were validated on the basis of experiments in a laboratory jaw crusher in the size reduction of a gold ore. Additionally, a sensitive analysis of the simulation model was carried out to investigate the effects of the closed side setting (CSS) and frequency on capacity, power, compressive force and reduction ratio.

2. Materials and Methods

2.1. Materials

The material used in this study is a gold-bearing ore from the San José mine in Íquira, Huila region, in Colombia. The ore is an intrusive igneous rock from the Ibagué batholith, with phaneritic texture, coarse-to-medium grain sizes, intermediate felsic composition, being predominantly composed of granodiorites. Table 1 presents the mineralogical composition of the Íquira gold ore determined by optical microscopy, which shows that it is composed mainly of quartz and feldspar as main gangue minerals, pyrite, carbonates with a smaller percentage of other metallic minerals such as galena, hematite and arsenopyrite [20].

Table 1. Mineral composition of the gold ore from Íquira-Huila

Mineral	Percent (%)
Gangue minerals (quartz and feldspar)	62
Pyrite	23
Carbonates	7
Chalcopyrite	4
Hematite	2
Sphalerite	1
Galena	1

Figure 1 shows a snapshot of the Run of Mine sample collected for testing. The particles were classified into three narrow sizes for testing: 63/53, 31.5/26.5 and 16.0/13.2 mm.

(a) (b)

Figure 1. Run of Mine sample of the Íquira-Huila gold ore (**a**) and particles classified in narrow sizes (**b**).

2.2. Laboratory-Scale Jaw Crushing Tests

Figure 2 shows a snapshot of the Otsuka Iron Works Ltd. laboratory-scale jaw crusher used in the present study, which has the universal design [2]. Table 2 summarizes the main operational parameters of the machine, used in the experimental tests. Each batch crushing test was conducted using about 6 kg of sample with particles contained either in narrow sizes or with a distribution of sizes.

The jaw crusher throughput for each experimental test was measured using an integrating load cell coupled to an Arduino UNO data acquisition device. The net power was calculated from measurements with an amperage multimeter, whereas the product particle size distributions were measured using a $\sqrt{2}$ series of sieves in a laboratory sieve shaker.

Table 2. Operational parameters used in the laboratory-scale jaw crushing experimental tests

Parameters	Units	Value
Nominal throughput	kg/h	300
Power	kW	2.2
Main shaft frequency	rpm	400
Swing jaw stroke frequency	Hz	6.0
Feed (top) opening	mm × mm	140×90
Discharge opening	mm × mm	140×7.5 (closed)
Swing jaw throw	mm	10

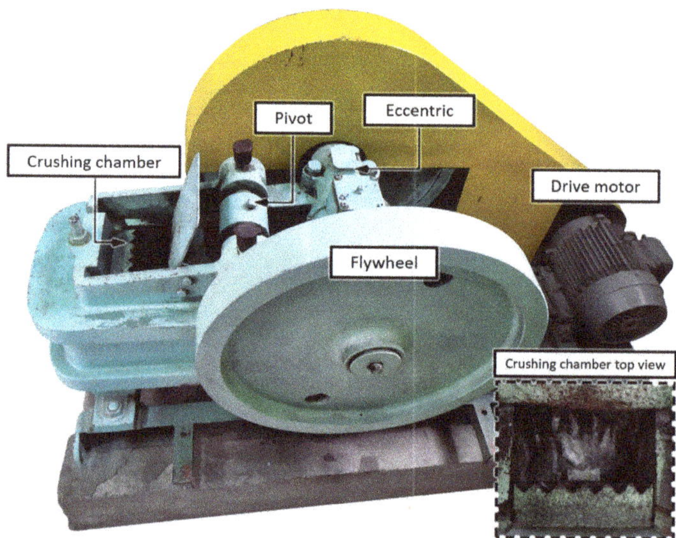

Figure 2. Laboratory-scale jaw crusher at the Colombian Geological survey, with the insert showing the top view of the crushing chamber.

Figure 3 shows the jaw crusher scheme indicating the main components of the machine and the operating variables such as the discharge opening and the displacement of the swing jaw.

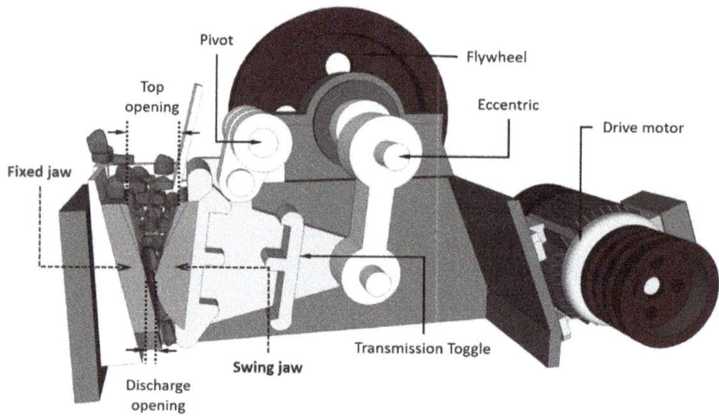

Figure 3. Scheme of the jaw crusher showing the main components of the machine.

2.3. DEM Particle Replacement Model Parameter Calibration

The particle replacement model (PRM) used in this work may be used to describe fragmentation of individual particles under compression or impact. It consists of instantaneously replacing a spherical particle by progeny fragments every time a critical condition for failure is met. Spheres that make up the progeny fragments are allowed to overlap each other at the instant of replacement and a relaxation factor of the repulsive force was applied to prevent them from explosively repelling each other given the large contact forces involved. The version used in the present work is a custom PRM implemented in the software EDEM®, using an Advanced Programming Interface (API). An extended description of the model is presented in the work of Barrios et al. (2020) [19].

The PRM formulation used in the present work has been successfully used previously in DEM simulations describing breakage of individual and beds of particles by impact [17], as well as the interaction of particles and complex geometries in High-Pressure Grinding Rolls [21].

The calibration of the PRM parameters was conducted from uniaxial compression tests of individual unconfined irregular particles using a universal press manufacture by Maekawa with a capacity of 400 kN, applying a methodology similar to that used by Qian et al. [22]. Displacement during compression was measured using a digital camera, and the force was measured using a load cell with a maximum capacity of 5 kN, coupled to an Arduino UNO data acquisition device connected to a laptop (Figure 4). Each test consisted of compressing 30 particles contained in each narrow size range.

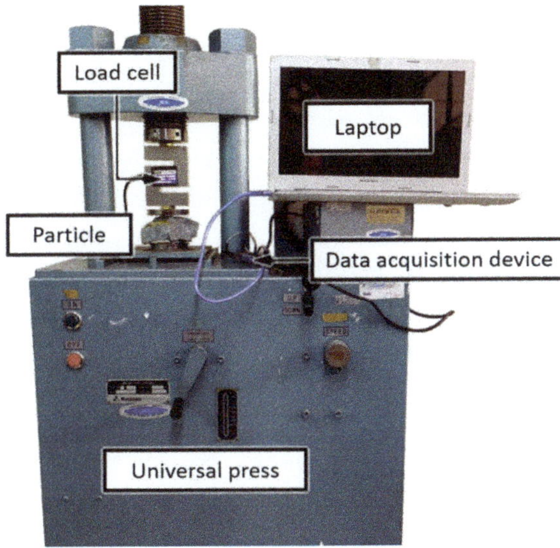

Figure 4. Uniaxial compression system used in unconfined testing of particles.

2.4. Jaw Crusher DEM Simulations

Tables 3 and 4 show individual and contact parameters of the materials used in the DEM simulations. These DEM parameters were chosen based on the work of Rodriguez et al. [23], and validated on the basis of repose angle results from the EDEM® software database (Generic EDEM Material Model Database—GEMM).

Table 3. Individual material parameters used in the DEM simulations

Parameters	Units	Steel	Ore
Density	kg/m^3	7800	2700
Shear modulus	Pa	7×10^9	1×10^8
Poisson's ratio	-	0.30	0.25

Table 4. Contact parameters used in the DEM simulations

Parameters	Units	Steel–Ore	Ore–Ore
Coefficient of restitution	-	0.37	0.35
Coefficient of friction	-	0.20	0.34
Coefficient of rolling friction	-	0.10	0.25

Material and contact parameters in Tables 3 and 4 were used both in calibration of the PRM parameters and in the DEM simulations of the jaw crusher.

EDEM® 2019 was used in the DEM simulations of the laboratory-scale jaw crusher. The CAD model of the jaw crusher geometry was designed using the software SketchUp (Boulder, CO, USA), based on the Otsuka Iron Works manual, as well as direct measurements of the discharge opening and the jaw wear surfaces. Table 5 shows the ranges of operating variables used in the DEM simulations of the jaw crusher.

The throughput of the simulated jaw crusher was obtained using the "flow sensor" feature of the EDEM® post processing module. The compressive force and the power on the swing and fixed jaws were calculated extracting the force and the torque on each element of the geometry. Finally, the particle size distribution of the product was calculated based on the mass of the particles produced by the PRM.

Table 5. Operating conditions adopted in the DEM simulations of the jaw crusher

Parameter	Units	Range	
Discharge opening	mm	2.5–7.5–12.5	
Swing jaw stroke frequency	Hz	0.5–3.0–6.0–9.0	
Feed particle narrow sizes	mm	63.0/52.0	100%
		31.5/26.5	100%
		16.0/13.2	100%
Feed particle size distribution	mm	63.0/52.0	33.3%
		31.5/26.5	33.3%
		16.0/13.2	33.3%

3. Results

3.1. Calibration of Particle Replacement Model Parameters

Figure 5 shows the comparison between the experimental and the DEM simulation set-up of the single particle compression test, used to calibrate the PRM parameters of mother particle fragmentation and breakage force.

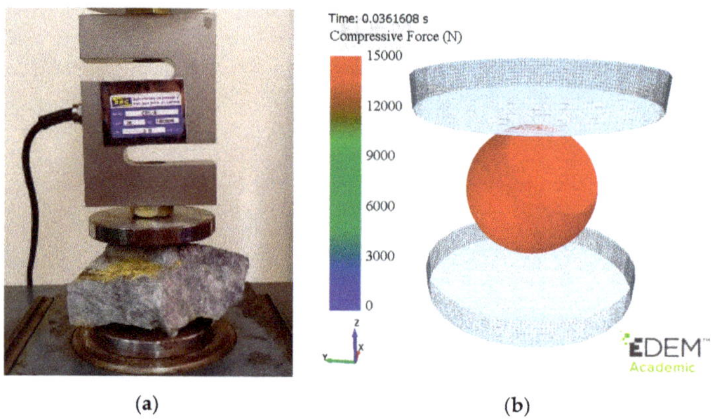

(a)   (b)

Figure 5. Experimental (**a**) and DEM simulation (**b**) set up of the single particle uniaxial compression test.

Figure 6a shows a comparison between the experimental and simulated force-deformation profiles of a single particle obtained from a compression test. In the figure, some of the limitations of the PRM, already discussed by Jimenez-Herrera et al. [17], in describing the fine details of the breakage process, are evident. Figure 6b shows the cumulative distribution of specific fracture energies for each size class. From the experimental distribution the median values of breakage force that is used to calibrate the threshold breakage force parameter of the DEM Particle Replacement Model, above which mother particles break and are replaced by daughter particles, is calculated.

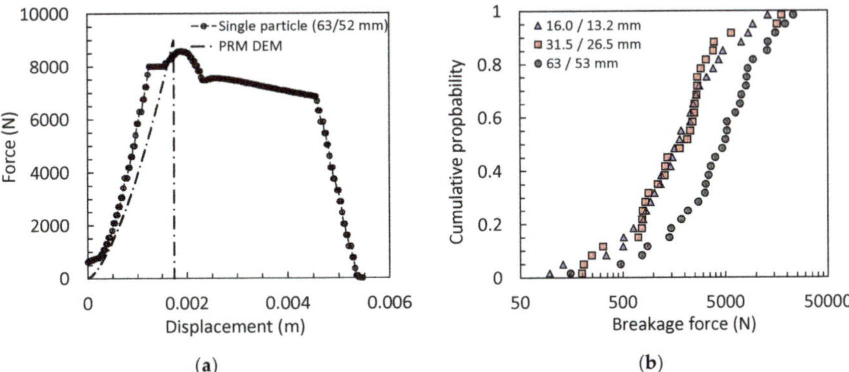

Figure 6. Force-deformation profile of a single particle compression test (experiment and simulation) (**a**) and distribution of breakage forces for three size classes (**b**).

Figure 7a shows the experimental product size distributions from the jaw crushing tests conducted using the different feeds, including the three different narrow size classes and the particle size distribution. Figure 7b compares results from modeling and calibration of the DEM PRM spherical daughter particle distribution based on the experimental results of the jaw crushing test using a distributed feed. It was fitted using a primary distribution in which every breakage event results in a generation of daughter particle contained in three size classes: 1 particle with a size ratio equal to 0.595 of the mother particle, 8 particles with size ratio equal to 0.354 of the mother particle and the last with 52 particles with size ratio equal to 0.210 of the mother particle. In addition, the value of the relaxation factor b_L equal to 0.0524 of the particle replacement model [19] was selected, which is responsible for capping the normal force calculated using the overlap of the daughter particles, so as to prevent the appearance of extremely high velocities of the fragments from breakage which would make simulations unrealistic.

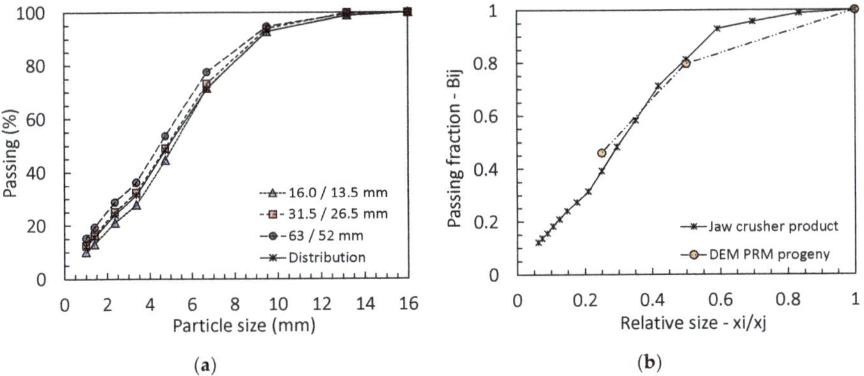

Figure 7. Product size distributions from laboratory jaw crushing tests with different feeds (**a**) and DEM PRM daughter spheres distribution modeled (**b**).

3.2. DEM Simulations of the Laboratory Jaw Crusher

Figure 8a is a snapshot of the DEM simulation of particles being crushed in the virtual jaw crusher. In this perspective, a view of the breakage of particles, modeled using the Particle Replacement Model, may be observed. On the other hand, Figure 8b shows qualitative results on the velocity profile of the particles along the crushing chamber. In this figure, the higher velocities of the particles appear closer to the discharge, given the greater freedom of the fragments to move downwards in the chamber as they become progressively finer. Such a result shows the capabilities of the DEM model to represent the dynamics of the particles that are expected to appear in real jaw crushers.

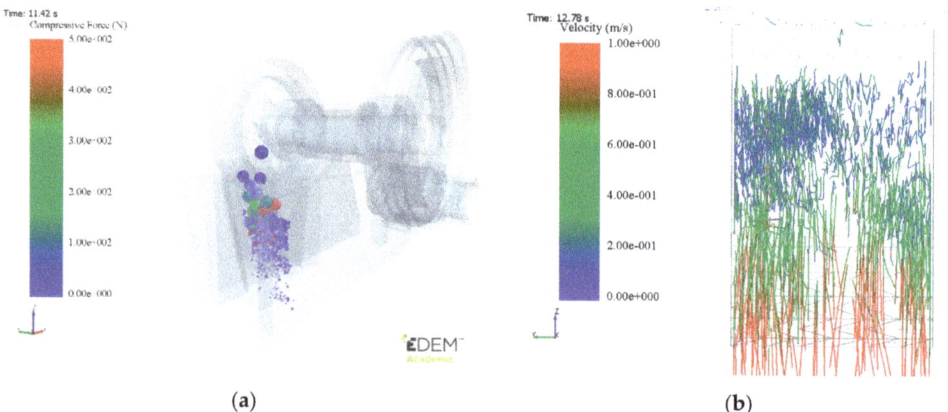

(a)   (b)

Figure 8. Perspective view of DEM simulation of the laboratory jaw crusher (**a**) and particle velocities profile through the crushing chamber (**b**).

Figure 9 shows a comparison between the experimental and simulated breakage of a single particle contained in the 63/53 mm range inside the jaw crusher chamber. The DEM simulation coupled to the PRM is capable to provide a valid qualitative representation of fragmentation observed in the experiment.

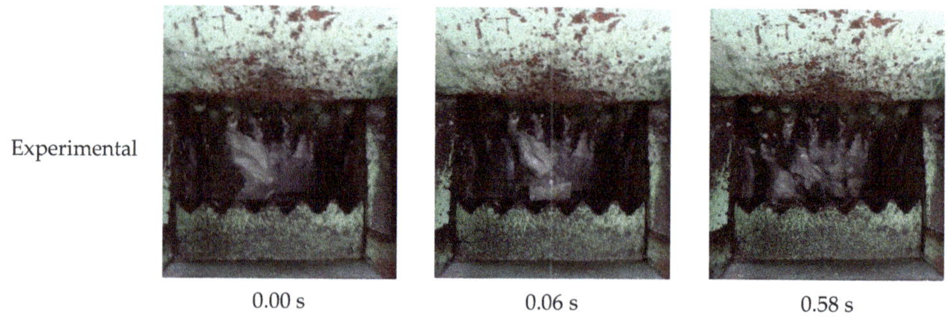

Experimental   0.00 s   0.06 s   0.58 s

Figure 9. *Cont.*

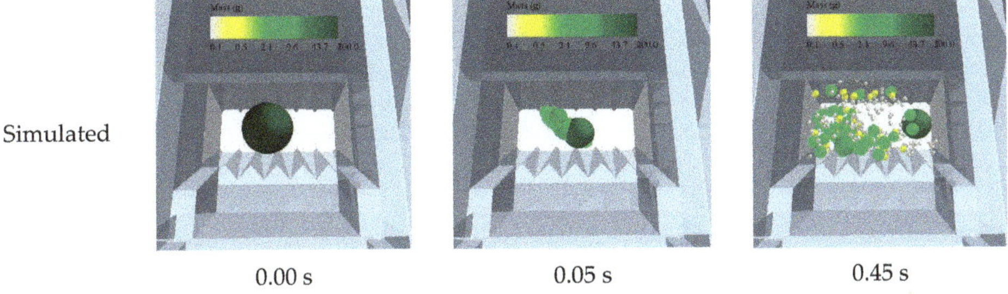

Figure 9. Comparison of experimental and simulated breakage of a single 63/52-mm particle in the jaw crusher chamber.

Simulation results are shown in Figure 10 for different feed particle sizes, showing the qualitative results of the particle bed compression within the crushing chamber. The compressive force on the particles is represented by the colored lateral bar. Particle visualization shows the plate zone where the particles are broken according to their sizes and the differences in the stress field exerted on the plates according to particle size. As expected, given the V-shape of the jaw plates, high stresses are experienced at the highest positions along the plate when coarser particles are fed and immediately get in contact with them. However, in order to simulate more precisely the behavior of the ore and conclude about the real stress patterns on the plates, it is necessary to take into account the decompression features of the PRM in the instant in time immediately after particle breakage. Such an improved description could be reached by an improved calibration of the relaxation parameter of the Particle Replacement Model, as presented by Barrios et al. [19].

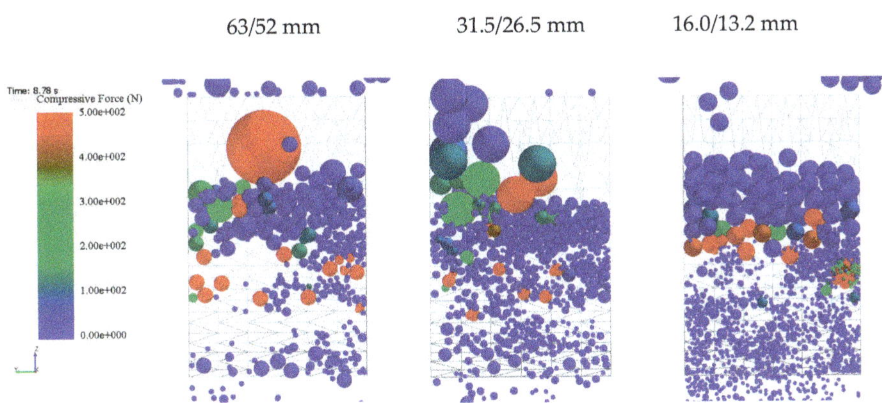

Figure 10. Particle bed compression inside the crushing chamber for different particle size classes fed to the crusher.

An equivalent analysis of the forces exerted on the plates can be seen from direct evaluation of the stress intensity on the plate geometry mesh (Figure 11). The figure is a snapshot of the DEM simulation that shows qualitatively the variation of the compressive force on the geometry of the swing and fixed plates. From the simulation it is observed that the area that is subjected to the highest stresses is located closest to the discharge opening, for all feed sizes. This is consistent with observations by Lindqvist and Evertsson [24], who studied the wear of the plates of a jaw crusher from both experiments and analytical models on the basis of measurement of pressure and forces exerted on the fixed and swing jaws. Figure 11 also shows that, when the crusher was fed with the coarser particles (63/52 mm), the plates were subjected to higher compressive forces.

Validation of the DEM simulation was carried out by comparing it to the experimentally measured values of throughput, power, and product size distribution. For instance, the mean net power demanded in experiments in which the crusher was fed with 63/52 mm material was 0.34 kW, whereas simulations yielded a mean value of 0.32 kW, thus demonstrating the very good agreement.

Figure 12 shows the variation of measured and the predicted net power of the jaw crusher, with good agreement between simulations and experiments. In the experiments, the highest powers were obtained when crushing the coarsest feed particles (63/52 mm). The fluctuation in the values were also found to be smaller for finer feeds. The peaks in power intensity represent the instants in which the machine squeezes and fractures either individual particles or assemblies of particles as they move down the crushing chamber, whereas the low values represent lack of particles being compressed by the crusher plates in those moments in time.

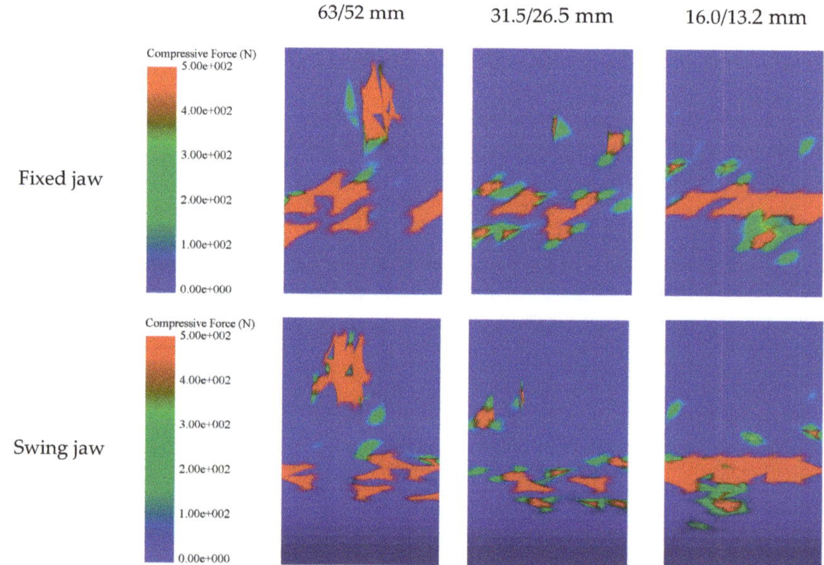

Figure 11. Snapshot of the compressive force on the geometries (fixed and swing jaws) for different feed particle sizes.

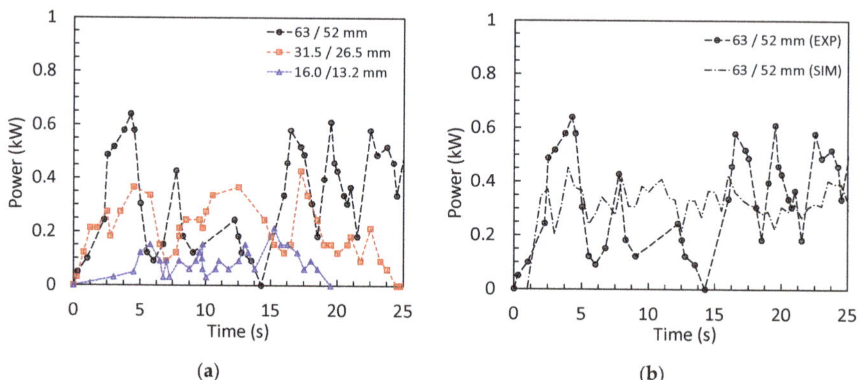

Figure 12. Measured (**a**) and comparison of measured and simulated (**b**) net power of the laboratory jaw crusher.

Figure 13 shows the variation of the throughput during the tests, which demonstrates that an increase in feed size resulted in a reduction in crusher throughput. The figure also shows the good agreement between the DEM simulation results and the measurements using the integrating load cell.

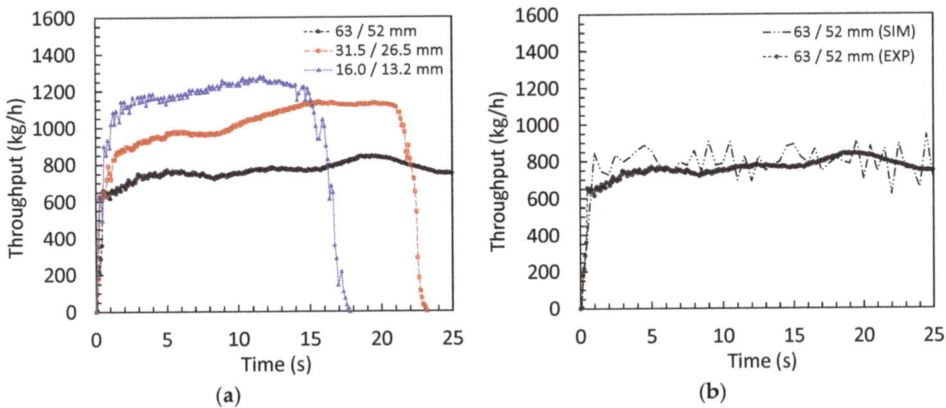

Figure 13. Experimental (**a**) and DEM simulations and experiment (**b**) of the laboratory jaw crusher throughput.

A more detailed comparison between experiments and simulation results is presented in Figure 14. The results show the good agreement between the experiments and the DEM simulations, in which the simulations captured very well the trends observed in the experiments regarding the effect of feed particle size on throughput and net power. Indeed, the reduction in throughput as particle size increased was well described by the DEM simulation (Figure 14a), as well as the increase in power required to crush the coarser feed particles (Figure 14b). In addition, Figure 14c compares measured and predicted product size distributions for the case in which the crusher was fed with a particle size distribution (Table 5). It shows very good agreement with the experimental results. Nevertheless, these are limited to the minimum sphere size simulated of 2.36 mm, given the additional computational cost associated to simulating finer particles.

Figure 14d shows the simulated compressive force for different feed particle sizes. The model shows that forces increase with feed particle size. This trend agrees with results from single particle experiments, as demonstrated by Tavares and King [25], as well as results from the present work (Figure 6). Unfortunately, it was not possible to validate this in the present work due to lack of proper instrumentation in the jaw crusher, such as load cells or others sensors, installed on the plates to register the changes in force, as carried out in earlier studies [24,26].

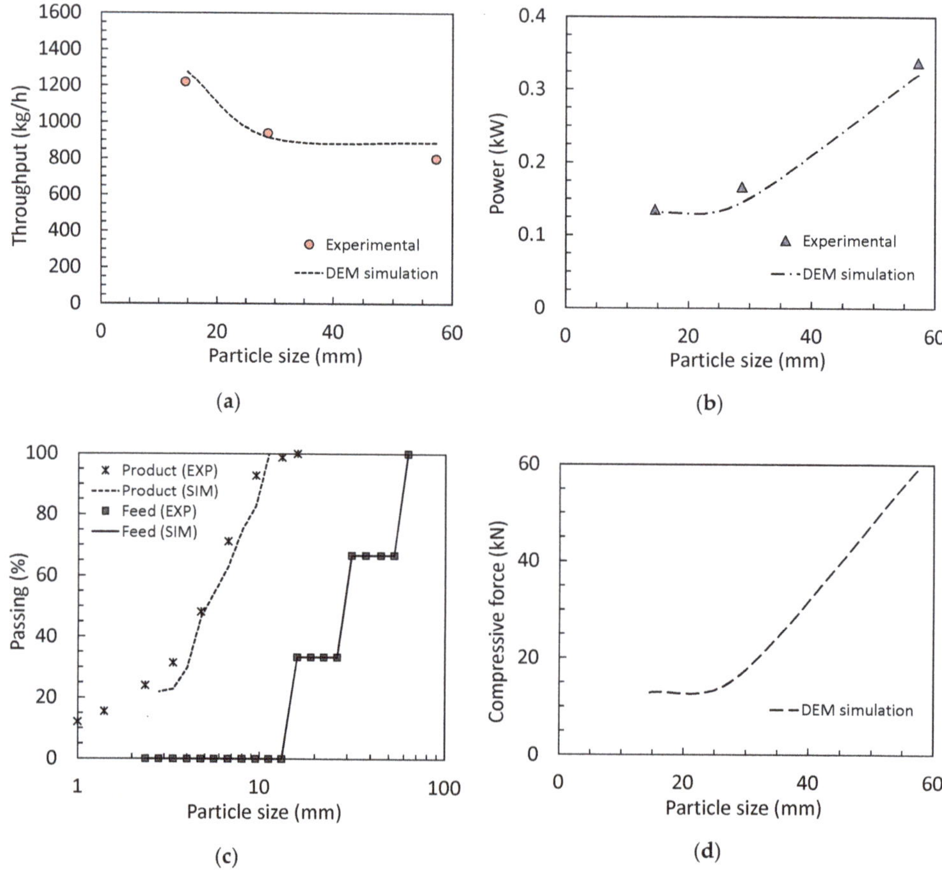

Figure 14. Comparison between experiments and DEM simulations for the jaw crusher throughput (**a**), power (**b**) and product size distribution (**c**); simulated trend on the compressive force (**d**). Figures a, b and d presented as a function of mean feed particle size.

3.3. Sensitivity Analysis of the Jaw Crusher DEM Model

Figure 15 shows the product particle size distributions for different stroke frequencies (Figure 15a) and closed side settings (Figure 15b) from the laboratory jaw crusher DEM simulations. Figure 15a shows a finer product for higher stroke frequencies for the same CCS, whereas Figure 15b shows a finer product for smaller CSS for a fixed value of frequency. A similar behavior of the product size distribution with frequency and closed side setting was found in earlier studies [6,14].

Time series showing the variation of the crusher throughput in the simulations as a function of frequency are presented in Figure 16. These data were extracted from the "particle mass flow sensor" of EDEM and filtered using a statistical moving mean (continuous line). The non-filtered data correspond to the markers. These data were extracted from the virtual experiments that considered a feed composed of multiple sizes operating in partially choke and non-choke feed condition during 7.5 s.

Results show the dynamics of the jaw crusher DEM model and the throughput response to the disturbances given by the stroke frequency of the swing plate. As observed in Figure 16, the DEM simulations respond to the changes in frequency. Additionally, more stable behavior with less scattered results from operation at the frequency of 9.0 Hz.

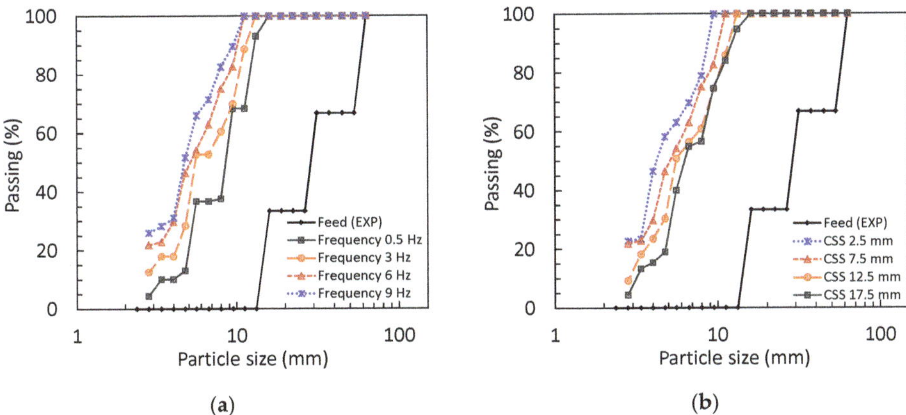

(a)   **(b)**

Figure 15. Product particle size distributions for different stroke frequencies and a constant CSS of 7.5 mm (**a**) and for different closed side settings for a fixed frequency of 6 Hz (**b**) in DEM simulations of the laboratory jaw crusher for a feed particle size distribution (Table 5).

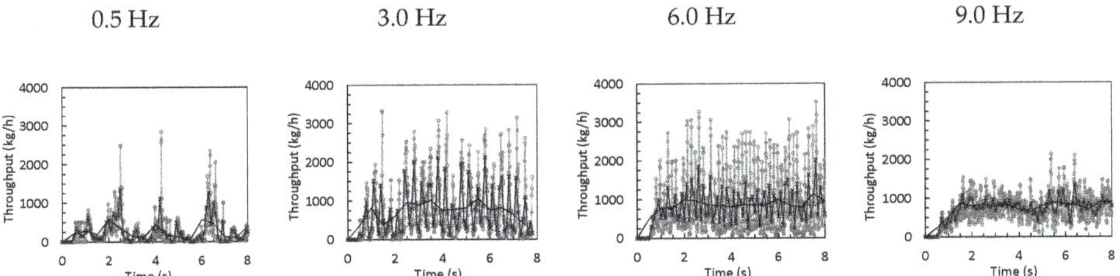

Figure 16. DEM jaw crusher simulation results showing raw throughput data for different stroke frequencies. Crusher fed with a range of sizes (63–13 mm) and CSS constant at 7.5 mm.

Figure 17 presents results on the sensitivity analyzes of the jaw crusher DEM model with the embedded Particle Replacement Model. It shows the variation of crusher capacity, compressive force, power and reduction ratio as a function of stroke frequency and closed side setting.

From Figure 17a, it is evident that throughput increases with closed side setting. This has been observed in several studies in the literature [6]. Regarding frequency, capacity initially increases, reaches a maximum value and then drops at higher frequencies. This trend is less pronounced for intermediate values of CSS, whereas it disappears at the largest values of CSS, in which throughput increases monotonically with frequency. As such, it is possible to conclude that high values of CSS combined with high frequencies apparently allow reaching the highest throughputs. The variation of throughput as a function of frequency was also described by Rose and English [10] using an analytical model, whereas the relationship among these variables has already been object of simulations by Johansson et al. [14] who found similar trends.

The effect of CSS and frequency on crusher power is shown in Figure 17b. It allows to conclude that power demanded by the jaw crusher increases significantly with frequency as well as with a reduction in closed side setting. A similar general trend was found elsewhere [14].

A further examination on the effectiveness of the crusher can be extracted from analyzing the reduction ratio (R_{80}). It expresses the ratio between the feed (F_{80}) and the product (P_{80}) 80% passing sizes of an operation [27]. Experimentally, it is typically desirable to reach the maximum values of reduction ratio for a given specific energy consumption. From the DEM model (Figure 17c) it is possible to conclude that the highest values of R_{80} are reached both with the smallest CSS as well as the highest frequency. The same trend was also experimentally found by Fladvad and Onnela [6].

Figure 17d presents the resulting compressive force on the plates as a function of frequency and CSS. The highest compressive forces are achieved with smaller values of CSS. The results also show a modest increase in compressive forces with frequency.

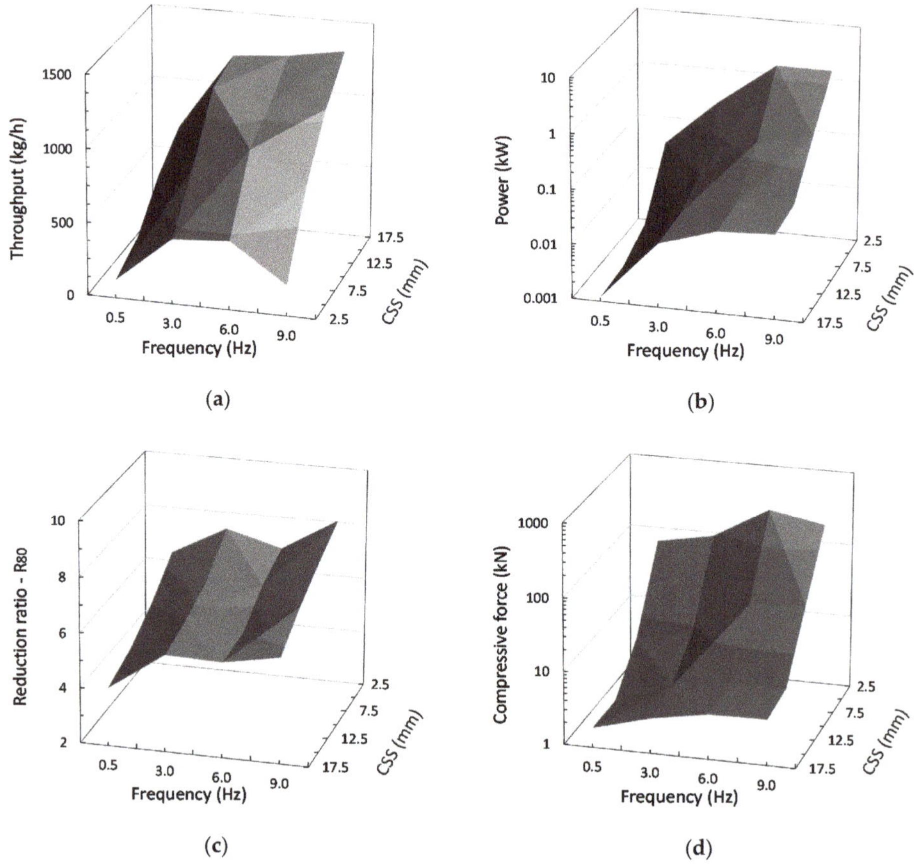

Figure 17. Response surface analysis of the jaw crusher DEM simulations showing throughput (**a**), power (**b**), reduction ratio (**c**) and compressive force (**d**) as a function of closed side setting (CSS) and stroke frequency.

4. Discussion

The DEM simulations results demonstrate the validity of properly calibrated simulations of the jaw crusher performance as a tool for improved design and optimization of industrial equipment. The approach of running simulations at a small (laboratory) scale allows the concept to be proved and the technology to be validated. The following step could be its application in the design of a customized jaw crusher with the aim of improving crusher performance. This could represent, for instance, setting the crusher frequency and stroke amplitude to values that maximize the machine performance for a given feed ore. Other potential applications include analyzes of jaw crusher operational response to fluctuations in particle size distribution in the feed, ore competence, as well as feed segregation. On the other hand, such a tool could be used by equipment manufacturers to customize the machine design in terms of jaw wear surface and crusher chamber geometry.

In particular, the use of PRM is an efficient solution to represent, in a realistic way, the performance of the jaw crusher in terms of throughput, specific energy consumption, compressive force, and product particle size distribution. This model offers some computational benefits in comparison with other DEM breakage methods, such as the Bonded-Particle Model [17]. Among its greatest advantages is its reduced computational effort, ability to describe different loading conditions, ease in model parameter estimation and calibration, besides reasonably good resolution in describing the product size down to relatively fine sizes.

5. Conclusions

Simulations of a laboratory jaw crusher with a Particle Replacement Model embedded in DEM were able to reproduce the experimental performance of a laboratory jaw crusher in size reduction of a gold ore from the department of Huila–Colombia, in terms of power, throughput, and product size distribution. Very good agreement was observed between measured and predicted results obtained for different feed sizes.

The DEM model demonstrated high sensitivity to changes in CSS and swing jaw frequency, allowing to show some trends in throughput, power, reduction ratio, and compressive force. The trends observed were generally in good agreement with the literature.

Simulations of the jaw crusher incorporating a description of particle breakage can be a useful tool in design and optimization of the machine, since DEM may be an excellent tool to predict machine performance, including assessing measures that are hard to obtain experimentally, such as the stress gradient on the jaw plates.

Author Contributions: Conceptualization, G.K.B. and N.J.-H.; methodology, G.K.B.; software, G.K.B.; validation, G.B., N.J.-H. and S.N.F.-T.; formal analysis, G.K.B. and N.J.-H.; investigation, G.K.B. and N.J.-H.; resources, G.K.B., N.J.-H. and L.M.T.; data curation, G.K.B.; writing—original draft preparation, G.K.B., N.J.-H., S.N.F.-T. and L.M.T.; writing—review and editing, G.K.B., N.J.-H. and L.M.T.; visualization, G.K.B. and N.J.-H.; supervision, G.K.B. and N.J.-H.; project administration, G.K.B. and L.M.T.; funding acquisition, G.K.B., N.J.-H., S.N.F.-T. and L.M.T. All authors have read and agreed to the published version of the manuscript.

Funding: One of the authors (L.M.T.) would also like to thank funding from the Brazilian Agencies CNPq (grant number 310293/2017-0) and FAPERJ (grant number E-26/202.574/2019).

Acknowledgments: The authors wish to thank the Servicio Geológico Colombiano, based on Cali for the support by means of its infrastructure, equipment and laboratories. The authors also thank the EDEM Company for providing the EDEM software by means of "Take EDEM with you" program.

Conflicts of Interest: The authors declare no conflict of interest.

References

1. Mular, A.L.; Halbe, D.N.; Barratt, D.J. *Mineral Processing Plant Design, Practice, and Control: Proceedings*; SME: Englewood, CO, USA, 2002; Volume 1, ISBN 0873352238.
2. Wills, B.A.; Napier-munn, T. Preface to 7th Edition. In *Wills' Mineral Processing Technology*; Butterworth-Heinemann: Oxford, UK, 2006; ISBN 0750644508.
3. Gupta, A.; Yan, D.S. *Mineral Processing Design and Operations: An Introduction*; Elsevier: Amsterdam, The Netherlands, 2016; ISBN 0444635920.
4. Olaleye, B.M. Influence of some rock strength properties on jaw crusher performance in granite quarry. *Min. Sci. Technol.* **2010**, *20*, 204–208. [CrossRef]
5. Naziemiec, Z.; Gawenda, T.; Tumidajski, T.; Saramak, D. The influence of transverse profile of crusher jaws on comminution effects. In Proceedings of the IMPC 2006—Proceedings of 23rd International Mineral Processing Congress, Istanbul, Turkey, 3–8 September 2006; pp. 69–74.
6. Fladvad, M.; Onnela, T. Influence of jaw crusher parameters on the quality of primary crushed aggregates. *Miner. Eng.* **2020**, *151*. [CrossRef]
7. Zeng, Y.; Zheng, M.; Forssberg, E. Monitoring jaw crushing parameters via vibration signal measurement. *Int. J. Miner. Process.* **1993**, *39*, 199–208. [CrossRef]

8. Olawale, J.O.; Ibitoye, S.A. Failure analysis of a crusher jaw. In *Handbook of Materials Failure Analysis*; Butterworth-Heinemann: Oxford, UK, 2018; pp. 187–207. ISBN 9780081019283.

9. Broman, J. Optimizing capacity and economy in jaw and gyratory crushers. *Eng. Min. J.* **1984**, *185*, 69–71.

10. Rose, H.E.; English, J.E. Theoretical analysis of the performance of jaw crushers. *IMM Trans.* **1967**, *76*, C32.

11. Andersen, J.S.; Napier-Munn, T.J. The influence of liner condition on cone crusher performance. *Miner. Eng.* **1990**, *3*, 105–116. [CrossRef]

12. Lynch, A.J. *Mineral Crushing and Grinding Circuits: Their Simulation, Optimisation, Design, and Control*; Elsevier Science Ltd: Amsterdam, The Netherlands, 1977; Volume 1.

13. Nikolov, S. A performance model for impact crushers. *Miner. Eng.* **2002**, *15*, 715–721. [CrossRef]

14. Johansson, M.; Bengtsson, M.; Evertsson, M.; Hulthén, E. A fundamental model of an industrial-scale jaw crusher. *Miner. Eng.* **2017**, *105*, 69–78. [CrossRef]

15. Fusheng, M.; Hui, L.; Xingxue, L.; Hongzhi, X. Jaw crusher based on discrete element method. *Appl. Mech. Mater.* **2013**, *312*, 101–105. [CrossRef]

16. Legendre, D.; Zevenhoven, R. Assessing the energy efficiency of a jaw crusher. *Energy* **2014**, *74*, 119–130. [CrossRef]

17. Jiménez-Herrera, N.; Barrios, G.K.P.; Tavares, L.M. Comparison of breakage models in DEM in simulating impact on particle beds. *Adv. Powder Technol.* **2018**, *29*, 692–706. [CrossRef]

18. Cleary, P.W.; Sinnott, M.D. Simulation of particle flows and breakage in crushers using DEM: Part 1—Compression crushers. *Miner. Eng.* **2015**, *74*, 178–197. [CrossRef]

19. Barrios, G.K.P.; Jiménez-Herrera, N.; Tavares, L.M. Simulation of particle bed breakage by slow compression and impact using a DEM particle replacement model. *Adv. Powder Technol.* **2020**. [CrossRef]

20. SGC. Guía Metodológica Para el Mejoramiento Productivo del Beneficio de Oro Sin el Uso de Mercurio: Íquira (Huila). 2018. Available online: http://srvags.sgc.gov.co/Archivos_Geoportal/Geologia/Guia-metodologica-Iquira-Huila.pdf (accessed on 30 March 2020).

21. Barrios, G.K.P.; Tavares, L.M. A preliminary model of high pressure roll grinding using the discrete element method and multi-body dynamics coupling. *Int. J. Miner. Process.* **2016**, *156*, 32–42. [CrossRef]

22. Qian, G.; Lei, W.S.; Yu, Z.; Berto, F. Statistical size scaling of breakage strength of irregularly-shaped particles. *Theor. Appl. Fract. Mech.* **2019**, *102*, 51–58. [CrossRef]

23. Rodriguez, V.A.; de Carvalho, R.M.; Tavares, L.M. Insights into advanced ball mill modelling through discrete element simulations. *Miner. Eng.* **2018**, *127*, 48–60. [CrossRef]

24. Lindqvist, M.; Evertsson, C.M. Linear wear in jaw crushers. *Miner. Eng.* **2003**, *16*, 1–12. [CrossRef]

25. Tavares, L.M.; King, R.P. Single-particle fracture under impact loading. *Int. J. Miner. Process.* **1998**, *54*, 1–28. [CrossRef]

26. Terva, J.; Kuokkala, V.; Valtonen, K.; Siitonen, P. Effects of compression and sliding on the wear and energy consumption in mineral crushing. *Wear* **2018**, *398–399*, 116–126. [CrossRef]

27. Taggart, A.F. *Handbook of Mineral Dressing*; Wiley: Hoboken, NJ, USA, 1945; Volume 1.

Article

An LSTM Approach for SAG Mill Operational Relative-Hardness Prediction

Sebastian Avalos [1,*], Willy Kracht [2,3] and Julian M. Ortiz [1]

[1] The Robert M. Buchan Department of Mining, Queen's University, Kingston, ON K7L 3N6, Canada; julian.ortiz@queensu.ca
[2] Department of Mining Engineering, Universidad de Chile, Santiago 8370448, Chile; wkracht@uchile.cl
[3] Advanced Mining Technology Center, AMTC, Universidad de Chile, Santiago 8370451, Chile
* Correspondence: sebastian.avalos@queensu.ca

Received: 1 August 2020; Accepted: 18 August 2020; Published: 20 August 2020

Abstract: Ore hardness plays a critical role in comminution circuits. Ore hardness is usually characterized at sample support in order to populate geometallurgical block models. However, the required attributes are not always available and suffer for lack of temporal resolution. We propose an operational relative-hardness definition and the use of real-time operational data to train a Long Short-Term Memory, a deep neural network architecture, to forecast the upcoming operational relative-hardness. We applied the proposed methodology on two SAG mill datasets, of one year period each. Results show accuracies above 80% on both SAG mills at a short upcoming period of times and around 1% of misclassifications between soft and hard characterization. The proposed application can be extended to any crushing and grinding equipment to forecast categorical attributes that are relevant to downstream processes.

Keywords: semi-autogenous grinding mill; operational hardness; energy consumption; mining; deep learning; long short-term memory

1. Introduction

In mining operations, the primary energy consumer is the comminution system, responsible for more than half of the entire mine consumption [1]. From all pieces of equipment that integrate the comminution circuit, the semi-autogenous grinding mill (SAG) is perhaps the most important in the system. With an aspect ratio of 2:1 (diameter to length), these mills combine impact, attrition and abrasion to reduce the ore size. SAG mills are located at the beginning of the comminution circuits, after a primary crushing stage. Although there are small SAG mills, their size usually ranges from 9.8 × 4.3 to 12.8 × 7.6 m, with a nominal energy demand of 8.2 and 26 MW, respectively [2], which make SAG mills the most relevant energy consumer within the concentrator. Modelling their consumption behaviour supports the operational control and energy demand-side management [3].

Most theoretical and empirical models [4–6] demand input feed characteristics, such as hardness, size distribution and inflow rate, SAG characteristics, such as sizing and product size distribution, and operational variables such as bearing pressure, water addition and grinding charge level. Although they are suitable to provide adequate design guidelines, they lack accurate in-situ inference since most assume steady-state and isolation from up and downstream processes. In response, model predictive control, SAG MPC [7], combines those methods with real-time operational information. However, expert knowledge is required to model the SAG mill dynamics properly.

From a geometallurgical perspective, the integration of new predictive methods that account for space and time relationships over real-time attributes has been defined as a fundamental challenge [8,9] in mining operations, particularly in an integrated system such as comminution. In response, data-driven approaches have been proposed ranging from support vector machines [10] and gene expression programming [11] to hybrid models that combine genetic algorithms and neural networks [12] and recurrent neural networks [13]. As data-driven methods are sensitive to the context (available information) and representation (information workflow), the authors have studied the use of several machine learning and deep learning methods in modelling the SAG energy consumption behaviour based only on operational variables [14].

The energy consumed by a SAG mill is related to several factors such as expert operator decisions, charge volume, charge specific gravity and the hardness of the feed material. Knowing the output hardness material becomes relevant for the downstream stage in the primary grinding circuit. Ore hardness can be characterized at sample support by combining the logged geological properties and the result of standardized comminution tests. They can be used to predict the hardness of each block sent to the process. However, these attributes are not always available. In response, a qualitative characterization of the ore hardness processed at time t, relative to the operational hardness of the ore processed at time $t + 1$ can be done using only operational variables rather than a set of mineralogical characterizations. This qualitative characterization is referred and here used as operational relative-hardness (ORH).

We take advantage of previous works [14] by knowing that the Long Short-Term Memory (LSTM) [15] outperforms other machine learning and deep learning techniques on inferring the SAG mill energy consumption. Therefore, Section 2 presents the ORH and LSTM models, Section 3 establishes the SAG mill experimental framework, the results of which are presented in Section 4, and conclusions are drawn in Section 5.

2. Model

2.1. Operational Relative-Hardness Criteria

From the several operational parameters that can be captured and associated to SAG mill operations, we consider the energy consumption (EC) and feed tonnage (FT) to build our operational relative-hardness criteria.

Let us assume that data $\{EC,FT\}_t$ is collected over a period of time T using a Δt discretization. By considering the one-step forward time difference of energy consumption ($\Delta EC_t = EC_{t+1} - EC_t$) and feed tonnage ($\Delta FT_t = FT_{t+1} - FT_t$), a qualitative assessment of the operational relative-hardness can be done. For instance, if the energy consumption is increasing and the feed tonnage is constant, it can be interpreted as an increase in ore hardness relative to the previous period. Similarly, if the feed tonnage is constant and the energy decreases, a decrease in ore hardness relative to the previous period can be assumed. Particularly, when both ΔEC_t and ΔFT_t show the same behaviour, the SAG can be either processing ore with medium operational relative-hardness or being filled up or emptied. To avoid misclassification in this last case, the operational relative-hardness is labelled as undefined. Table 1 summarizes the nine combinations of states and the associated operational relative-hardness.

The qualitative labelling of ΔEC_t and ΔFT_t as increasing, constant or decreasing can be established based on their global distribution over the period T as:

$$\Delta EC_t = \begin{cases} \text{Increasing} & \text{if } \Delta EC_t > \lambda \cdot \sigma_{\Delta EC} \\ \text{Constant} & \text{if } |\Delta EC_t| \leq \lambda \cdot \sigma_{\Delta EC} \\ \text{Decreasing} & \text{if } \Delta EC_t < -\lambda \cdot \sigma_{\Delta EC} \end{cases} \qquad \Delta FT_t = \begin{cases} \text{Increasing} & \text{if } \Delta FT_t > \lambda \cdot \sigma_{\Delta FT} \\ \text{Constant} & \text{if } |\Delta FT_t| \leq \lambda \cdot \sigma_{\Delta FT} \\ \text{Decreasing} & \text{if } \Delta FT_t < -\lambda \cdot \sigma_{\Delta FT} \end{cases} \qquad (1)$$

where $\sigma_{\Delta EC}$ and $\sigma_{\Delta FT}$ represent the standard deviations over the period T of EC and FT, respectively, and λ is a scalar value that modulates the labelling distribution. Note that (i) a λ value above 1.5 would make the entire definition meaningless since most values would remain as constant, and (ii) the λ value definition is an external model parameter and can be guided either subjectively or via statistical meaning.

Table 1. Operational relative-hardness criteria based on one time-step difference of energy consumption and feed tonnage.

Energy Consumption	Feed Tonnage	Operational Relative-Hardness
Constant	Decreasing	Hard
Increasing	Constant	Hard
Increasing	Decreasing	Hard
Decreasing	Decreasing	Undefined
Increasing	Increasing	Undefined
Constant	Constant	Undefined
Constant	Increasing	Soft
Decreasing	Constant	Soft
Decreasing	Increasing	Soft

2.2. Long Short-Term Memory

The Long Short-Term Memory (LSTM) [15] neural network architecture belongs to the family of recurrent neural networks in Deep Learning [16]. They are suitable to capture short and long term relationships in temporal datasets. Internally, LSTM applies several combinations of affine transformations, element-wise multiplications and non-linear transfer functions, for which the building blocks are:

- x_t: input vector at time t. Dimension $(m, 1)$.
- $\mathbf{W}_f, \mathbf{W}_i, \mathbf{W}_c, \mathbf{W}_o$: weight matrices for x_t. Dimensions (n_H, m).
- h_t: hidden state at time t. Dimension $(m, 1)$.
- $\mathbf{U}_f, \mathbf{U}_i, \mathbf{U}_c, \mathbf{U}_o$: weight matrices for h_{t-1}. Dimensions (n_H, m).
- $\mathbf{b}_f, \mathbf{b}_i, \mathbf{b}_c, \mathbf{b}_o$: bias vectors. Dimensions $(n_H, 1)$.
- $\mathbf{V}$: weight matrix for h_t as output. Dimension (K, m).
- c: bias vector for output. Dimension $(K, 1)$.

where m is the number of variables as input, K is the number of output variables, and n_H is the number of hidden units. Let $\tau \in \mathbb{N}$ be a temporal window. At each time $t \in \{1, ..., \tau\}$, the LSTM receives the input x_t, the previous hidden state h_{t-1} and previous memory cell c_{t-1}. The forget gate $f_t = \sigma(\mathbf{W}_f x_t + \mathbf{U}_f h_{t-1} + \mathbf{b}_f)$ is the permissive barrier of the information carried by x_t. The input gate $i_t = \sigma(\mathbf{W}_i x_t + \mathbf{U}_i h_{t-1} + \mathbf{b}_i)$ decides the relevance of the information carried by x_t. Note that both f_t and i_t use sigmoid $\sigma(x) = (1 + e^{-x})^{-1}$ as the activation function over a linear combination of x_t and h_{t-1}.

By passing the combination of x_t and h_{t-1} through a Tanh function, a candidate memory cell $\tilde{c}_t = Tanh(\mathbf{W}_c x_t + \mathbf{U}_c h_{t-1} + \mathbf{b}_c)$ is computed. The final memory cell $c_t = f_t \odot c_{t-1} + i_t \odot \tilde{c}_t$ is computed as a sum of (i) what to forget from the past memory cell as an element-wise multiplication ($\odot$) between f_t and c_{t-1}, and (ii) what to learn from the candidate memory cell as an element-wise multiplication ($\odot$) between i_t and $\tilde{c}_t$.

Similar to i_t and f_t the output gate $o_t = \sigma(\mathbf{W}_o x_t + \mathbf{U}_o h_{t-1} + \mathbf{b}_o)$ passes through a sigmoid function a linear combination between x_t and h_{t-1}. It controls the information passing from the current memory cell c_t to the final hidden state $h_t = Tanh(c_t) \odot o_t$ as an element-wise multiplication between o_t and $Tanh(c_t)$. At the final step τ, the output is computed as $y_\tau = (\mathbf{V}h_\tau + c)$. When dealing with more than one categorical prediction ($K > 1$), as in the present work for ORH forecasting, a softmax function is applied over y_τ to obtain the normalized probability distribution, and the category k has a probability of $\hat{p}(k) = \dfrac{\exp(y_{\tau,k})}{\sum_{c=1}^{K} \exp(y_{\tau,c})}$.

An illustrative scheme of the internal connection at time step t inside an LSTM is shown in Figure 1 (left). The ORH prediction has three categories (hard, soft and undefined) and the probability is computed at the last unit, at time step τ, as shown in the unrolled LSTM in Figure 1 (right).

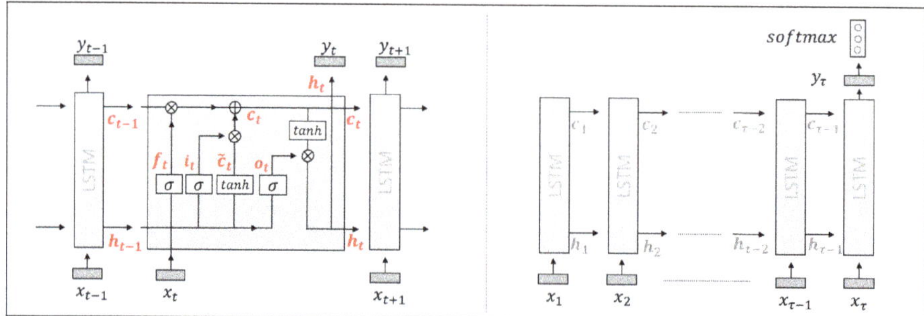

Figure 1. Schemes. Information flow inside Long Short-Term Memory (LSTM) (**left**) and unrolled LSTM where the output is computed at the last recurrence (**right**).

3. Experiment

3.1. Dataset

We used two datasets containing operational data for two independent SAG mills every half hour over a total time of 340 and 331 days, respectively. Each one of the SAG mills receives fresh feed and is connected in an open circuit configuration (SABC-B) where the pebble crusher product is sent to ball mills. At each time t, the dataset contains Feed tonnage (FT) (ton/h), Energy consumption (EC) (kWh), Bearing pressure (BPr) (psi) and Spindle speed (SSp) (rpm). They are split into two main subsets (a validation dataset is not considered since the optimum LSTM architecture to train is drawn from previous work [14]): training and testing (Table 2). This is an arbitrary division, and we seek to have a proportion of ~50/50, respectively.

Table 2. Summary statistics over training testing dataset on semi-autogenous grinding mill (SAG) mills.

SAG Mill 1				Training | Testing Dataset						
Variable	**Min**		**Mean**		**Max**		**St Dev**		**Count**	
Feed Tonnage (ton/h)	0	0	911	884	2111	1953	497	480	8170	8170
Energy Consumption (kWh)	0	0	9927	8920	12,248	10,809	1245	959	8170	8170
Bearing Pressure (psi)	0	0	12.7	11.9	13.7	13.7	2.2	2.2	8170	8170
Spindle Speed (rpm)	0	0	9.2	9.1	10.3	10.7	0.7	0.7	8170	8170
SAG Mill 2				**Training | Testing Dataset**						
Variable	**Min**		**Mean**		**Max**		**St Dev**		**Count**	
Feed Tonnage (ton/h)	0	0	2077	2073	3477	3452	1136	1134	7953	7952
Energy Consumption (kWh)	0	0	16,709	17,445	19,688	19,533	1504	1462	7953	7952
Bearing Pressure (psi)	0	0	13.8	14.8	18.3	18.3	3.5	3.8	7953	7952
Spindle Speed (rpm)	0	0	9.1	8.9	10.0	9.9	0.6	0.6	7953	7952

As it can be seen in Table 2, the predictive methods are trained with the first 50% and tested with the upcoming 50%, without being fed with the previous 50% of historical data.

Note that the comminution properties of the ore, such as $a \times b$ or BWi, are not included in the datasets; therefore, the relationship between forecasted ORH and comminution properties is not explored in this work. The results herein presented, however, serve as a basis to examine such a relationship if those properties were known.

3.2. Assumptions

SAG mills are fundamental pieces in comminution circuits. As no information regarding downstream/upstream processes is available, recognizing bottlenecks in the dataset becomes subjective. We assume that SAG mills will potentially show changes from steady-state to under capacity and vice versa along with the dataset. Thus, stationarity of all operational variable distributions is assumed throughout this work, including the ore grindability. It means that the entire dataset belongs to a known and planned combination of ore characteristics (geometallurgical units). By doing so, we limit the applicability of the present models beyond the temporal dataset without a proper training process.

As explained in the problem statement section, we make use of the temporal average over energy consumption and feed tonnage as input for operational hardness prediction. Thus, we assume an additivity property over those variables as their units are kWh and ton/h, respectively, over constant temporal discretization so averaging adjacent data points is mathematically consistent.

In the operation from which the datasets were obtained, the SAG mill liners are replaced every 5–7 months. Since the datasets cover almost a year, we can ensure that the liners were replaced in each SAG mill at least once during the tested period, which may alter the relationship between energy consumption and other operational variables, inducing a discontinuity in the temporal plots. However, since in this work the temporal window for ORH evaluation is eight hours, the local discontinuity associated with liners replacement is not expected to affect the forecast at that time frame. The ORH is related to what was happening in the corresponding mill within the last few hours, and not to the mill behaviour prior to the last replacement of liners.

3.3. Problem Statement

The aim is to forecast the operational relative-hardness. To do so, we need to label the datasets with the associated ORH category at data point. We know from Equation (1) that the ORH labelling process requires as input (i) the one-step forward differences on energy consumption (ΔEC_t) and feed tonnage (ΔFT_t), and (ii) a lambda (λ) value. In addition, we are interested in forecasting the ORH at different time supports.

Since the information is collected every 30 min, the upcoming energy consumption EC_{t+1} and feed tonnage FT_{t+1} at 0.5 h support are denoted simply as EC_{t+1} and FT_{t+1} in reference to $EC_{t+1}^{(0.5\,h)}$ and $FT_{t+1}^{(0.5\,h)}$, respectively. An upcoming EC and FT at 1 h support, $EC_{t+1}^{(1\,h)}$ and $FT_{t+1}^{(1\,h)}$, are computed by averaging the next two energy consumption, EC_{t+1} and EC_{t+2}, and the two feed tonnage, FT_{t+1} and FT_{t+2}. Similarly, by averaging the upcoming ECs and FTs, different supports can be computed. Let **s** be the time support in hours, which represents the average over a temporal interval of a given duration, then $EC_{t+1}^{(sh)}$ and $FT_{t+1}^{(sh)}$ are calculated as:

$$EC_{t+1}^{(sh)} = \frac{EC_{t+1} + \ldots + EC_{t+2s}}{2s} \qquad FT_{t+1}^{(sh)} = \frac{FT_{t+1} + \ldots + FT_{t+2s}}{2s} \tag{2}$$

In this experiment, three different supports (sh) are considered: 0.5, 2 and 8 h.

Figure 2 illustrates the ORH criteria using a half-hour time support on SAG mill 1 dataset. From the daily graph of $EC_t^{(0.5\,h)}$ and $FT_t^{(0.5\,h)}$ at the top, the graph of $\Delta EC_t^{(0.5\,h)}$ and $\Delta FT_t^{(0.5\,h)}$ are extracted and presented at the centre and bottom, respectively. Three different bands, corresponding to λ: 0.5, 1.0 and 1.5, are shown. The values that are above the band are considered as increasing, the ones below it are considered as decreasing and inside as undefined (relatively constant). The corresponding categories for EC and FT are used to define the operational relative-hardness (as in Table 1). It can be seen that, when λ increases, the proportions of hard and soft instances decrease. Since λ is an arbitrary parameter, a sensitivity analysis is performed in the range $[0.5, 1.5]$ to capture its influence on the resulting LSTM accuracy to suitably learn to predict the ORH at the different time supports.

At each time t the input variables considered to predict $ORH_{t+1}^{(sh)}$ are FT_t, BPr_t and SSp_t. To account for trends, and since FT and SSp are operational decisions, the differences $FT_{t+1} - FT_t$ and $SSp_{t+1} - SSp_t$ are also considered as inputs. Therefore, the dataset of predictors and output $\{\mathbf{X}, \mathbf{Y}\} \in \mathbb{R}^5 \times \mathbb{R}$, at each time support sh, has samples $\{\mathbf{x}_t, \mathbf{y}_t\} \in \{\mathbf{X}, \mathbf{Y}\}$ made by $\mathbf{x}_t = \{FT_t, BPr_t, SSp_t, FT_{t+1} - FT_t, SSp_{t+1} - SSp_t\}$ and $\mathbf{y}_t = \{ORH_{t+1}^{(sh)}\}$. We also tried several other combinations of input variables, but all led to results with lower quality. A temporal window of the previous four hours (previous eight consecutive data points) are used as input for training and testing the LSTM models.

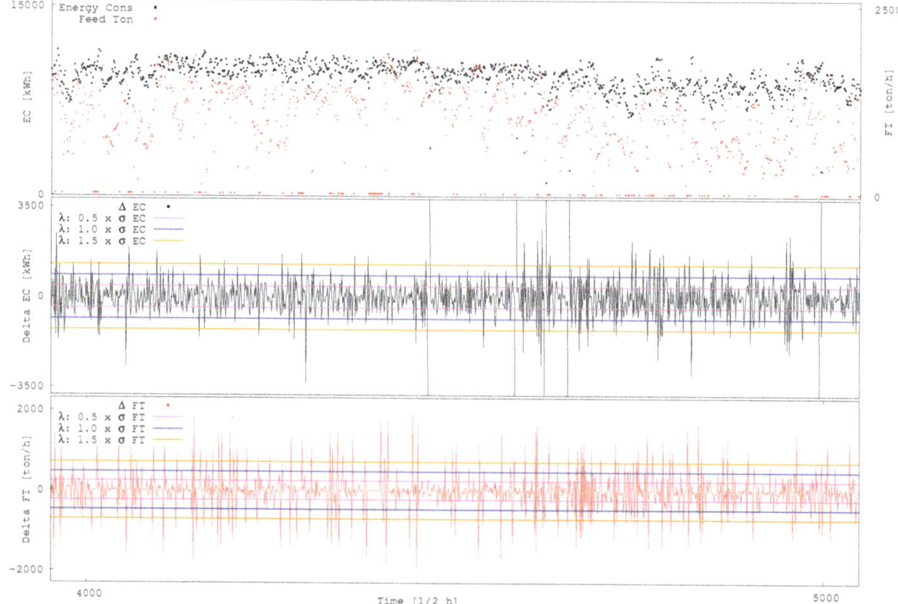

Figure 2. SAG mill 1. Graphic representation of the relative-hardness inference criteria at 0.5 h time support. Daily graphs of energy consumption and feed tonnage (**top**), delta of energy consumption (**centre**), and delta of feed tonnage (**bottom**).

3.4. Preprocessing Dataset

A preprocessing step is performed over the raw datasets to make them suitable for deep neural network training and inference processes. The aim is to make all input attributes fall into certain regions of the non-linear transfer functions via normalization and to be properly coded in categories via one-hot encoding. Thus, we normalize the entire raw dataset with the mean and standard deviation of the training dataset.

Let $\mathbf{x}_t^{(var)} \in \mathbf{x}_t$ be one of the five input variables (*var*) at time t, its normalized expression is computed as $\mathbf{x}_t^{(var)} = \dfrac{var_t - m_{var}}{s_{var}}$, where m_{var} and s_{var} represent the mean and standard deviation of *var* in the training dataset. We normalize the first three attributes of $\mathbf{x}_t$, FT_t, BPr_t and SSp_t while for last two attributes, the differences between the original values $FT_{t+1} - FT_t$ and $SSp_{t+1} - SSp_t$, are replaced by the differences between the normalized values of FT and SSp.

The known operational relative-hardness at time t ($\mathbf{y}_t$) is one-hot encoding such that soft, undefined and hard are encoded as $[1, 0, 0]$, $[0, 1, 0]$ and $[0, 0, 1]$, respectively.

3.5. Optimal LSTM Architecture

From the training dataset, sequence $\{\mathbf{x}_1, .., \mathbf{x}_\tau\}$ of length τ are extracted to train the LSTM model in order to forecast the operational relative-hardness at next time step $\tau + 1$, at different time supports. The chosen length is four hours (τ: 8).

The external hyper-parameter to be optimized on any LSTM architecture is the number of hidden units, n_H. Based on a previous work [14], the optimum number of hidden units was found and here used. They are displayed in Table 3.

Adam Optimizer is used to train the LSTM with hyper-parameters $\epsilon = 1 \times e^{-8}$, $\beta_1 = 0.9$ and $\beta_2 = 0.999$ as recommended by [17].

Table 3. Optimal number of hidden units in the LSTM architecture at different time supports [14].

LSTM	SAG Mill 1			SAG Mill 2		
Time support	$ORH^{(0.5\,h)}$	$ORH^{(2\,h)}$	$ORH^{(8\,h)}$	$ORH^{(0.5\,h)}$	$ORH^{(2\,h)}$	$ORH^{(8\,h)}$
Model (n_H)	280	240	516	596	576	488

4. Results

Directly from the datasets, the real operational relative-hardness ORH_R is calculated from Equation (1), varying λ in the set $(0.5, 0.6, ..., 1.4, 1.5)$ at each time t and for each time support. On the other hand, a probability vector with soft, undefined and hard ORH states is predicted. By taking the highest probability, the predicted ORH_P is obtained. Then, a confusion matrix, filled with the number of instances of pairs (RH_R, RH_P), is built for each time support and each λ value. Table 4 summarizes and presents the cases of λ: 0.5, 1.0 and 1.5, and supports 0.5, 2 and 8 h over the SAG mill 1, while the Table 5 summarizes the same results over the SAG mill 2.

Table 4. SAG mill 1. Confusion matrices (number of instances) of operational relative-hardness (ORH) predictions using λ: 0.5, 1.0 and 1.5 at 0.5, 2 and 8 h time supports.

0.5 h, $\lambda = 0.5$

Real \ Prediction	Soft	Und	Hard	Total
Soft	1515	591	16	1816
Und	295	3179	390	4325
Hard	6	555	1606	2012
Accurate →				6300

0.5 h, $\lambda = 1.0$

Real \ Prediction	Soft	Und	Hard	Total
Soft	957	268	1	1199
Und	242	5255	137	5885
Hard	0	362	931	1069
Accurate →				7143

0.5 h, $\lambda = 1.5$

Real \ Prediction	Soft	Und	Hard	Total
Soft	622	140	2	777
Und	151	6298	130	6577
Hard	4	139	667	799
Accurate →				7587

2.0 h, $\lambda = 0.5$

Real \ Prediction	Soft	Und	Hard	Total
Soft	1204	922	288	2165
Und	731	1659	904	3495
Hard	230	914	1301	2493
Accurate →				4164

2.0 h, $\lambda = 1.0$

Real \ Prediction	Soft	Und	Hard	Total
Soft	406	1120	15	582
Und	160	4793	102	7261
Hard	16	1348	193	310
Accurate →				5392

2.0 h, $\lambda = 1.5$

Real \ Prediction	Soft	Und	Hard	Total
Soft	290	609	2	463
Und	170	6172	66	7521
Hard	3	740	101	169
Accurate →				6563

8.0 h, $\lambda = 0.5$

Real \ Prediction	Soft	Und	Hard	Total
Soft	1277	805	370	2105
Und	626	1457	1083	3076
Hard	202	814	1519	2972
Accurate →				4253

8.0 h, $\lambda = 1.0$

Real \ Prediction	Soft	Und	Hard	Total
Soft	683	873	33	1030
Und	324	4135	504	6061
Hard	23	1053	525	1062
Accurate →				5343

8.0 h, $\lambda = 1.5$

Real \ Prediction	Soft	Und	Hard	Total
Soft	308	579	4	598
Und	286	6038	119	7273
Hard	4	656	159	282
Accurate →				6505

Table 5. SAG mill 2. Confusion matrices (number of instances) of ORH predictions using λ: 0.5, 1.0 and 1.5 at 0.5, 2 and 8 h time supports.

0.5 h, $\lambda = 0.5$

Real \ Prediction	Soft	Und	Hard	Total
Soft	1704	434	13	2039
Und	330	2718	416	3600
Hard	5	448	1882	2311
Accurate →				6304

0.5 h, $\lambda = 1.0$

Real \ Prediction	Soft	Und	Hard	Total
Soft	1085	334	2	1266
Und	180	4485	360	5119
Hard	1	300	1203	1565
Accurate →				6773

0.5 h, $\lambda = 1.5$

Real \ Prediction	Soft	Und	Hard	Total
Soft	640	274	8	753
Und	111	5916	91	6469
Hard	2	279	629	728
Accurate →				7185

2.0 h, $\lambda = 0.5$

Real \ Prediction	Soft	Und	Hard	Total
Soft	1026	1049	149	1614
Und	460	2224	720	4491
Hard	128	1218	976	1845
Accurate →				4226

2.0 h, $\lambda = 1.0$

Real \ Prediction	Soft	Und	Hard	Total
Soft	676	768	47	1119
Und	418	4178	395	6012
Hard	25	1066	395	819
Accurate →				5231

2.0 h, $\lambda = 1.5$

Real \ Prediction	Soft	Und	Hard	Total
Soft	338	593	12	567
Und	228	5721	133	7101
Hard	1	787	137	282
Accurate →				6196

8.0 h, $\lambda = 0.5$

Real \ Prediction	Soft	Und	Hard	Total
Soft	917	1151	196	1368
Und	361	2052	896	4285
Hard	90	1082	1205	2297
Accurate →				4174

8.0 h, $\lambda = 1.0$

Real \ Prediction	Soft	Und	Hard	Total
Soft	789	735	29	1169
Und	358	4118	353	6001
Hard	22	1148	398	780
Accurate →				5305

8.0 h, $\lambda = 1.5$

Real \ Prediction	Soft	Und	Hard	Total
Soft	325	641	10	606
Und	273	5660	133	6991
Hard	8	690	210	353
Accurate →				6195

The accuracy of the model prediction, ORH_P, defined as the percentage of right predictions is computed as:

$$ORH_{Accuracy} = \frac{\#(\mathbf{soft}_R, \mathbf{soft}_P) + \#(\mathbf{und}_R, \mathbf{und}_P) + \#(\mathbf{hard}_R, \mathbf{hard}_P)}{\#Total} \cdot 100 \qquad (3)$$

and it represents the percentage of elements in the confusion matrix diagonal. The relative percentage of predictions of each class (rows) is shown in Table 6 for SAG mill 1 and in Table 7 for SAG mill 2.

As shown in Tables 6 and 7 at 0.5 h time support, the LSTM is able to predict with enough confidence the ORH regardless the value of λ. Nevertheless, as λ increases, the number of instances of soft and hard ORH decreases improving the final accuracy since the higher the value of λ, the more data points are classified as undefined. Particularly, for 0.5 h time support, increasing λ from 0.5 to 1.5 makes real undefined points increase from 4325 to 6577 (from 53.0% to 80.7%) in SAG mill 1 and from 3600 to 6469 (from 45.3% to 81.4%) in SAG mill 2. Therefore, increasing λ improves accuracy, but the price is resolution. On the other hand, the number of extreme cases ($\mathbf{soft}_R$, $\mathbf{hard}_P$) and ($\mathbf{hard}_R$, $\mathbf{soft}_P$) is close to zero. This is a great result, since predicting soft hardness when it is actually hard (or vice versa) may induce bad short term decisions on how to operate the SAG mill, along with other downstream decisions.

Table 6. SAG mill 1. Confusion matrices (percentage) of ORH prediction using λ: 0.5, 1.0 and 1.5 at 0.5, 2 and 8 h time supports.

0.5 h $\lambda = 0.5$	Prediction Soft	Und	Hard	Total
Real Soft	71.4	27.9	0.8	22.3
Und	7.6	82.3	10.1	53.0
Hard	0.3	25.6	74.1	24.7
	Accurate →			77.3
2.0 h $\lambda = 0.5$	Prediction Soft	Und	Hard	Total
Real Soft	49.9	38.2	11.9	26.6
Und	22.2	50.4	27.4	42.9
Hard	9.4	37.4	53.2	30.6
	Accurate →			51.1
8.0 h $\lambda = 0.5$	Prediction Soft	Und	Hard	Total
Real Soft	52.1	32.8	15.1	25.8
Und	19.8	46.0	34.2	37.7
Hard	8.0	32.1	59.9	36.5
	Accurate →			52.2
0.5 h $\lambda = 1.0$	Prediction Soft	Und	Hard	Total
Real Soft	78.1	21.9	0.1	14.7
Und	4.3	93.3	2.4	72.2
Hard	0.0	28.0	72.0	13.1
	Accurate →			87.6
2.0 h $\lambda = 1.0$	Prediction Soft	Und	Hard	Total
Real Soft	26.3	72.7	1.0	7.1
Und	3.2	94.8	2.0	89.1
Hard	1.0	86.6	12.4	3.8
	Accurate →			66.1
8.0 h $\lambda = 1.0$	Prediction Soft	Und	Hard	Total
Real Soft	43.0	54.9	2.1	12.6
Und	6.5	83.3	10.2	74.3
Hard	1.4	65.8	32.8	13.0
	Accurate →			65.5
0.5 h $\lambda = 1.5$	Prediction Soft	Und	Hard	Total
Real Soft	81.4	18.3	0.3	9.5
Und	2.3	95.7	2.0	80.7
Hard	0.5	17.2	82.3	9.8
	Accurate →			93.1
2.0 h $\lambda = 1.5$	Prediction Soft	Und	Hard	Total
Real Soft	32.2	67.6	0.2	5.7
Und	2.7	96.3	1.0	92.2
Hard	0.4	87.7	12.0	2.1
	Accurate →			80.5
8.0 h $\lambda = 1.5$	Prediction Soft	Und	Hard	Total
Real Soft	34.6	65.0	0.4	7.3
Und	4.4	93.7	1.8	89.2
Hard	0.5	80.1	19.4	3.5
	Accurate →			79.8

Table 7. SAG mill 2. Confusion matrices (percentage) of ORH prediction using λ: 0.5, 1.0 and 1.5 at 0.5, 2 and 8 h time supports.

0.5 h $\lambda = 0.5$	Prediction Soft	Und	Hard	Total
Real Soft	79.2	20.2	0.6	25.6
Und	9.5	78.5	12.0	45.3
Hard	0.2	19.2	80.6	29.1
	Accurate →			79.3
2.0 h $\lambda = 0.5$	Prediction Soft	Und	Hard	Total
Real Soft	46.1	47.2	6.7	20.3
Und	13.5	65.3	21.2	56.5
Hard	5.5	52.5	42.0	23.2
	Accurate →			53.2
8.0 h $\lambda = 0.5$	Prediction Soft	Und	Hard	Total
Real Soft	40.5	50.8	8.7	17.2
Und	10.9	62.0	27.1	53.9
Hard	3.8	45.5	50.7	28.9
	Accurate →			52.5
0.5 h $\lambda = 1.0$	Prediction Soft	Und	Hard	Total
Real Soft	76.4	23.5	0.1	15.9
Und	3.6	89.3	7.2	64.4
Hard	0.1	19.9	80.0	19.7
	Accurate →			85.2
2.0 h $\lambda = 1.0$	Prediction Soft	Und	Hard	Total
Real Soft	45.3	51.5	3.2	14.1
Und	8.4	83.7	7.9	75.6
Hard	1.7	72.6	25.7	10.3
	Accurate →			65.8
8.0 h $\lambda = 1.0$	Prediction Soft	Und	Hard	Total
Real Soft	50.8	47.3	1.9	14.7
Und	7.4	85.3	7.3	75.5
Hard	1.4	73.2	25.4	9.8
	Accurate →			66.7
0.5 h $\lambda = 1.5$	Prediction Soft	Und	Hard	Total
Real Soft	69.4	29.7	0.9	9.5
Und	1.8	96.7	1.5	81.4
Hard	0.2	30.7	69.1	9.2
	Accurate →			90.4
2.0 h $\lambda = 1.5$	Prediction Soft	Und	Hard	Total
Real Soft	35.8	62.9	1.3	7.1
Und	3.7	94.1	2.2	89.3
Hard	0.1	85.1	14.8	3.5
	Accurate →			77.9
8.0 h $\lambda = 1.5$	Prediction Soft	Und	Hard	Total
Real Soft	33.3	65.7	1.0	7.6
Und	4.5	93.3	2.2	87.9
Hard	0.9	76.0	23.1	4.4
	Accurate →			77.9

The percentage of extreme cases $(($soft$_R$, hard$_P)$ and $($hard$_R$, soft$_P))$ using λ: 0.5 increases when moving from 0.5 to 8 h time support, on both SAG mills. However, they decrease to a value close to zero when increasing λ from 0.5 to 1.5, at all time supports. However, LSTM loses accuracy in terms of predicting the relevant cases ($\mathbf{soft}_R$, $\mathbf{soft}_P$) and ($\mathbf{hard}_R$, $\mathbf{hard}_P$) as soon as the time support increases, on both SAG mills.

The accuracy graph (Figure 3) shows the λ sensitivity at all time supports on both SAG mills. The lower accuracy is 51% and is achieved at 2 h time supports with λ: 0.5 on SAG mill 1. Its accuracy increases to 66% with λ: 1.0 and 81% with λ: 1.5. The best results are achieved at 0.5 h time support (same support as the original data) where 77%, 88% and 93% of accuracy are obtained with λ: 0.5, 1.0

and 1.5, respectively on SAG mill 1, and 79%, 85% and 90% of accuracy with λ: 0.5, 1.0 and 1.5 on SAG mill 2.

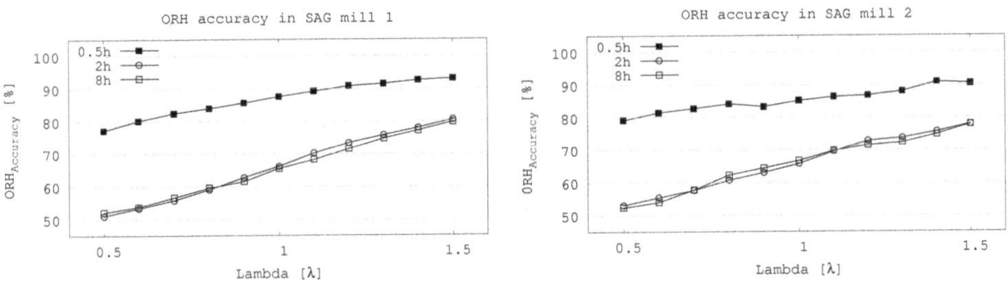

Figure 3. Accuracy of operational relative-hardness prediction at different time support as function of lambda (λ) on both SAG mills.

5. Conclusions

This work proposes the use of Long Short-Term Memory networks to forecast relative operational hardness in two SAG mills using operational data. We have presented the internal architecture of the deep networks, how to deal with raw operational datasets, and qualitative criteria to estimate the operational hardness of processing material inside the SAG mill based on the consumed energy, feed tonnage and a statistical distribution using a lambda value. Particularly, Long Short-Term Memory models have been trained to predict the operational relative-hardness based only on low-cost and fast acquiring operational information (feed tonnage, spindle speed and bearing pressure).

The LSTM network shows great results on predicting the relative operational hardness at 30 min time support. On SAG mill 1, using a lambda value of 0.5, the obtained accuracy was 77.3% while increasing the lambda to 1.5 led to an increase in accuracy of 93.1%. Similar results were found on the second SAG mill. As the time support increases to two and eight hours, the accuracy drops to around 52% using a lambda value of 0.5 and 78% with a lambda value of 1.5, on both SAG mills.

The inaccuracy of LSTM, when predicting extreme cases such as soft hardness when it is hard and vice-versa, is pretty low. Extreme misclassification is close to 1% at 0.5 h time support on both SAGs regardless of the lambda value. Although it increases to around 20% when increasing the time support using a lambda value of 0.5, it rapidly decreases to around 1% as lambda increases.

Lastly, the proposed application can be extended to any crushing and grinding equipment, under a similar context of real-data acquisition in order to forecast categorical attributes that are relevant to downstream processes.

Author Contributions: Conceptualization, S.A. and W.K.; methodology, S.A.; codes, S.A.; validation, S.A., W.K. and J.M.O.; formal analysis, S.A.; investigation, S.A.; resources, W.K.; data curation, S.A.; writing—original draft preparation, S.A.; visualization, S.A.; supervision, W.K. and J.M.O.; project administration, W.K.; funding acquisition, W.K. and J.M.O. All authors have read and agreed to the published version of the manuscript.

Funding: This research was funded by the Natural Sciences and Engineering Council of Canada (NSERC) grant number RGPIN-2017-04200 and RGPAS-2017-507956, and by the Chilean National Commission for Scientific and Technological Research (CONICYT), through CONICYT/PIA Project AFB180004, and the CONICYT/FONDAP Project 15110019.

Conflicts of Interest: The authors declare no conflict of interest.

Abbreviations

The following abbreviations are used in this manuscript:

LSTM	Long Short-Term Memory
ORH	Operational relative-hardness
FT	Feed tonnage
BPr	Bearing pressure
SSp	Spindle speed
SAG	Semi-autogenous grinding

References

1. Cochilco. *Actualización de Información sobre el Consumo de Energía asociado a la Minería del Cobre al año 2012*; COCHILCO: Santiago, Chile, 2013.
2. Jones, S.M.; Fresko, M. Autogenous and semiautogenous mills 2010 update. In Proceedings of the Fifth International Conference on Autogenous and Semiautogenous Grinding Technology, Vancouver, BC, Canada, 25–28 September 2011.
3. Ortiz, J.M.; Kracht, W.; Pamparana, G.; Haas, J. Optimization of a SAG mill energy system: Integrating rock hardness, solar irradiation, climate change, and demand-side management. *Math. Geosci.* **2020**, *52*, 355–379. [CrossRef]
4. Jnr, W.V.; Morrell, S. The development of a dynamic model for autogenous and semi-autogenous grinding. *Miner. Eng.* **1995**, *8*, 1285–1297.
5. Morrell, S. A new autogenous and semi-autogenous mill model for scale-up, design and optimisation. *Miner. Eng.* **2004**, *17*, 437–445. [CrossRef]
6. Silva, M.; Casali, A. Modelling SAG milling power and specific energy consumption including the feed percentage of intermediate size particles. *Miner. Eng.* **2015**, *70*, 156–161. [CrossRef]
7. Salazar, J.L.; Valdés-González, H.; Vyhmesiter, E.; Cubillos, F. Model predictive control of semiautogenous mills (sag). *Miner. Eng.* **2014**, *64*, 92–96. [CrossRef]
8. Ortiz, J.; Kracht, W.; Townley, B.; Lois, P.; Cardenas, E.; Miranda, R.; Alvarez, M. Workflows in geometallurgical prediction: Challenges and outlook. In Proceedings of the 17th Annual Conference of the International Association for Mathematical Geosciences IAMG, Freiberg, Germany, 5–13 September 2015.
9. Van den Boogaart, K.; Tolosana-Delgado, R. Predictive Geometallurgy: An Interdisciplinary Key Challenge for Mathematical Geosciences. In *Handbook of Mathematical Geosciences*; Springer: Berlin, Germany, 2018; pp. 673–686.
10. Curilem, M.; Acuña, G.; Cubillos, F.; Vyhmeister, E. Neural networks and support vector machine models applied to energy consumption optimization in semiautogeneous grinding. *Chem. Eng. Trans.* **2011**, *25*, 761–766.
11. Hoseinian, F.S.; Faradonbeh, R.S.; Abdollahzadeh, A.; Rezai, B.; Soltani-Mohammadi, S. Semi-autogenous mill power model development using gene expression programming. *Powder Technol.* **2017**, *308*, 61–69. [CrossRef]
12. Hoseinian, F.S.; Abdollahzadeh, A.; Rezai, B. Semi-autogenous mill power prediction by a hybrid neural genetic algorithm. *J. Cent. South Univ.* **2018**, *25*, 151–158. [CrossRef]
13. Inapakurthi, R.K.; Miriyala, S.S.; Mitra, K. Recurrent Neural Networks based Modelling of Industrial Grinding Operation. *Chem. Eng. Sci.* **2020**, *219*, 115585. [CrossRef]
14. Avalos, S.; Kracht, W.; Ortiz, J.M. Machine learning and deep learning methods in mining operations: A data-driven SAG mill energy consumption prediction application. *Min. Metall. Explor.* **2020**, *37*, 1–16. [CrossRef]
15. Hochreiter, S.; Schmidhuber, J. Long short-term memory. *Neural Comput.* **1997**, *9*, 1735–1780. [CrossRef] [PubMed]
16. Goodfellow, I.; Bengio, Y.; Courville, A.; Bengio, Y. *Deep Learning*; MIT Press: Cambridge, MA, USA, 2016; Volume 1.
17. Kingma, D.P.; Ba, J. Adam: A method for stochastic optimization. *arXiv* **2014**, arXiv:1412.6980.

Article

Industrial Vertical Stirred Mills Screw Liner Wear Profile Compared to Discrete Element Method Simulations

Priscila M. Esteves [1,*], Douglas B. Mazzinghy [1], Roberto Galéry [1] and Luís C. R. Machado [2]

[1] Programa de Pós-Graduação em Engenharia Metalúrgica, Materiais e de Minas (PPGEM), Universidade Federal de Minas Gerais (UFMG), Belo Horizonte 31270-901, Brazil; dmazzinghy@demin.ufmg.br (D.B.M.); rgalery@demin.ufmg.br (R.G.)

[2] Anglo American-Minas-Rio Project, Minas Gerais 35860-000, Brazil; luis.machado@angloamerican.com

* Correspondence: pmxesteves@gmail.com

Abstract: Vertical stirred mills have been widely applied in the minerals industry, due to its greater efficiency in comparison with conventional tumbling mills. In this context, the agitator liner wear plays an important role in maintenance planning and operational costs. In this paper, we use the discrete element method (DEM) wear simulation to evaluate the screw liner wear. Three different mill rotational velocities are evaluated in the simulation, according to different scale-up procedures. The wear profile, wear measurement, power consumption, and particle contact information are used for obtaining a better understanding of the wear behavior and its effects on grinding mechanisms. Data from a vertical stirred mill screw liner wear measurement obtained in a full-scale mill are used to correlate with simulation results. The results indicate a relative agreement with industrial measurement in most of the liner lifecycle, when using a proper mill velocity scale-up.

Keywords: Vertimill; Tower Mill; liner wear; fine grinding; discrete element method

Citation: Esteves, P.M.; Mazzinghy, D.B.; Galéry, R.; Machado, L.C.R. Industrial Vertical Stirred Mills Screw Liner Wear Profile Compared to Discrete Element Method Simulations. *Minerals* **2021**, *11*, 397. https://doi.org/10.3390/min11040397

Academic Editor: Luís Marcelo M. Tavares

Received: 5 August 2020
Accepted: 21 September 2020
Published: 10 April 2021

Publisher's Note: MDPI stays neutral with regard to jurisdictional claims in published maps and institutional affiliations.

1. Introduction

The reduction of mineral deposits created the necessity for extracting more complex minerals with reduced ore grades and grindability [1]. Consequently, there is an arising need for processing reduced particle size that can propitiate the liberation of valuable minerals for post concentration stages. In this context, additional power is required to achieve a finer grain size. Therefore, grinding and regrinding applications are increasing in importance.

At the same time, grinding is energy inefficient, and this is the main reason why it is pointed as responsible for around 34% to 44% of the energy required in a mineral processing plant [2,3]. The numbers are even more alarming if we consider the situation in a global context. In the 1980s, grinding was responsible for around 3% to 4% of total world electrical energy consumption [2], and more recently, this number was indicated as close to 1.8% [4]. In fact, Napier-Munn went beyond and argued that real energy consumption in grinding is even greater than those numbers, once consumption of liners and grinding media represents a large amount of extra energy. Regardless of which is the correct number, it is conclusive that the grinding process is responsible for a significant portion of the world's power consumption.

Given the rising importance of the representative energy consumption of grinding equipment, it is very important to develop the process of improved energy efficiency. In relation to this issue, Shi (2009) [5] demonstrated that the application of vertical stirred mills presents the energy saving around 30% when compared to tumbling ball mills used for coarse grinding. By applying various methodologies, other researches confirmed the obtained results and reinforced this better energy efficiency behavior for fine-grinding and ultra-fine grinding applications. This behavior is the main reason why vertical stirred mills are pointing as trend equipment for fine-grinding and ultra-fine-grinding applications.

Several approaches for vertical stirred mills modeling can be found in literature, among which stand out the mechanistic approach, Discrete Element Method (DEM), empirical models, and finally, the Population Balance Model (PBM). However, the understanding of the screw liner wear is not well developed yet.

As indicated by Esteves et al. [6], operational costs of vertical stirred mills are divided into electrical energy (50%), balls (40%), and liners (10%). Liner wear intensifies grinding media consumption, increases mill filling, and consequently, affects electrical power consumption. Consequently, liner wear affects all operational costs components, such as maintenance practices and equipment reliability. As highlighted by Allen and Noriega [7], the understanding of liner wear life can also provide information that allows best practices for maintenance schedules that can minimize spare parts and labor costs. Consequently, a better understanding of wear behavior, such as monitoring methods can precisely provide its measurement and/or prediction is a topic of great need in the mineral processing industry. In this paper, DEM simulations are used to predict the screw liner wear behavior of a vertical stirred mill.

Computational simulations based on DEM made many contributions to the science of comminution since its first application by Mishra and Rajamani [8], for the simulation and modeling of tumbling ball mills. According to Weerasekara et al. [9], since the last two decades, the DEM became an essential tool to help design, optimization, and modeling of comminution devices. More recently, there is also an increasing interest in applying the method for the prediction of liners and lifters wear, such as its effects on load behavior.

As there is no available information about how to determine and quantify wear for vertical stirred mills, the approach is based on what is presented for other grinding equipment. The understanding of liner and lifters wear, especially in tumbling mills, is not a new subject in the area, and several papers are found addressing this issue.

In the specific case of wear evaluation, Cleary [10] proposed a method that used DEM to predict liner wear rates and distribution in a 5.0 m diameter ball mill. With a similar approach, Cleary and Owen [11] evaluated wear in a 3D slice of a SAG mill, and finally, Cleary, Sinnott, and Morrison [12] performed an evaluation and comparison between wear in tower and pin mills. Kalala and Moys [13] used DEM to estimate adhesion, abrasion, and impact wear in dry ball mills with further validation with industrial wear measurement. Later, Kalala, Bwalya, and Moys [14] validated the use of DEM for wear prediction in mill liners by comparing normal and tangential forces between experimental data and DEM simulations. More recently, Cleary and Owen [15] simulated liner wear evolution in a Hicom Mill and was able to convert DEM abrasion measurements in a wear rate measurement that was calibrated with experimental data. Boemer and Ponthot [16] established a generic wear prediction procedure applying DEM to a 3D ball mill and validated the results with experimental data obtained in a 5.8 m diameter industrial cement ball. Finally, Xu et al. [17] obtained a numerical prediction of wear in SAG mills using DEM simulations that were quantitatively validated by experiment data available in other researches. In fact, the use of DEM for wear prediction in SAG mills reached such a high level that has recently been used to evaluate operational strategies to account for liner wear.

For the specific case of vertical stirred mills, DEM was first applied by Cleary, Sinnott, and Morrison [12] to investigate the relative performance of stirred mills with two different agitator designs. The paper analysis media flow, energy absorption, flow structures, wear, mixing, and transport efficiency, and allows a good and wide understanding of the performance of vertical stirred mills. For the case of the screw agitator, it can be seen that media motion is a simultaneously lifting and circulating movement inside the equipment. The simulations indicate that the collisional energy associated with media motion is mostly dominated by shear. In relation to media-media and media-liner interactions, shear energy represents more than three times the impact energy. Based on shear energy absorptions, the author infers that the equipment wear is dominated by abrasion and is more intensive on the outer radial edge of the screw. In a slightly different approach, Morrison, Cleary, and Sinnott [18] used DEM to compare the energy efficiency between the ball and vertical stirred mills in pilot-scale and proposed that the higher efficiency associated with the stirred mill can be explained by analyzing energy spectra associated with collisions frequencies inside the mill. Therefore, the higher energy efficiency of the vertical mill can be explained by the presence of a great number of contacts of low energy. Sinott, Cleary, and Morrison [19] applied DEM to understand how flow and energy are affected by media shape in stirred mills, concluding that grinding performance tends to significantly deteriorate when using non-spherical media. By analyzing rates of shear power absorption, the author also proposes that the increase in non-sphericity of the media can significantly intensify the screws' wear.

In a different approach, Sinott, Cleary, and Morrison [20] evaluated slurry transport inside the mill using the SPH (Smoothed Particle Hydrodynamics) method and based on DEM simulations. Allen and Noriega [7] applied DEM with SPH to understand the screw liner wear. The results demonstrate that shear power is more intensive at the screw outside edges and at the bottom of the mill. This explains why the screw wears from the outside to inside and from the bottom to the top, such as have been seen in the industry. In addition, results indicate that shear power is exponentially related to screw diameter, and consequently, worn screws tend to wear slower as they present smaller diameters. Based on this description, it is possible to note that although DEM is widely applied in the understanding and evaluation of vertical stirred mills performance, the liner wear topic is still superficially approached.

To summarize, the use of DEM to evaluate liner wear in tumbling mills came from a quantitative evaluation and evolved another level. Nowadays, DEM can predict and quantify the wear in SAG and Ball mills, being used in the optimization of materials and operational parameters. Unfortunately, for the case of the vertical stirred mills, the studies are still preliminary and qualitative. Solving this issue requires more in-depth studies and validation with experimental data.

In this paper, the simulation results were compared with industrial measurements performed in Minas Rio Project, an iron ore beneficiation plant of Anglo American, which is in Minas Gerais State, in Brazil. An empirical evaluation of wear is an important tool for validation and calibration of modeling techniques, such as performed by some researches [13,15–17,21,22] for several types of grinding equipment. However, there are few available information about how to determine and quantify wear for vertical stirred mills. Based on that, the approach is based on what is presented for other mills.

The understanding of liner and lifters wear in tumbling mills is not a new subject in the area, and several papers are found addressing this topic. In the case of tumbling mills, the wear profile is commonly estimated in two-dimensional methods, taking the assumption that wear has a uniform profile. More recently, with the advent of new technologies, it is easier to perform three-dimensional measurements, and there are also highly sophisticated devices available [16]. Three-dimensional measurements are being widely used to calculate wear on the surface of liners in grinding equipment, and the technique was successfully used for SAG mills [17,23], and also for a Hicom Mill [15].

2. Gravity Induced Stirred Mill

Vertical stirred mills are grinding equipment applied for comminution, especially for fine grinding in regrinding applications where feed material is under 1mm. The main parts of the equipment are the grinding chamber, internal liner, screw impeller, screw liner, motor engine, ball feeder, slurry feeder, and the discharge, as shown in Figure 1.

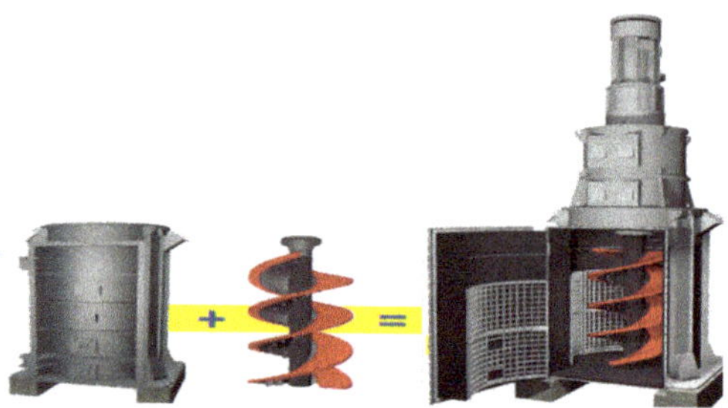

Figure 1. Vertical stirred mill components [24].

Figure 2 emphasizes the screw impeller, such as its liner parts. As the screw impeller is responsible for media motion and is currently in contact with media and slurry, it suffers intensive wear. Due to the wear, liner parts required periodic replacement. The liner is divided into several parts, allowing the substitution to be performed differently, according to the wear pattern. The number of parts depends on the equipment supplier and on the equipment size. The liner parts are attached to the screw and protect it from wear.

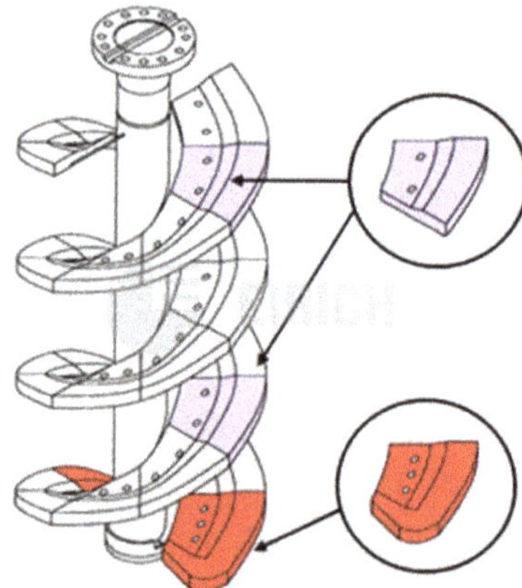

Figure 2. Vertical stirred mill screw liner parts [24].

Figure 3 shows the expected wear shape of the base liner part. From that, it is possible to note that the liner does not suffer homogeneous wear. In this sense, it is expected that the bottom part of the liner suffers more intensive wear at the edges.

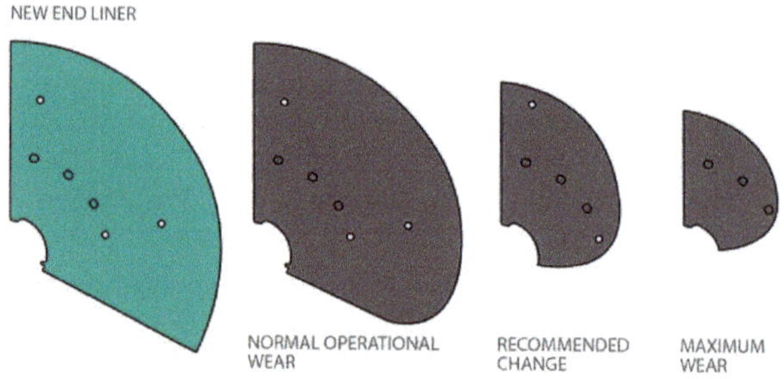

Figure 3. Wear pattern for the bottom liner [25].

As the liner wears, the total surface area decreases, and consequently, both media motion and power draw decrease. In the operational context, the liner wear compensation is performed by adding additional grinding media to keep constant power draw. In this sense, the liner wear measurement can be estimated and accomplished by measuring mill filling. Although this provides an idea about wear conditions, visual inspections of the liner are necessary. For this inspection, it is necessary to completely empty the mill, in a very effort and time-consuming inspection.

Figure 4 shows the screw liner after completely emptying the mill. It can be seen the difference between a new and an old liner of the VTM-1500. The figure on the left side shows a liner with intensive wear, and the right side shows the replaced liner. From the figure, it can be inferred that wear predominates at the bottom and edges of the screw, causing a great decrease in the screw area.

Figure 4. Left: Screw liner with intensive wear at the bottom. Right: New screw liner [6].

Based on what was presented, the aim of this paper is to gather what has already been discussed on the subject and also to propose the use of DEM for evaluating the liner wear during a complete liner lifecycle.

3. Methodology

3.1. Dem Model Setup

DEM simulations were performed using the Rocky 4.2.2 software (ESSS, Florianópolis, Brazil). A 1:10 scale version of the Metso Vertimill VTM-1500 (Metso, Helsinki, Finland) was simulated. Table 1 and Figure 5 shows geometry dimensions and screw design.

Table 1. Dimensions and operation parameters of VTM-1500 and its scaled-down version considered in the discrete element method (DEM) simulations.

Mill	VTM-1500	Scaled Down (Simulated)
Scale	1/1	1/10
Screw Diameter (mm)	3300	330
Ball load (kg)	80,000	80
Mill filling	80%	80%

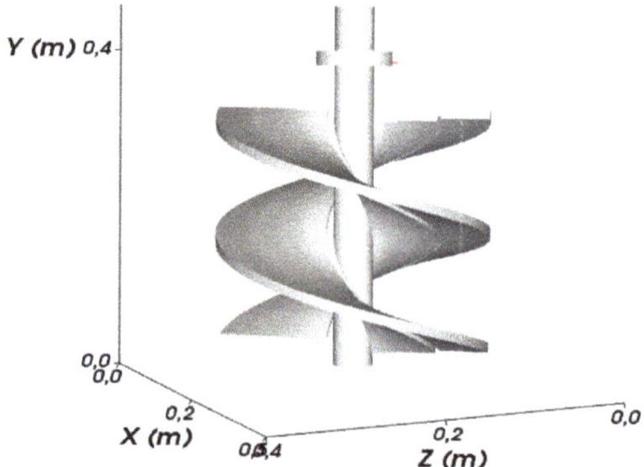

Figure 5. Vertimill 1:10 scale of the VTM-1500.

The contact models used in the simulations were Linear Hysteresis for normal forces and Elastic Coulomb for tangential forces [26]. Table 2 summarizes the contact model parameters and materials properties setup used for the simulation.

Table 2. Material and contact parameters used in the DEM simulations.

Variable	Value
Young's Modulus (Pa)	1×10^{11}
Density (kg/m^3)	7850
Poisson's Ratio	0.3
Coefficient of Static Friction (particle-particle)	0.7
Coefficient of Dynamic Friction (particle-particle)	0.7
Coefficient of Static Friction (particle-geometry)	0.3
Coefficient of Dynamic Friction (particle-geometry)	0.3
Restitution coefficient	0.3

The simulation was performed only considering grinding media in the grinding environments, in the absence of slurry and ore particles. This can be acceptable for the process where the ore contributions for breakage are negligible, such as in the vertical stirred mills, where the breakage mechanisms are mainly created by the energy involved in grinding media collisions. However, it is also known that grinding media interaction is affected by ore and slurry properties. Consequently, both solids concentration and particle size distribution can affect power consumption and grinding performance. To approximate the slurry effect in the grinding environment, the shear modulus was reduced, thus making contact between balls softer, in comparison to steel-steel contact (Steel: 200 Gpa). This approximates to the ore and slurry interactions behavior.

In relation to the mill rotational speed, three different velocities were simulated: 87 rpm, 130 rpm, and 190 rpm. The 87 rpm was obtained based on a model for velocity scale-up, as proposed by Mazzinghy et al. [27] and adapted by Esteves et al. [28]. The model is based on an equipment dataset and consists of a more recent approach for velocity scale-up. The 190 rpm is obtained considering a fixed tip speed of 3.5 m/s for the mill agitator. This is the usual method for velocity scale-up and was widely applied for reduced scale equipment test work and simulation. The 130 rpm was obtained as an intermediary velocity in between the two scale-up approaches.

3.2. Dem Outputs

During the simulation, the interaction between particles and geometry is individually described by first physical principles, according to the contact model defined. This information is saved during each simulation step and is further used to generate the simulation outputs, such as: particle energy spectra, particle trajectory, particle absolute translational velocity, power consumption, and wear design.

3.3. Wear Model

The Archard's wear law, together with the DEM outputs, were used to quantify liner boundary wear. The wear quantification is realized step by step, and this information is used to currently update the liner shape during the simulation. The Archard's wear law is presented by Equation (1) [26]:

$$A.\,dh = K.\,dw \tag{1}$$

where A is the surface area of a boundary element (m^2), h is the loss in depth (m), w is the shear work (J), and K is the wear rate (m^3/J).

The incremental loss in depth consists of the amount of wear generated in the liner surface. This information is used to currently update the liner shape and volume. The shear work consists of the DEM shear outputs. In this sense, the DEM shear stress results for the interaction between the liner and grinding media are used to continuously quantify the boundary wear. The wear factor is defined as the relationship between the contact energy and the amount of lost surface. A wear factor of 1×10^{-6} m^3/J was established for the simulation. This indicated the amount of volume that is lost according to the energy amount involved at each contact between grinding media and liner. The increase of these parameters can extremely accelerate the wear ratio. Although the use of greater values can accelerate the obtaining of a wear surface, it can also generate unwanted damages on the surface that are not in accordance with reality. Whereas the use of very small factors can bring the need for a very long simulation to obtain representatives wear information. In the present case, several values were tested, until the obtaining of a feasible shape for the wear. As a result, it is necessary to adapt to the simulation and real time.

3.4. Boundary Definition

The boundary definition of the geometrical mesh directly affects the computational cost to run the wear simulation, such as the wear results. Figure 6 shows the effect of triangle size in mesh refinement.

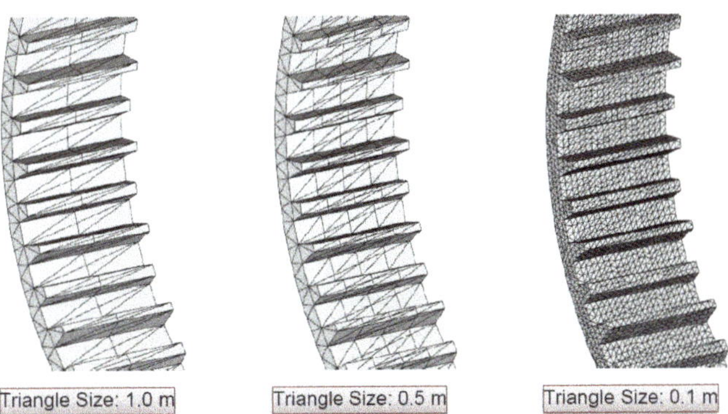

Figure 6. Boundary definition for the geometrical mesh of the wear parts [29]

The defining of greater triangle sizes decreases the total number of triangles and creates a bad refinement for the mesh, decreasing the computational simulation cost. Contrastingly, the existence of smaller triangle sizes highly increases the number of triangles, refining the geometrical mesh and causing a representative increase in the computational simulation cost. Figure 7 shows the effect of mesh refinement in the wear simulation results.

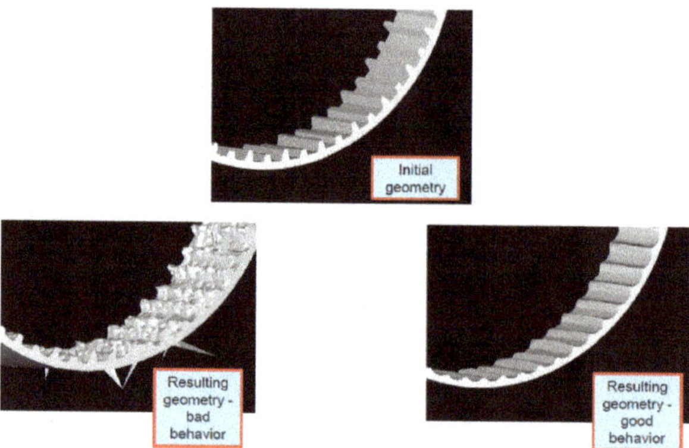

Figure 7. Effect of boundary definition and mesh refinement in the obtained wear pattern [29].

It is possible to note the final wear shape is affected by a refinement degree. In this sense, the existence of a more refined mesh brings a better wear resolution in the final geometry of the object. However, as finer meshes representatively increase computational cost, it is very important to investigate the sensitivity of mesh refinement degree, to obtain a reasonable balance between wear resolution and feasibility of computational cost [26].

3.5. Industrial Wear Measurement

The industrial measurements presented were obtained by Silva [30] from the regrinding circuit of the Minas-Rio project, which is an iron ore plant from Anglo American located in the southeast region of Brazil. The regrinding circuit is composed of two parallel lines, each one with eight VTM-1500. The regrinding product is specified with a P_{80} of 36 µm, mill capacity as 190 t/h per mill, and finally, specific energy is established as 5.9 kWh/t.

The process flowsheet is indicated in Figure 8 and shows that the regrinding circuit is located downstream of the flotation plant [27]. Thus, the regrinding plant is fed with concentrate material, which tends to present greater stability in relation to density and solids concentrate. This creates the perfect space to use this data as input for simulation validation.

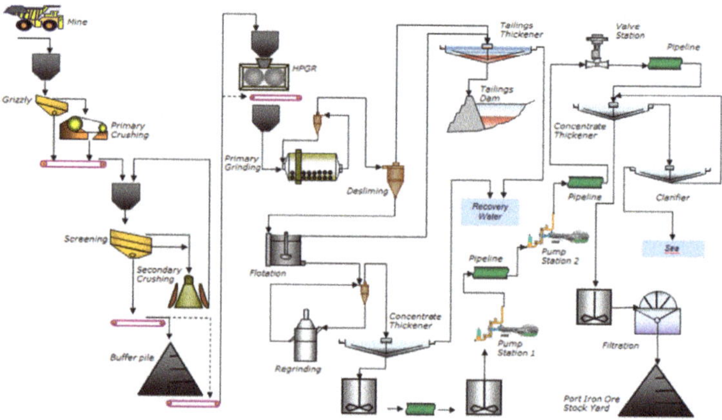

Figure 8. Minas-Rio process flowsheet [27].

The wear measurements were performed for the base and intermediate liner parts at three different liner lifetimes: 0 h, 1000 h, 2000 h and 3000 h. The measurement was performed using a three-dimensional laser scanning technique. Performing the measurement required to complete stop and unload the mill, as shown in Figure 9.

Figure 9. Scanning procedure to obtain a three-dimensional measurement of liner parts [30].

After unloading the mill, a laser device and a reception sensor are used to capture data that is used to generate a three-dimensional geometry of the worn parts. The generated geometry is then compared with the new liner model, in a coordinate system.

The gaps associated with geometrical positions are used to generate a dimensional report and a deviation map, which are later converted into wear quantification. The wear measurements obtained on the VTM-1500 will be presented and used for the comparison to DEM wear simulations.

3.6. Wear Model

The industrial wear measurements were compared with the DEM simulation results. For that, three-dimensional geometries of the agitator screw were exported at different simulation times. Each geometry was sliced in different liner parts, such as the industrial VTM-1500 design. The Meshlab software was used to calculate the volume of base and intermediate worn liners, thus generating a relationship between liner volume and simulation time.

This result was then compared with industrial wear measurement of the full scale VTM-1500. A liner fitting was established to correlate DEM simulation time (in seconds) and operational time (in hours). This scale-up factor was obtained for each simulation, at different agitator velocities. Finally, the scale-up factor and wear simulation are combined, to predict base and intermediate liners wear. The prediction can then be compared with industrial measurement. The explained process is shown in Figure 10.

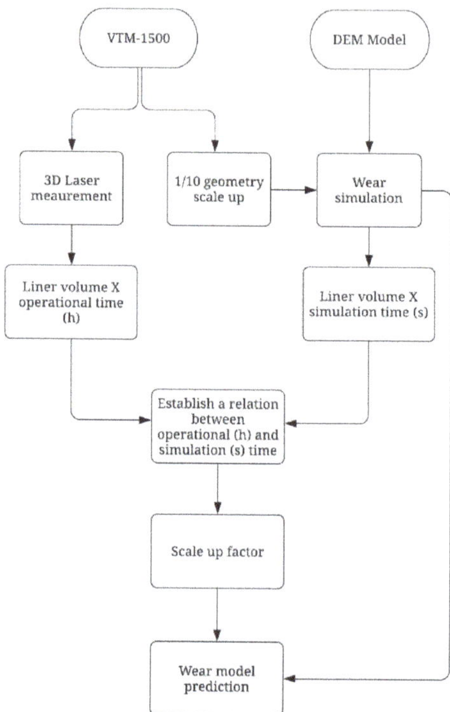

Figure 10. A flowchart explaining the wear model development based on the DEM model and VTM-1500 equipment.

4. Results

Results are divided into four sections: Steady-state simulation results, wear simulation results, VTM-1500 wear measurements, and finally, the DEM wear model prediction. The steady-state simulation results are shown in the initial section, as well as the present results that were obtained at the beginning of the simulation, after a minimum of three mill revolutions and before the beginning of the wear simulation. This is defined as steady-state simulation and aims to understand the mill behavior before wear. The second section presents results obtained during the wear simulation, after wear parameters were settled. This comprises of evaluating the simulation outputs, such as wear quantification. The VTM-1500 wear measurement consists of the next stage and is focused on presenting and adapting industrial wear measurements presented by Silva [30]. Finally, the last section comprises the correlation between industrial and simulated wear patterns, such as the presentation of the wear model prediction. This is performed for the three different mill agitator velocities.

4.1. Steady-State Simulation

4.1.1. Particle Trajectory

Figure 11 shows the particle trajectory colored as a function of particle absolute translational velocity, for the three agitator velocities. A red box was constructed in the surrounding area of the particle trajectory, for the 190 rpm condition. By placing the same box over the particle trajectory of the mills operating at 130 rpm and 87 rpm, it noticeable that the top part of the box is not filled with trajectory streaming lines. This means that the media motion height range is reduced when reducing mill velocity. In other words, it indicates that by increasing mill velocity, the upper parts of the mill were activated.

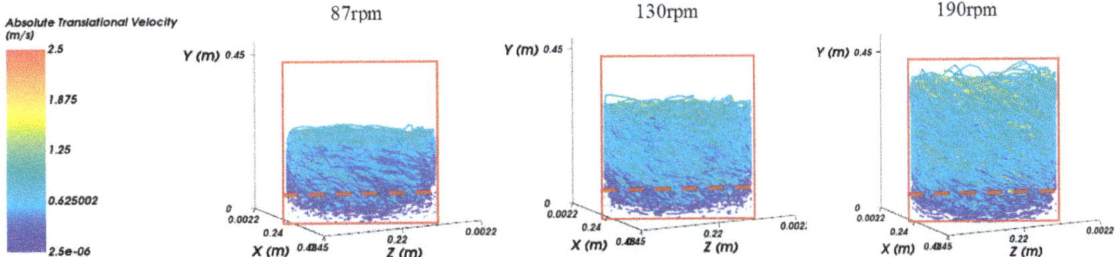

Figure 11. Particle trajectory at different mill rotational velocities.

A dashed red line is placed at the bottom part of the mill. It is visible that streaming lines are also reduced in this bottom area. Because of the reduced motion, this area is identified as "dead zone". The velocity reduction, and thus, the dead zone effect is intensified for reduced mill velocities. This is in accordance with the velocity behavior at the mill top zone. Based on that, it was verified that agitator rotational velocity has a direct effect on the particle trajectory distribution, over the complete height of the mill.

4.1.2. Particle Velocity and Power

Figure 12 shows the particle average translational velocity and simulation power as a function of mill rotational speed. It is conclusive that both particle velocity and power increases with mill speed. This demonstrates that the agitator rotational velocity influences the overall mill power and grinding media kinetic energy, thus having a direct relation in the energy that can be transferred into breakage mechanisms.

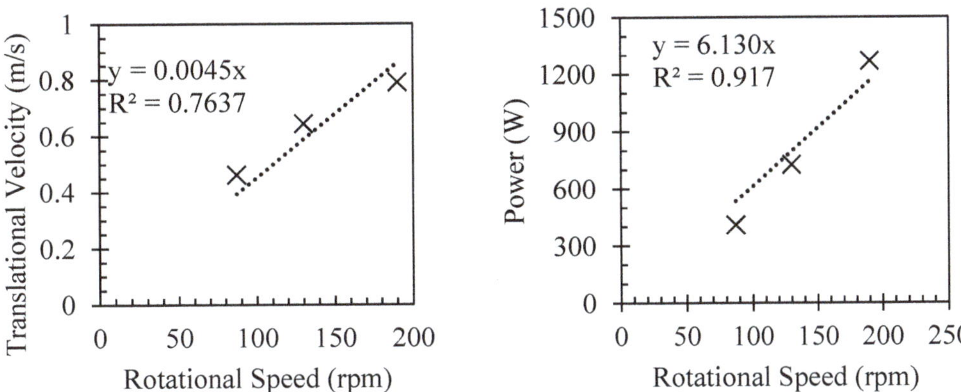

Figure 12. Left: The relationship between agitator speed and particle average translational velocity (m/s). Right: The relationship between agitator speed and simulation power (W).

4.1.3. Particle Spectrum

The collision frequency distribution of particle specific energy was plotted for normal and tangential particle contact in Figure 13. This represents the collisions statistics for specific energy that are applied to particles per time unit. The x axis indicates the specific energy of contact, which means the amount of energy that is transferred per mass of grinding media. The y axis indicates the number of contacts per unit of time, at each specific energy level.

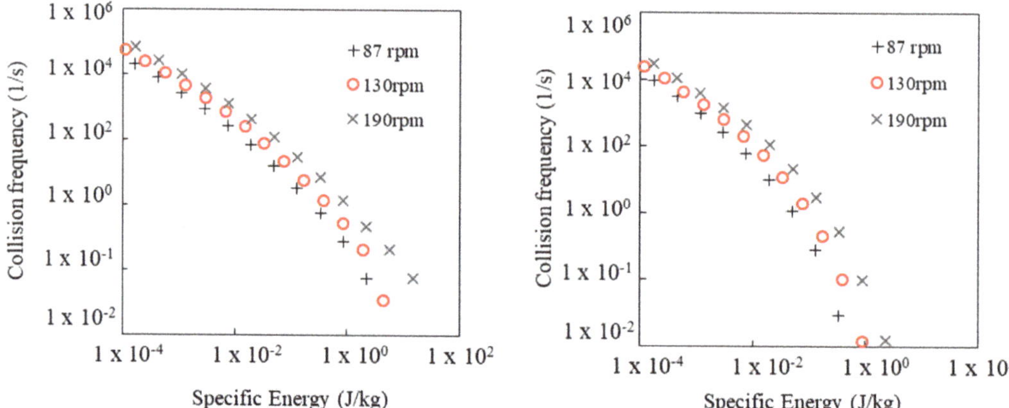

Figure 13. Particle collision spectra for normal (left) and shear (right) contact at different operational velocities.

The wider specific energy distribution is observed for normal contact, in comparison to the shear contact distribution. This means that normal contacts can have a higher specific energy value than shear contacts. In relation to the agitator rotational velocity, higher velocities are associated with a small increase in both collisions' frequency and specific energy range. This can be observed for both normal and tangential contacts.

By looking at the relationship between energy levels and collision frequency, it is also possible to correlate with breakage behavior. In this sense, contacts with greater values of specific energy have a higher tendency to overcome the minimum specific energy required for a certain particle to break. This means that contacts of higher specific energy values have a higher probability of causing breakage in a single or fewer contacts. However, this can also lead to a non-energy efficient behavior once this increases the chance to apply more energy than necessary to cause particle breakage. In contrast, smaller values of specific energy might initially lead to particle weakness, instead of direct breakage. Because of that, it can be necessary for a larger amount of low specific energy collisions cause breakage. At the same time, by applying a greater amount of small specific energy contacts increases the chance to apply the minimum amount of energy that is required for particle breakage.

4.2. Wear Simulation

The wear simulation started after at least three revolutions at steady-state conditions, thus, allowing the wear simulation to be performed under stable conditions.

4.2.1. Particle Trajectory and Velocity

Figure 14 shows the streaming lines associated with particle absolute translational velocity.

A red box surrounds the grinding media trajectory for the new liner condition, shown in the left side of the image. Comparing the left and right side of the picture, it can be noted that red boxes on the right side miss the streaming lines both at the top and bottom parts of the mill. This indicates that media motion is negligible in those regions. In the bottom part of the mill, it is noted that the dead zone effect is emphasized by liner wear.

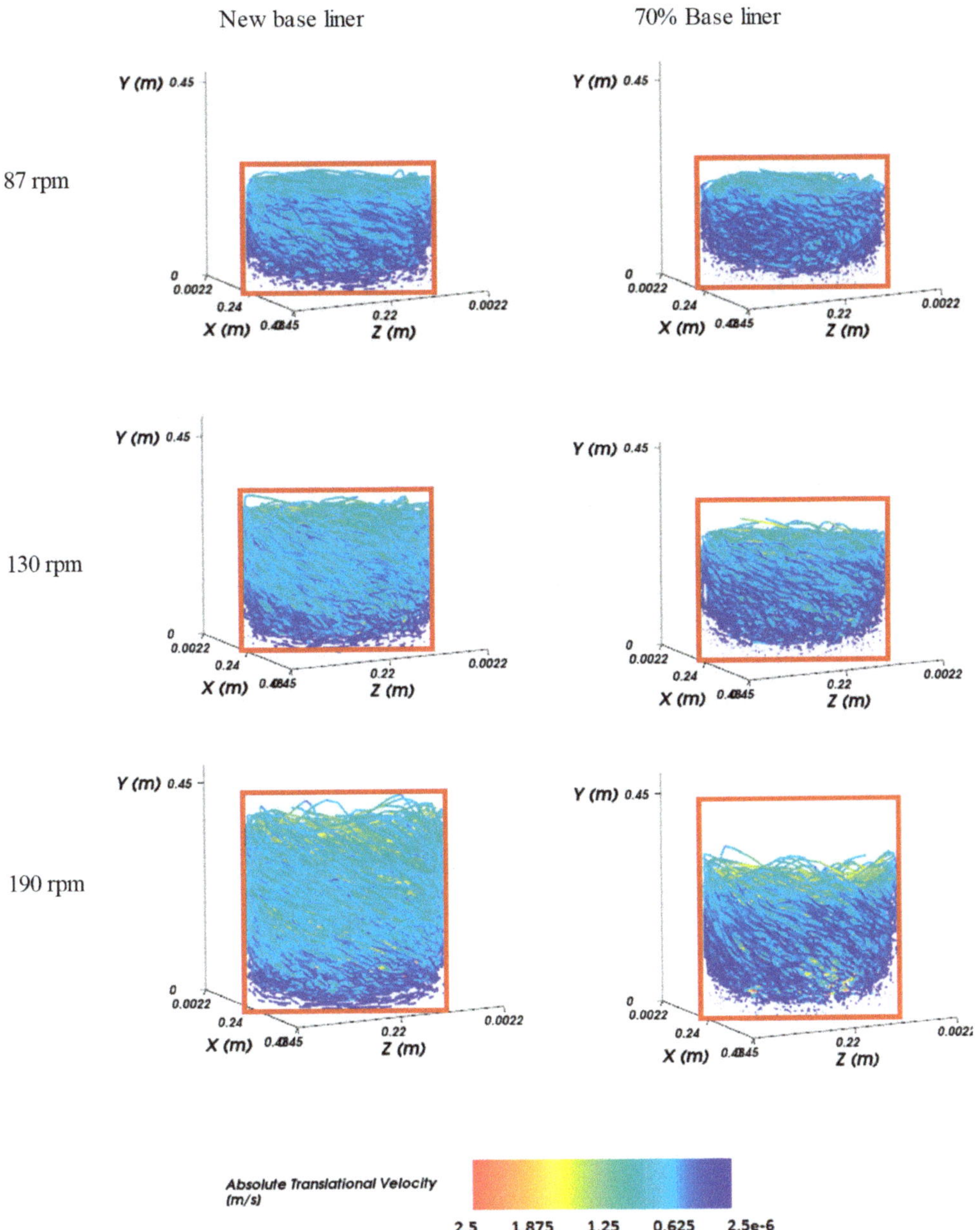

Figure 14. Particle trajectory and absolute translational velocity for new (left) and worn liner (right) conditions, at different mill velocities.

Figure 15 shows the relationship between particle absolute translational velocity and the base liner wear. It is possible to note that particle velocity reduces with wear, due to the reduction in the liner surface area. This reduction is more aggressive for higher agitator velocities. This means that a more representative decrease in particle velocity is obtained when operating at higher agitator velocity. Based on the extrapolated data, it can be expected that particle velocity will be independent on agitator velocity when the base liner is approximately 40% to 50% worn.

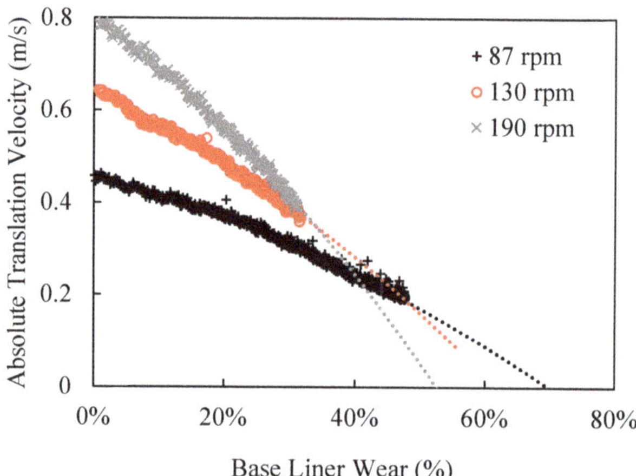

Figure 15. The relationship between average particle velocity and the base liner wear intensity.

4.2.2. Power

Figure 16 shows the relationship between base liner volume and simulation power. It is possible to note that power reduction is intensified at higher rotational velocities. In the operational context, the power is kept constant during the liner lifecycle by increasing mill filling.

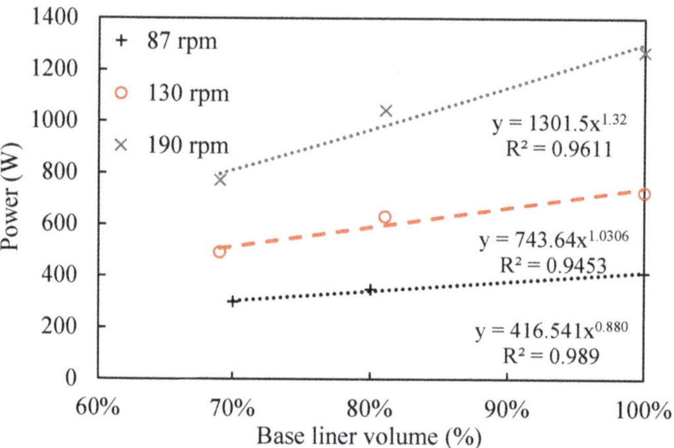

Figure 16. The relationship between simulation power and base liner volume.

4.2.3. Particle Spectrum

Figure 17 shows normal and shear energy spectrums for the three operational velocities, at different wear stages. By comparing shear (right) and normal (left) collision spectra, it is possible to note that normal contacts are more affected by wear. In this sense, a greater reduction in the collision frequency was observed for normal contacts, in comparison to shear. This reduction is emphasized at greater specific energies, especially above 1×10^{-1} J/kg. Based on the idea that collisions of greater specific energies can lead to an inefficient use of energy, it can be suggested that wear scenarios can be related to a better energy use behavior. This indicates that the agitator wear pattern affects mostly high energy contact, while maintaining low energy contacts that present a better energy efficiency behavior. However, it is important to emphasize that the collision frequency and energy reduction will probably generate a coarser product, and thus, the better energy use behavior will be mainly caused by the overall power consumption reduction. Recently, Oliveira [31] applied a mechanistical model to predict product size distribution by using the particle contact spectra obtained with DEM simulations. The simulation considered grinding media as the mill charge, in the absence of ore and slurry. By taking into consideration different proportions of shear power involved in inter particle collision, the model successfully predicted product size distribution of laboratory-scale experiment, for different solids concentration ratios. In this sense, this model can be applied to quantify the wear effect on product size distribution.

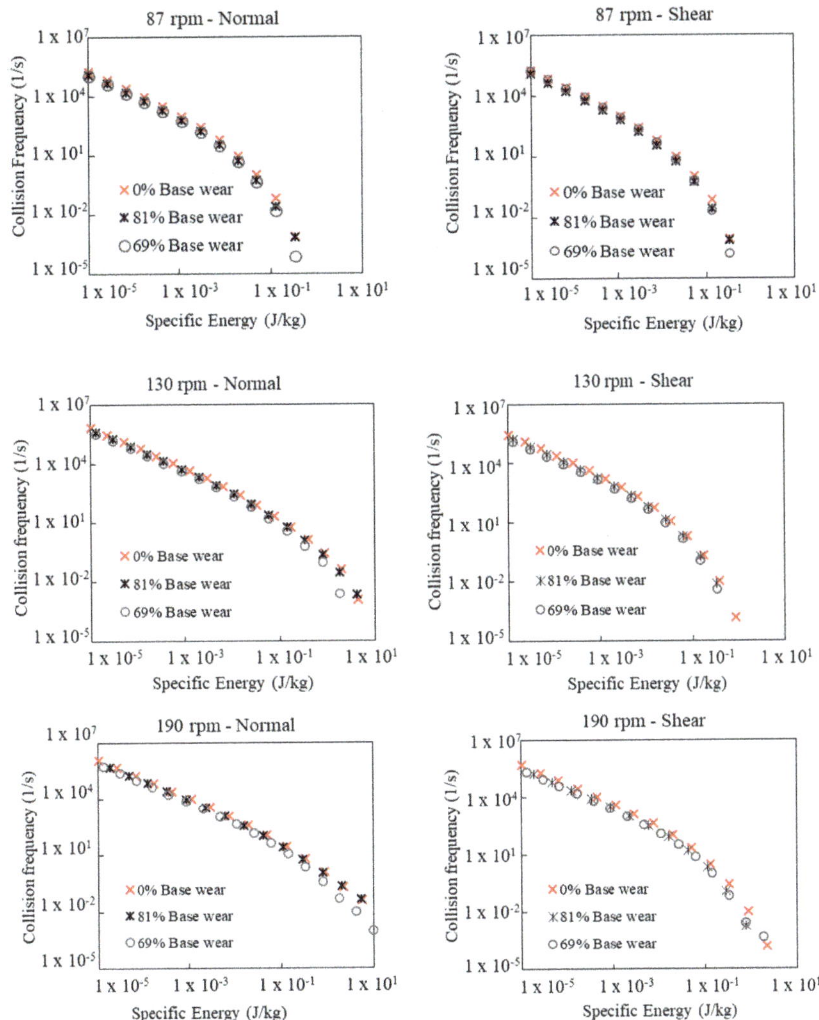

Figure 17. Particle collision spectra at different liner conditions and rotational velocities.

In the operational context, it is important to evaluate the effect on the target product size, such as mineral liberation, to prevent mineral losses in further beneficiation processes. In this sense, a wider evaluation is necessary to guarantee the overall process efficiency, which includes a proper understanding of the effect on the downstream process. As an alternative, various operational conditions can be approached to guarantee the obtaining of the target mineral product size, while maximizing energy efficiency. In the specific case of the Mins-Rio project, the product requirement is in relation to fines generation, for pipeline transport requirements. In this sense, it is necessary to evaluate the effect on fines generation to guarantee the product adequacy.

Comparing the collision spectra behavior at different mill rotational velocities it can be noted that higher rotational velocities intensified the overall collision frequency reduction, such as reduced the maximum value of specific energy (J/kg). This explains the greater power reduction observed for this rotational velocity.

4.2.4. Wear Volume

The relationship between the liner volume and simulation time is shown in Figure 18 for the base and intermediate liner parts. It can be noted that wear is more intensive in the base liner, resulting in a more representative volume reduction. The increase in the agitator velocity also intensifies wear. This effect is more representative of the intermediate liner part. This is a consequence of a wider distribution of particle trajectory when operating at higher rotational velocity, as shown in Figure 18.

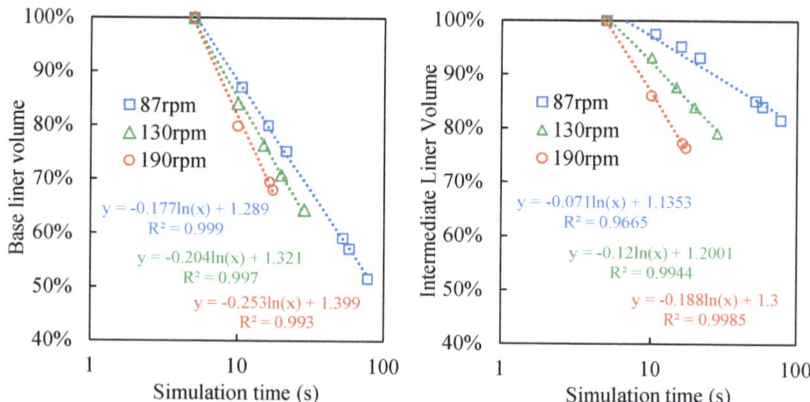

Figure 18. Base and intermediate liner volume during simulation time.

4.3. VTM-1500 Wear Measurement

Based on the three-dimensional scanning of VTM-1500 liner parts, [30] presented the relationship between operational hours and liner mass, separately for the base and intermediate liner parts.

The absolute mass value was converted as liner volume percentage, and are presented in Figure 19, in relation to the operational time. The wear ratio is defined as the relationship between wear percentage and operational time.

The liner volume comparison between the two liner parts reinforces that wear is more aggressive in the base liner. In this sense, at the end of the lifecycle, or 3000 h, the intermediate liner reached approximately 67% of the initial volume, while the base liner reached 35.5% of its initial volume. Because of that, the base liner requires sooner and more frequent replacement. Moreover, the base liner wear ratio significantly increases after 2000 h.

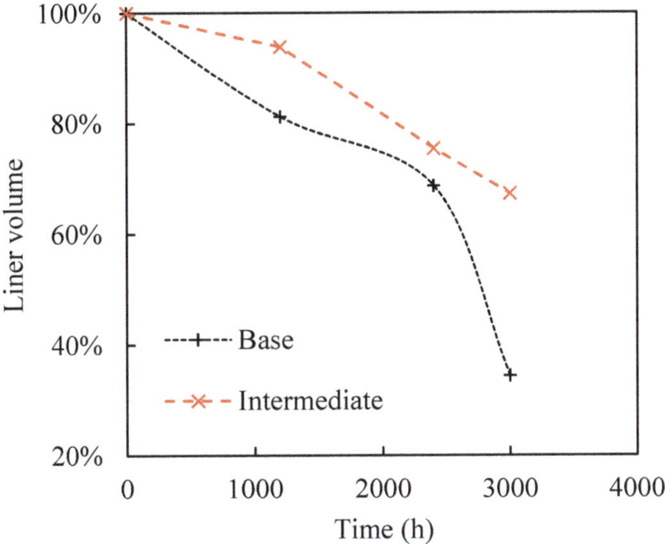

Figure 19. Base and intermediate liner volume reduction during liner operational time (in hours).

4.4. DEM Wear Modeling

Figure 20 compares the simulated geometry and industrial liner designs under several wear conditions. The visual comparison between wear patterns shows that DEM provided a remarkably similar wear design. The next stage would be to quantify the wear relation. For that, a time relation was established in order to obtain a scale-up factor that correlates simulation seconds and operational hours. The scale-up factors were calculated for the base and intermediate liners, at each operational velocity. The final scale-up factor, per velocity, was obtained as an average value in between the liner and intermediate factors.

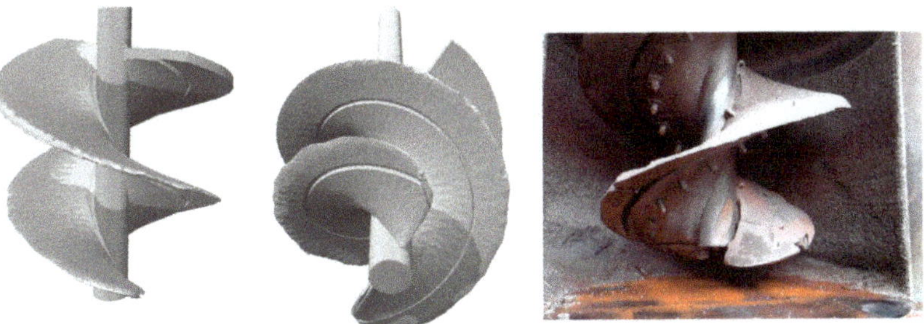

Figure 20. Comparison between 3D worn screw after DEM simulation and industrial worn liner.

Figure 21 shows the predicted liner volume, based on the obtained DEM model. By comparing the intermediate and base liner predictions, a better agreement was achieved when using the 87 rpm as agitator velocity. In relation to the base liner, the results compared favorably during most of the operational time, except for the last liner measurement, which corresponds to approximately 3000 h. This is probably because the last liner measurement presents an aggressive behavior, which representatively differs from the trend.

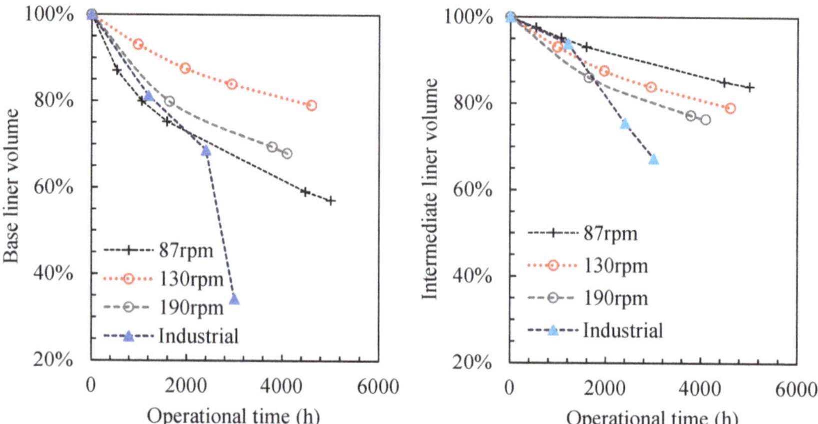

Figure 21. Model predicted and measured values of base and intermediate liner volume during operational time (h).

For the intermediate liner, the behavior was well predicted up to approximately 1500 h. This is probably due to the fact of filling increases during the liner lifetime. The filling increasing is performed during the liner lifetime with the aim to keep constant power by activating upper parts of the mill. Because of that, added grinding media gets in contact with upper parts of the intermediate liner, thus resulting in additional wear. In contrast, the liner volume predictions for 130 rpm and 190 rpm did not fit very well when evaluating base and intermediate liner parts.

To summarize, it was noted that the wear prediction presented a relative agreement with industrial measurements in the first half of the liner lifecycle. Differently, the agreement was not very effective for the second half of the lifecycle, especially for the intermediate liner part. In relation to the liner part, the disagreement can be explained by the wear compensation in the industrial context. In order to keep constant power consumption, additional grinding media is added to the mill, thus activating the upper parts of the intermediate and generating extra wear. This filling compensation was not taken into account by the DEM simulation, and then it can be expected a lower wear rate and the end of the lifecycle for the intermediate liner part. In relation to the base liner, the differences after 3000 h can be explained by a possible variation in the material properties after several wears, which should be confirmed by liner material properties evaluation. In this sense, the intensive wear could substantially affect liner material resistance properties, thus reducing wear resistance for very worn conditions. This effect was not taken into account by the DEM simulations, resulting in a large difference for the base liner prediction after 3000 h.

5. Conclusions

Once the vertical stirred mill screw liner is responsible for media movement, and consequently, grinding, it is very important to perform accurate monitoring of wear progress. However, it is not possible to directly install a sensor for wear measurement. The paper addressed this issue by performing DEM simulations, with the aim to provide a better understanding of screw liner wear behavior and effects. The simulation was performed for a 1:10 reduced scale geometry from the Metso Vertimill VTM-1500 model, to reduce simulation effort.

The simulations applied different scale-up methodologies for agitator rotational velocity. In this sense, three different mill velocities were simulated. Firstly, a reduced velocity was obtained based on a dataset model. This consists of a more recent approach for velocity scale-up in laboratory-scale equipment. Moreover, a direct scale-up factor based on the geometry reduction was applied, thus resulting in a higher operational velocity of 190 rpm. Finally, an intermediary velocity was tested to provide a better understanding of the effects of operating in between the lower and maximum mill velocities.

The wear simulation results qualitatively showed that the wear profile obtained using DEM presents a relative similarity to the wear design observed in the full-scale equipment at the beginning of the liner lifecycle. The worn agitator geometries were exported, at different simulation times, to provide a volume quantification of wear. By comparing worn geometries obtained from DEM simulations and industrial measurement, a time correlation was proposed. In this sense, a scale-up factor was obtained to correlate simulation time, in seconds, and operational time, in hours. This scale-up factor was then applied to simulation results to predict wear in the operational context, for the base and intermediate liner parts. By comparing the obtained wear prediction and industrial measurement, a better agreement was encountered for the 87 rpm mill velocity. Based on that, the recommendation would be to apply the developed model under different operational conditions, to evaluate different strategies for wear compensation, such as changing mill speed and grinding media filling.

Additionally, DEM outputs were evaluated to provide a better understanding of wear effects. From that, it was noted that wear introduced a significant decrease in simulation power and particle average translational velocity. In relation to particle trajectory, the particle velocity reduction was intensified in the top and bottom parts of the mill, thus resulting in null particle motion in those areas. In the bottom part of the mill, this is known as the dead zone effect. The particle collision spectra indicated that wear affects more intensively normal contacts of greater specific energy. Although this might cause a decrease in fines generation, this can lead to an improvement in energy usage. This is because the predominance of low energy contacts is related to an increase in energy efficiency, once this reduces the occurrence of high-intensity contacts, which apply greater energy than required for particle breakage.

To summarize, the obtained results indicate that liner wear affects particle breakage and energy consumption. A better understanding and proper quantification about its effects can play a key role in the development of strategies for optimizing operational procedures to keep grinding efficiency and to reduce wear rates. In this sense, the recommendations would be to evaluate the relationship between operational conditions and liner wear, in the industrial context.

The model did not consider particle breakage, as ore particles itself are not included in the simulation. Grinding media wear also not considered. Finally, the current work did not consider the slurry (fluid), so as there would be a recommendation to apply CFD (computational fluid dynamics) or SPH (smoothed particle hydrodynamics) to describe the mill charge with more details.

Author Contributions: P.M.E. wrote the paper and carried out all the experimentation, D.B.M. planned the experimental part and reviewed the paper, R.G. reviewed the paper and L.C.R.M. provided the technical support on the site of Minas-Rio operation. All authors have read and agreed to the published version of the manuscript.

Funding: The authors thank CAPES and Anglo American for the financial support.

Acknowledgments: The authors acknowledge the license of ROCKY Software provided by ESSS and its technical support during the simulations performed. Thanks to Anglo American for the support and permission to publish the results from Vertimill™ laser wear measurement. Thanks to peer reviewers for the comments that significantly contributed to the paper.

Conflicts of Interest: The authors declare no conflict of interest.

References

1. Hasan, M.M. Process Modelling of Gravity Induced Stirred Mills. Ph.D. Thesis, University of Queensland, Brisbane, Australia, 2016.
2. DOE. *Comminution and Energy Consumption: Report of the Committee on Comminution and Energy Consumption*; National Materials Advisory Board, Commission on Sociotechnical Systems: Washington, DC, USA, 1981; p. 283.

3. Marsden, J. Energy Efficiency and Copper Hydrometallurgy. In *Hydrometallurgy 2008: Proceedings of the Sixth International Symposium*; SME: Englewood, CO, USA, 2008.

4. Napier-Munn, T. Is progress in energy-efficient comminution doomed? *Miner. Eng.* **2015**, *73*, 1–6. [CrossRef]

5. Shi, F.; Morrison, R.; Cervellin, A.; Burns, F.; Musa, F. Comparison of energy efficiency between ball mills and stirred mills in coarse grinding. *Miner. Eng.* **2009**, *22*, 673–680. [CrossRef]

6. Esteves, P.M.; Mazzinghy, D.B.; Galéry, R.; Filho, B.C.; Silva, J.F.L.; Russo, J.F.C. Predictive modelling of vertical stirred mills liner wear using vibration signature analysis. In Proceedings of the 11th International Comminution Symposium, Cape Town, South Africa, 16–19 April 2018.

7. Allen, J.; Noriega, E.R. Screw liner replacement in a VERTIMILL®grinding mil—Determining best practice. In Proceedings of the 8th International Mining Plant Maintenance Meeting, Antofagasta, Chile, 7–9 September 2011.

8. Mishra, B.K.; Rajamani, R.K. *Analysis of Media Motion in Industrial Ball Mills. Comminution: Theory and Practice*; Society for Mining, Metallurgy and Exploration, Inc.: Littleton, CO, USA, 1992; pp. 426–440.

9. Weerasekara, N.; Powell, M.; Cleary, P.; Tavares, L.; Evertsson, M.; Morrison, R.D.; Quist, J.; Carvalho, R.M. The contribution of DEM to the science of comminution. *Powder Technol.* **2013**, *248*, 3–24. [CrossRef]

10. Cleary, P. Predicting charge motion, power draw, segregation and wear in ball mills using Discret Element Methods. *Miner. Eng.* **1998**, *11*, 1061–1080. [CrossRef]

11. Cleary, P.W.; Owen, P. Development of models relating charge shape and power draw to SAG mill operating parameters and their use in devising mill operating strategies to account for liner wear. *Miner. Eng.* **2018**, *117*, 42–62. [CrossRef]

12. Cleary, P.W.; Sinnott, M.; Morrison, R. Analysis of stirred mill performance using DEM simulation: Part 2—Coherent flow structures, liner stress and wear, mixing and transport. *Miner. Eng.* **2006**, *19*, 1551–1572. [CrossRef]

13. Kalala, M.H. Discrete element method modelling of liner wear in dry ball milling. *J. S. Afr. Inst. Min. Metall.* **2004**, *104*, 597–602.

14. Kalala, J.T.; Bwalya, M.M.; Moys, M.H. Discrete element method (DEM) modelling of evolving mill liner profiles due to wear. *Part I DEM validation. Miner. Eng.* **2005**, *18*, 1386–1391.

15. Cleary, P.W.; Owen, P. Effect of liner design on performance of a HICOM (R) mill over the predicted liner life cycle. *Int. J. Miner. Process.* **2010**, *134*, 11–22. [CrossRef]

16. Boemer, D.; Ponthot, J.P. A generic wear prediction procedure based on the discrete element method for ball mill liners in the cement industry. *Miner. Eng.* **2017**, *109*, 55–79. [CrossRef]

17. Xu, L.; Luo, K.; Zhao, Y. Numerical prediction of wear in SAG mills based on DEM simulations. *Powder Technol.* **2018**, *39*, 353–363. [CrossRef]

18. Morrison, R.D.; Cleary, P.W.; Sinnott, M.D. Using DEM to compare energy efficiency of pilot scale ball mill and tower mills. *Miner. Eng.* **2009**, *22*, 665–672. [CrossRef]

19. Sinnott, M.D.; Cleary, P.W.; Morrison, R.D. Is media shape important for grinding performance in stirred mills? *Miner. Eng.* **2011**, *24*, 138–151. [CrossRef]

20. Sinnott, M.D.; Cleary, P.W.; Morrison, R.D. Slurry flow in a tower mill. *Miner. Eng.* **2011**, *24*, 152–159. [CrossRef]

21. Radziszewski, P.; Moore, A. Understanding the effect of pressure profile on stirred mill impeller wear. *Miner. Eng.* **2017**, *103*, 54–59. [CrossRef]

22. Toor, P.; Bird, M.; Perkins, T.; Powell, M.; Franke, J. The influence of liner wear on milling efficiency. In Proceedings of the METPLANT 2011—Metallurgical Plant Design and Operating Strategies, Perth, WA, USA, 8–9 August 2011; pp. 193–212.

23. Yahyaei, M.; Banisi, S.; Hadizadeh, M. Modification of SAG mill liner shape based on 3-D liner wear profile measurements. *Int. J. Miner. Process.* **2009**, *91*, 111–115. [CrossRef]

24. Eirich, Brochure: EIRICH Tower Mill-Vertical Agitated Media Mill. 2018. Available online: https://www.eirichusa.com/products/towermills/tower-mill-vertical-ag-med-mill (accessed on 27 July 2020).

25. Metso, Blog: When to Change Vertimill™ Liners and How to Do it safely! 3 October 2017. Available online: https://www.metso.com/blog/mining/blog-when-to-change-vertimill-liners-and-how-to-do-it-safely/ (accessed on 17 September 2019).

26. Rocky, Rocky Capabilities Chart. 2018. Available online: http://rocky-dem.com/index.php?pg=capability (accessed on 22 January 2018).

27. Mazzinghy, D.B.; Lichter, J.; Schneider, C.L.; Galéry, R.; Russo, J.F. Vertical stirred mill scale-up and simulation: Model validation by industrial samplings results. *Miner. Eng.* **2017**, *103*, 127–133. [CrossRef]

28. Esteves, P.; Mazzinghy, D.; Hilden, M.; Yahyaei, M.; Powell, M.; Galery, R. Qualitative evaluation of the grinding efficiency of a gravity induced stirred mill using the size specific energy approach. In Proceedings of the 16th European Symposium on Comminution and Classification (ESCC 2019), Leeds, UK, 2–4 September 2019.

29. Rocky, Workshop 4—SAG Mill (Wear and Particle Energy Spectra). 2017. Available online: https://www.bajajfinserv.in/customer-portal (accessed on 8 January 2018).

30. Silva, J. Effect of Process Parameters on Vertical Mill Power. Master's Thesis, Federal University of Minas Gerais, Belo Horizonte, Brazil, 2019. (In Portuguese)

31. Oliveira, A.L.; Rodriguez, V.A.; de Carvalho, R.M.; Powell, M.S.; Tavares, L.M. Mechanistic modelling and simulation of a batch vertical stirred mill. *Miner. Eng.* **2020**, *156*, 106487. [CrossRef]

Article

Effects of Ball Size on the Grinding Behavior of Talc Using a High-Energy Ball Mill

Hyun Na Kim [1,*], Jin Woo Kim [1], Min Sik Kim [1], Bum Han Lee [2] and Jin Cheul Kim [2]

[1] Department of Earth and Environmental Sciences, Kongju National University, Gongju 32588, Korea; ahop2000@kongju.ac.kr (J.W.K.); msms6365@smail.kongju.ac.kr (M.S.K.)

[2] Korea Institute of Geoscience and Mineral Resources, 124 Gwahak-ro, Yuseong-gu, Daejeon 34132, Korea; leebh@kigam.re.kr (B.H.L.); kjc76@kigam.re.kr (J.C.K.)

* Correspondence: hnkim@kongju.ac.kr; Tel.: +82-41-850-8513; Fax: +82-41-850-8953

Received: 26 September 2019; Accepted: 28 October 2019; Published: 31 October 2019

Abstract: The properties and preparation of talc have long been investigated due to its diverse industrial applications, which have expanded recently. However, its comminution behavior is not yet fully understood. Therefore, having better control of the particle size and properties of talc during manufacturing is required. In this study, we investigate the effect of the ball size in a high-energy ball mill on the comminution rate and particle size reduction. High-energy ball milling at 2000 rpm produces ultrafine talc particles with a surface area of 419.1 m^2/g and an estimated spherical diameter of 5.1 nm. Increasing the ball size from 0.1 mm to 2 mm increases the comminution rate and produces smaller talc particles. The delamination of (00l) layers is the main comminution behavior when using 1 mm and 2 mm balls, but both the delamination and rupture of (00l) layers occurs when using 0.1 mm balls. The aggregation behavior of ground talc is also affected by the ball size. Larger aggregations form in aqueous solution when ground with 0.1 mm balls than with 1 mm or 2 mm balls, which highlights the different hydro-phobicities of ground talc. The results indicate that optimizing the ball size facilitates the formation of talc particles of a suitable size, crystallinity, and aggregation properties.

Keywords: nanoscale talc; wet milling; high-energy ball milling; ball size; aggregation

---

1. Introduction

Talc ($Mg_3Si_4O_{10}(OH)_2$) is a phyllosilicate mineral with a T-O-T layer composed of tetrahedral silicon and octahedral magnesium, which share oxygen and are strongly bonded with each other. The weak van der Waals bonds between the T-O-T layers is the origin of softness of talc. The silicon in the siloxane sheet renders talc hydrophobic, since it cannot easily be substituted with aluminum [1]. Talc is physically easy to handle due to low hardness, does not react with acid due to its chemical stability, and has high adsorptivity, low plasticity, and low thermal/electrical conductivity. Because of these characteristics, talc is used as a coating, refractory, and additive in various industrial fields such as paper, paint, rubber, ceramic, refractory material, and polymer manufacturing. Chemical and pharmaceutical industries also require high-grade talc powders with high purity and uniform particle sizes. The grade and usage of talc is classified considering the purity, whiteness, particle size, and more. For example, talc with a particle size of 44 μm is used in ceramics and paints, 8–12 μm in paper, 7 μm in cosmetics, and 2 μm or smaller in rubber [2]. In addition, the nanoscale talc recently received attention for its application in areas such as improving the heat resistance of nanocomposites, and for use in waste water filter materials [3–5].

Given the importance of the above, various attempts have been made to produce ultrafine talc powders and better understand their physicochemical properties [6–11]. Various processes, such as planetary ball milling, tumbling milling, stirred ball milling, and disk milling, have been used to grind talc [10,12–17]. A reduction in particle size and the collapse of the crystalline structure of talc have been observed in mechanical grinding studies [8,10,12,13,15,18–20]. Table 1 summarize the particle size and/or surface area of ground talc with processed various milling methods. Delamination of the layered structure in the direction of the (00*l*) surface is the predominant mechanism of grinding, which, consequently, accompanies the breaking of individual plates and disordering in the talc crystal [8,10,15,21]. Further excessive grinding generally causes the collapse of the crystalline structure into the amorphous phase and the aggregation of ultrafine particles [13,14]. The degree of particle size reduction varies depending on the type of milling machine and milling conditions such as duration and rotating speed [11,21–26]. Besides the collapse of the crystalline structure, the physicochemical properties of talc such as thermal and dispersion behaviors, cation exchange capacity, wettability, and whiteness are also changed by mechanical grinding [6,10,12,14,27–29]. In accordance with the increasing demand for high-quality and ultrafine talc powder, the importance of comminution technology, capable of controlling crystallinity, particle shape, and particle size, is increasing.

Table 1. Specific surface area and equivalent spherical diameter of talc powder depending on the milling condition.

Milling Machine	Milling Type	Rotation Speed (rpm)	Grinding Time (h)	Ball Size (mm)	S.S.A [†] (m²/g)	ESD [‡] (nm)	D_{50} (μm)	References
Tumbling mill	dry	86	116.6	-	14.0	153	-	[13]
Planetary ball mill	dry	1350	0.5	16	10.5	204	-	[13]
Planetary ball mill	dry	-	0.5	-	109.7	20	-	[14]
Planetary ball mill	dry	-	5	-	28	76	-	[10]
Stirred ball mill	wet	2500	3.5	-	-	-	0.2	[17]
Stirred ball mill	dry	600	0.25	2.5–3.5	3.34	641	2.33	[16]
Disk mill	dry	70	6		17.3	128		[15]
High energy ball mill	wet	2000	12	2	419.1	5.1	0.37	This study
				1	365.1	5.9	0.44	
				0.1	171.7	12.5	7.96	

[†] S.S.A. = specific surface area. [‡] ESD = equivalent spherical diameter.

Planetary ball milling, which is one of the most frequently used lab-scale milling tools, has been applied to increase the grinding efficiency of talc. The collision impact and friction between powder samples and balls in a rotating jar grind the microscale talc into sub-microscale particles. The grinding efficiency of ball milling is greatly affected by the jar rotation speed, ball size, and ratio of balls and sample talc [22,24,25]. A faster rotation speed generally produces talc with a smaller particle size and lower crystallinity [11]. Recently developed high-energy ball mills reach rotation speeds of up to 2000 rpm and provide an opportunity to reduce the size of talc powder to the nanoscale level. The high-energy ball mill system consists of turn discs and jars. The oppositely rotating jar and turning disc produce highly energetic impact energy (Figure 1) [30]. However, while the ball size is an important factor in ball milling, the effect of the ball size on the comminution of talc has not yet been fully understood [31,32]. While the particle size is known to generally increase with an increasing ball size, in 2013, Shin et al. showed that particle size decreased and then increased as the ball size increased from 1 to 10 mm during the comminution of Al_2O_3 powder at low rotation speeds (50–153 rpm) [32]. The optimum ball size is dependent on the milling conditions and the properties of the mineral. Thus, to obtain ultrafine talc powder while minimizing the loss of crystallinity, a systematic approach is necessary to understand the effect of ball size in high-energy ball milling on the comminution of talc.

In this study, high-energy ball milling with high rotation speed of 2000 rpm is applied to obtain the nanoscale talc powder. The effect of the ball size for planetary ball milling on the grinding rate and behavior of talc is investigated with a varying ball size (2, 1, and 0.1 mm). Changes in the agglomeration characteristics of mechanochemical talc and the relationship with the ball size is also investigated.

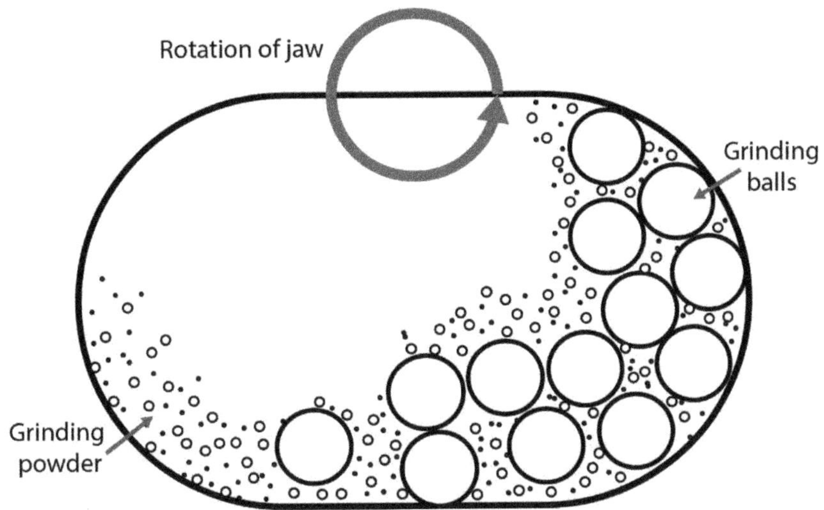

Figure 1. Schematic view of the motion of the balls and powder in a jar during high energy ball milling.

2. Materials and Methods

2.1. Sample Preparation

A commercial talc powder with high purity (>99%, XTA-400, KOCH corp., Shenyang, China) was used in this study. The mean particle size of an as-received powder is approximately 12 μm. The sample was mechanically deformed using a high-energy ball mill (Emax, Retsch, Haan, Germany), which generates ultrafine particles by combining impact and shear stress. Wet milling was performed by loading the talc powder, distilled water, and zirconia balls in the volume ratio of 2:3:4 into a 50 mL zirconia (ZrO_2) jar. Rotation speed was fixed at 2000 rpm and grinding time varied from 10 to 360 min. The time interval for grinding was selected by considering the trends in particle size and crystallinity of the talc following preliminary experiments. Zirconia balls of various sizes (2, 1, and 0.1 mm) were used. During grinding, the temperature of the milling jar was maintained at approximately 50 °C by a water-cooling system.

2.2. Characterization Methods

2.2.1. X-Ray Diffraction Analysis

X-ray diffraction (XRD) analysis was performed with the Rigaku SmartLab X-ray diffractometer (Rigaku, Tokyo, Japan) in the Korea Institute of Geoscience and Material Resources. Cu-Kα radiation was used with a tube voltage of 40 kV and a current of 30 mA. The scan range was 3–90° (2θ) with an interval of 0.02°.

2.2.2. BET (Brunauer-Emmett-Teller) Specific Surface Area Analysis

The specific surface area of the talc sample was measured using the specific surface area analyzer (Micromeritics ASAP-2420, Norcross, GA, USA) in the Jeonju Center of Korea Basic Science Institute. Before the experiment, the adsorbed water on the talc particle surface was removed at 300 °C for 2 h, and then the BET method involving nitrogen adsorption method at 77 K was used.

2.2.3. Laser Diffraction Particle Size Analysis

The particle size was determined using a laser diffraction particle size analyzer (Mastersizer3000, Malvern Panalytical, Malvern, UK) in the Korea Institute of Geoscience and Material Resources. The talc in an aqueous colloid solution was dispersed for 10 min using a 28 kHz ultrasonic dispersing machine.

2.2.4. Scanning Electron Microscopy

Scanning electron microscopy was performed using the high-resolution scanning electron microscope (HR FE-SEM, TESCAN-MIRA3, Brno, Czech Republic) at the Kongju National University. The sample was coated with platinum (Pt) using an ion-coating machine at an accelerated voltage of 10.0 kV.

3. Results

3.1. X-Ray Diffraction Analysis

Figure 2 shows the XRD patterns of talc ground for up to 360 min using different ball sizes, which shows the effect of ball size on the crystallinity of ground talc.

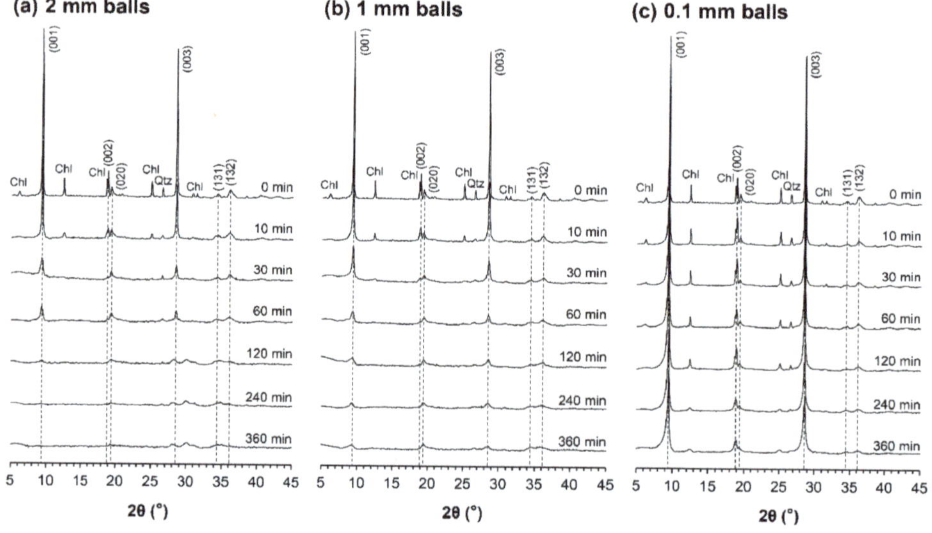

Figure 2. XRD patterns of talc powder upon grinding with (**a**) 2 mm, (**b**) 1 mm, and (**c**) 0.1 mm balls. Chl: chlorite, Qtz: quartz.

In the as-received talc, (001), (002), and (003) peaks of talc with a narrow width were observed at 2θ angles of 9.4°, 18.9°, and 28.6°, respectively [12]. The small peaks for chlorite and quartz were also observed, which were often found with talc as a result of carbonate alteration [33]. Upon grinding, the X-ray diffraction peak intensity gradually decreased while the peak width increased, which implied the decrease in crystallinity of ground talc. The degree and rate of reduction in peak intensity varies with the size of the balls used in grinding. When 2 mm or 1 mm balls were used for grinding, the peak intensity of the talc markedly decreased until 60 min of grinding. After 120 min of grinding, most diffraction peaks disappeared when the 2 mm balls were used. When 1 mm balls were used, small diffraction peaks were still observed, even though their intensities are very small compared with prior to grinding. A further change in the diffraction pattern was not observed for grinding by more than 120 min. These results indicated that ground talc for 120 min and more has a disordered or amorphous structure when 2 mm or 1 mm balls were used for grinding. When 0.1 mm balls were used, the decrease in peak intensity was relatively sluggish. The peak intensity consistently decreased even after 120 min of grinding, which was in contrast with the cases using 2 mm or 1 mm balls. The unique diffraction pattern of talc was still apparent even after 360 min of grinding, which indicated a crystalline structure of talc left after grinding.

To quantitatively compare the effects of the ball size on the crystallinity of ground talc, the change in the relative height of the talc (001) peak, according to grinding, was analyzed (Figure 3). When 2 mm balls were used, the peak height after 60 min of grinding decreased to approximately 9% compared with the height prior to grinding.

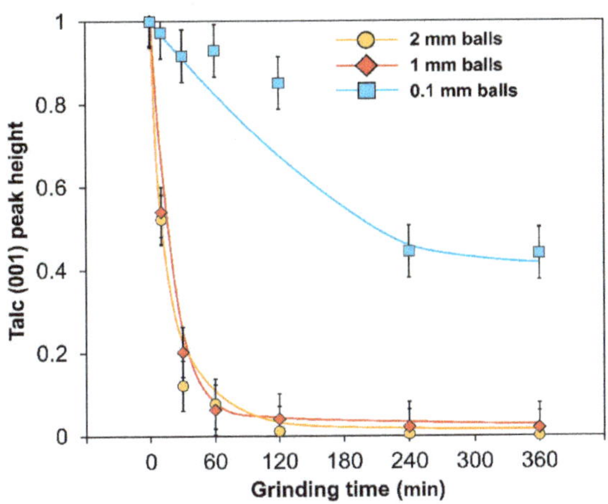

Figure 3. Changes in the relative intensities of (001) XRD peak of talc powder upon grinding with varying ball sizes for high-energy ball milling. The relative peak intensities are normalized with the (001) XRD peak of as-received talc.

Almost no peak was observed after 120 min of grinding. Similar results were obtained when 1 mm balls were used, but, after 120 min, the peak height decreased slightly to less than that obtained using 2 mm balls. When 0.1 mm balls were used, the peak height showed a slight decrease, and approximately 44% remained even after 360 min of grinding. These results indicate that the crystallinity of talc decreases more when 2 mm and 1 mm balls are used than when 0.1 mm balls are used for milling. As the ball size increases, greater size reduction of the talc particle and loss of crystallinity occur during grinding. This is due to the reduced kinetic energy of collision between rotating balls in the mill with 0.1 mm balls when compared with 2 mm and 1 mm balls [31,32].

Figure 4 shows the variation in relative peak intensities (I_t/I_0) of talc's main diffraction peaks such as (001), (132), (003), and (020) upon grinding, which shows the effect of the ball size on the direction in the fracture of talc. For 2 mm balls, the (001) and (003) lattice planes, which are perpendicular to the c-axis of talc, decreased faster than the (132) and (020) peaks. In particular, the decrease rate of (001) and (003) peaks was very high during the first 10 min and gradually slows at 60 min of grinding. The (001) and (003) peaks mostly disappeared after 120 min of grinding. In contrast, the decrease rate of the (132) and (020) peaks is relatively low and its peak intensities are approximately 20% after 360 min of grinding compared with that prior to grinding. The use of 1 mm balls yielded similar results to those obtained using 2 mm balls. The decrease rate of (001) and (003) peaks was very high in the early stage of grinding, and most of the peaks subsequently disappear. The (132) and (020) peak intensities are approximately 50% after 360 min of grinding when compared with that prior to grinding, which indicates less amorphization when using 1 mm balls than when using 2 mm balls.

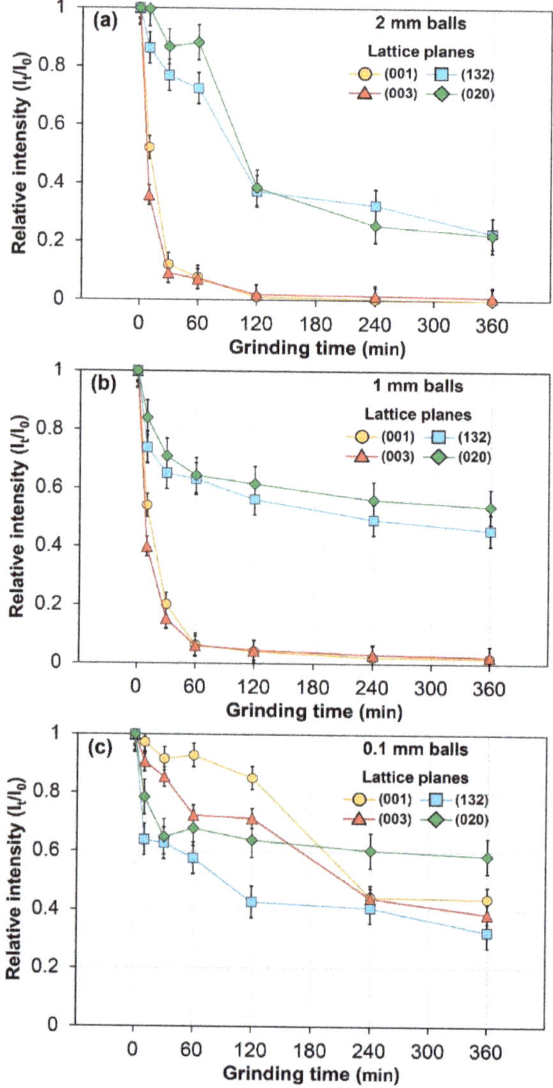

Figure 4. Effects of grinding time and ball size on the relative XRD intensities of lattice planes. The ball size for high-energy ball milling is (**a**) 2 mm, (**b**) 1 mm, and (**c**) 0.1 mm. The relative peak intensities are normalized with the XRD peak of as-received talc.

In contrast, when 0.1 mm balls were used, the decrease rate of (132) and (020) peaks was higher than that of the (001) and (003) peaks, especially during the first 60 min of grinding. After 360 min of grinding, all peaks except the (020) peak reached approximately 40% compared with that prior to grinding. As the ball size increases, the reduction of peak intensity prior to grinding increases for most peaks. The delamination of (00l) planes, in particular, are significantly affected by the ball size. When 2 mm or 1 mm balls were used, the peak intensity of (00l) planes reduced to less than 5% of its state prior to grinding, while it only reduced to 40% when using 0.1 mm balls. The relative intensities of (020) plane after 360 min of grinding were approximately 20%, 55%, and 60% when using 2 mm, 1 mm, and 0.1 mm balls, respectively. These results indicate that the size reduction mechanism of talc changes with varying ball size. The contribution of delamination toward the decrease in particle size is high when 2 mm or 1 mm balls are used but low when 0.1 mm balls are used. The particle size reduction of talc occurred in all directions when using 0.1 mm balls.

3.2. Analysis of Specific Surface Area

Figure 5 shows the variation in the specific surface area of talc according to the ball size used in grinding.

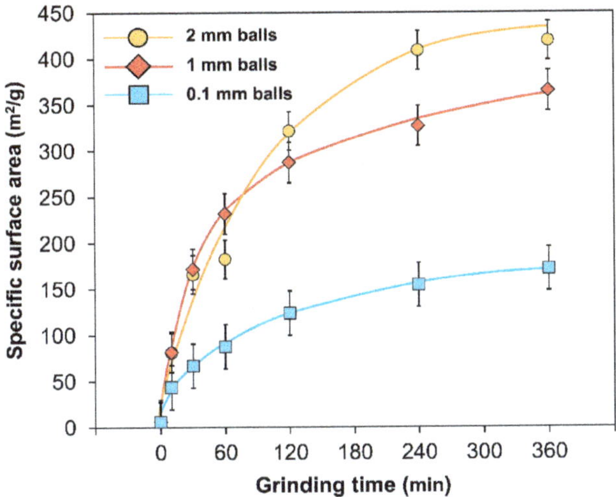

Figure 5. Variation of BET specific surface area of talc upon grinding with varying ball sizes for high-energy ball milling.

As the grinding progresses, the increase in the specific surface area of talc is initially abrupt and then becomes gradual. The overall tendency of the increase in the specific surface area is similar for all ball sizes. However, the degree of increase varied according to the ball size. When 2 mm or 1 mm balls were used, the tendency was very similar until 60 min of grinding. The specific surface area was slightly higher when 1 mm balls were used than when 2 mm balls were used. However, since the grinding time is increased above 60 min, the specific surface area of the particles ground using 2 mm balls increased more toward the end of the grinding period. The specific surface area was approximately 6.1 m²/g before grinding and increased to approximately 419.1 m²/g for 2 mm balls and approximately 365.1 m²/g for 1 mm balls after 360 min of grinding. The grinding efficiency when using 2 mm balls was slightly lower than when using 1 mm balls for the first 60 min of grinding, but prolonged grinding yielded a higher specific surface area when 2 mm balls were used. The 0.1 mm ball showed the lowest grinding efficiency, with a specific surface area of only 171.7 m²/g even after 360 min of grinding. These results indicate that the use of 1 mm and 2 mm balls in grinding talc could yield a larger specific surface area than the use of 0.1 mm balls could. In planetary ball milling, the proper ball size could maximize the grinding efficiency with a balance of aggregation and particle size reduction [32]. Our results indicate that the specific surface area eventually increases when the 2 mm balls are used for grinding, even though it is not clearly observed during the early stage of grinding.

The estimated spherical diameters (ESDs) were calculated using the equation $D = 6/\rho S$, where ρ is the density of talc (2.8 g/m³), S is the specific surface area (m²·g⁻¹), and the constant 6 is the shape factor assuming that the talc particles are spherical. The results are listed in Table 2.

Table 2. Changes in mineralogical parameters as a function of grinding time.

Talc	Ball Size (mm)	Grinding Time (min)						
		0	10	30	60	120	240	360
Relative	2		61.8	14.3	9.1	1.3	0.4	0.1
(001) Peak height	1	100	63.9	23.9	7.2	4.7	2.3	2.2
(%)	0.1		97.3	91.7	92.9	85.1	44.4	43.9
	2		81.1	165.8	182.5	321.7	408.8	419.1
S_{BET} (m^2/g)	1	6.1	81.5	171.6	231.8	287.3	326.9	365.1
	0.1		43.5	66.7	87.7	123.7	154.4	171.7
	2		26.4	12.9	11.7	6.7	5.2	5.1
E.S.D. (nm)	1	351.3	26.3	12.5	9.2	7.5	6.6	5.9
	0.1		49.3	32.1	24.4	17.3	13.9	12.5
Particle size	2		5.58	2.05	0.55	0.41	0.38	0.37
(D_{50}, µm)	1	12.8	3.25	0.71	0.52	0.45	0.41	0.44
	0.1		12.2	11.3	10.2	9.32	8.42	7.96

ESD, equivalent spherical diameter. $D = 6/\rho \cdot S$, where ρ is 2.8 g/cm^3.

The powder had ESD of approximately 350 nm prior to grinding, which decreased to 5.1, 5.9, and 12.5 nm after 360 min of grinding using 2 mm, 1 mm, and 0.1 mm balls, respectively. These are much lower ESDs of talc than previously reported, as shown in Table 1 [10,14]. Data from previous reports listed in Table 1 were generally acquired under dry conditions, unlike the wet milling performed in this study.

We also obtained ground talc series with a similar specific surface area, even though the ball size varied. Similar specific surface areas of 81.1, 81.5, and 87.7 m^2/g (approximately 85 m^2/g) were obtained after 10 min of grinding with 2 mm and 1 mm balls and after 60 min of grinding with 0.1 mm balls, respectively. Furthermore, similar specific surface areas of 165.8, 171.6, and 171.7 m^2/g (approximately 170 m^2/g) were obtained after 30 min of grinding with 2 mm and 1 mm balls and after 360 min of grinding with 0.1 mm balls, respectively. These two series with a similar specific surface area will be discussed below.

3.3. Analysis of Particle Size by the Laser Diffraction Method

Figure 6 shows the evolution of the talc's particle size upon grinding, as measured by the laser diffraction particle size analysis.

The as-received talc shows a single distribution with a mean particle size of approximately 12 µm. With an increase in grinding time, the particle size generally decreased. However, the minimum particle size and reduction rate varied according to the ball size. When 2 mm and 1 mm balls were used, the particle size rapidly decreased to less than 1 µm for the first 60 min of grinding and did not change much after 120 min. However, when 0.1 mm balls were used, the particle size gradually decreases until 360 min of grinding, but the changes are relatively insignificant compared with the changes when using 1 mm and 2 mm balls.

The distribution pattern of particle sizes also varies upon grinding. The particle size of as-received talc shows a monomodal distribution ranging from ~3 to ~50 µm with a center at ~10 µm. As the grinding proceeds, an increase in submicron particles appeared in addition to the existing predominately micron-sized particles of approximately ~10 µm, which forms a bimodal distribution. In this case, we refer to the two distribution groups as a microscale group and a sub-microscale group. When using 1 mm and 2 mm balls, the sub-microscale group appears even after 10 min of grinding and its population gradually increases until 120 min of grinding at the expense of the microscale group. The mean particle size of the microscale group gradually decreases from ~10 to ~2 µm as the grinding proceeds for up to 60 min. The microscale group is not observed after 120 min of grinding, which indicates that talc powders are fully pulverized into sub-microscale after 120 min of grinding. However, re-aggregation occurred in some samples, and, thus, small amounts of ~4 and ~20 μm particles were observed after 120 min of grinding. When using 0.1 mm balls, the population of the microscale group does not show a significant reduction even after 360 min of grinding and the population of the sub-microscale group is relatively insignificant when compared to the results when using 1 mm and 2 mm balls. We note that dispersion of the ground talc may be incomplete in aqueous solution and the sizes of the aggregated particles can be measured by the laser diffraction method.

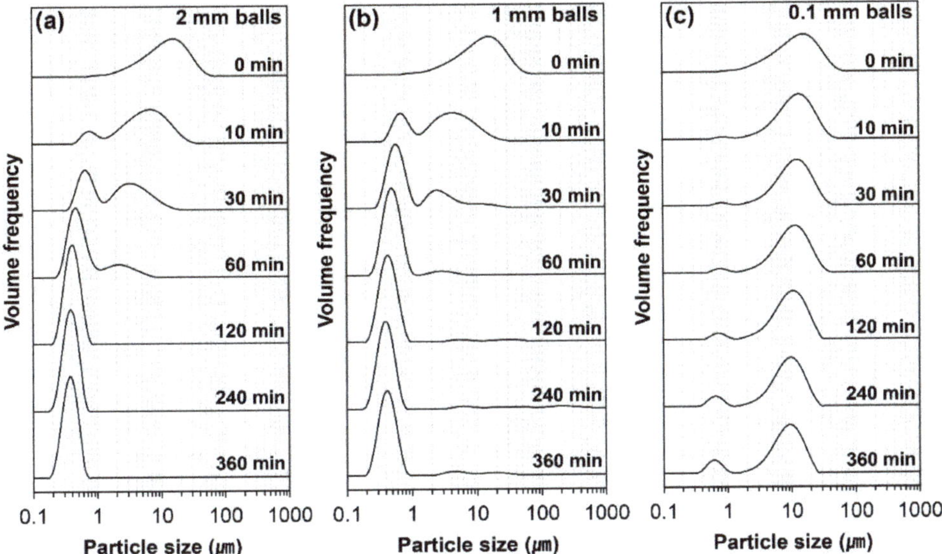

Figure 6. Changes in particle size of talc upon grinding, measured using laser diffraction particle size analysis. The ball size for high-energy ball milling is (**a**) 2 mm, (**b**) 1 mm, and (**c**) 0.1 mm.

Figure 7 shows the change in the D_{50}, which corresponds to 50% of the cumulative volume of the particle size distribution, upon grinding, according to the ball size.

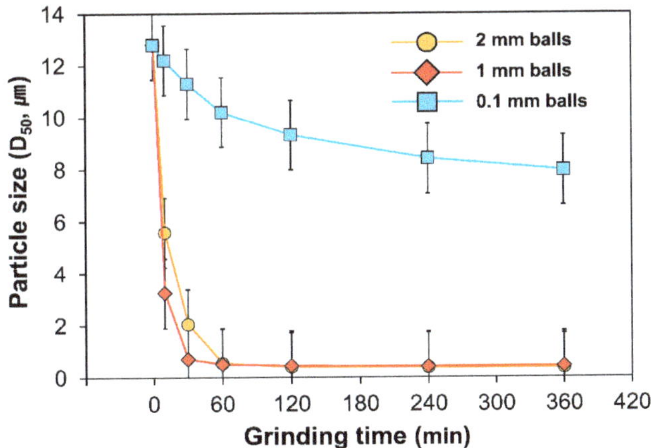

Figure 7. Change of particle size (D_{50}) of talc upon grinding.

When 2 mm balls were used, D_{50} decreased sharply to approximately 0.5 μm after 60 min of grinding, but did not show significant changes until 360 min. Similarly, when 1 mm balls were used, D_{50} decreased to approximately 0.7 μm in the first 30 min of grinding, but it did not change significantly after that. These results show that the particle size rarely decrease after approximately 60 min of grinding, which indicates that it reached the grinding limit under the grinding condition employed in this study. Until 60 min of grinding, the grinding efficiency of the 2 mm balls was slightly lower than that of the 1 mm balls. However, beyond this time frame, the D_{50} was smaller and the particle size distribution was narrower. In contrast, when 0.1 mm balls were used, the particle size steadily decreased until 360 min from the beginning of grinding. The final D_{50} was approximately 7.9 μm, which is approximately 7 μm larger than that obtained using 2 mm or 1 mm balls. These results indicate that, when 0.1 mm balls are used, the pulverization rate is slow and showed a lower grinding efficiency compared with the 2 mm and 1 mm balls.

3.4. Aggregation of Ground Talc

Laser diffraction particle size analysis and BET-specific surface area analysis are general methods to analyze the changes in particle size upon grinding. Though the calculated ESD based on the BET-specific surface area is much smaller than the results of particle size analysis because the shape of phyllosilicate talc and the increase in specific surface area generally indicates the decrease in particle size. According to the laser diffraction particle size analysis, the particle size of talc ground using 2 mm and 1 mm balls significantly decreased until 60 min of grinding, and further size reduction was not subsequently observed. On the contrary, BET-specific surface area continuously increases after 60 min of grinding. In addition, the significant difference in a specific surface area between talc powders ground using 2 mm and 1 mm balls is observed while the result of the particle size analysis shows only a trivial difference. These differences between the laser diffraction particle size and surface area analysis can appear when the ultrafine talc particles are aggregated by grinding [12]. In the BET-specific surface area analysis, there is a rare effect of aggregation because the N_2 gas is adsorbed to the particle surface. However, in the particle size analysis using laser diffraction, the size of aggregates can be measured rather than the individual particles. These results indicate that aggregation of ultrafine talc powder occurred as a result of grinding.

The aggregation of ground talc particles can be observed in SEM images of the talc particles grounded using 1 mm balls (Figure 8).

Before grinding, a unique laminar structure of talc is clearly observed with a size of tens of micrometers. With increasing grinding time, the particle size decreases to approximately 5 μm after 30 min, less than approximately 1 μm after 240 min, and approximately 200–300 nm after 360 min of grinding. The aggregation of ultrafine particles is clearly observed after 240 min of grinding, which is consistent with results of the laser diffraction particle size and BET-specific surface area analysis.

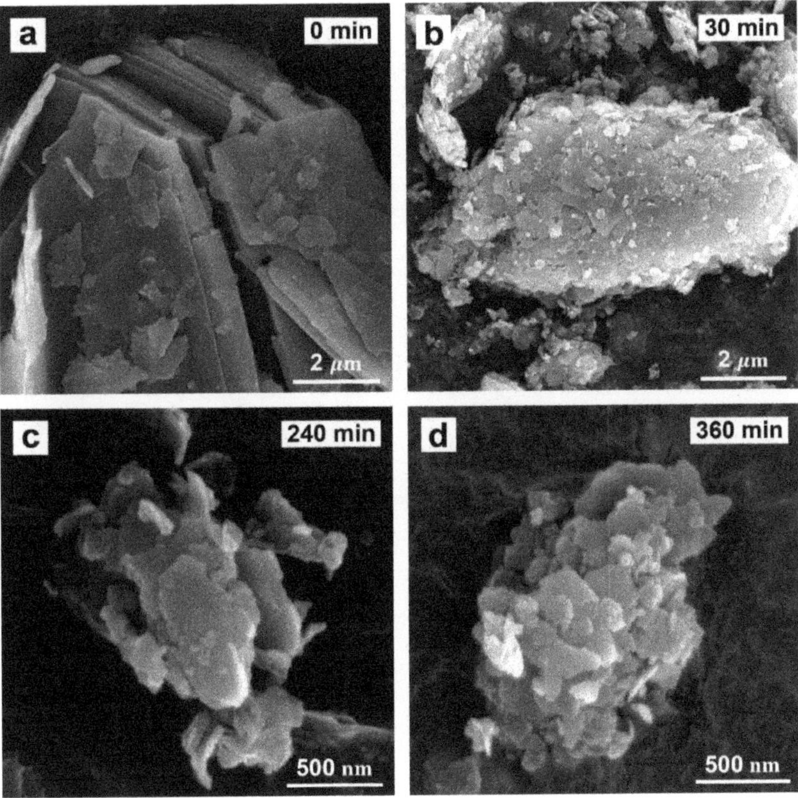

Figure 8. SEM images of ground talc for (**a**) 0 min, (**b**) 30 min, (**c**) 240 min, and (**d**) 360 min using 1 mm balls for high-energy ball milling.

In order to understand the effect of ball size on the aggregation of ground talc, Figure 9 compares the particle size analysis results of talc powders with similar specific surface areas. As described above, we obtain the ground talc series that have a similar specific surface area of talc particles that are approximately 85 m^2/g and 170 m^2/g. The laser diffraction particle size of the ground talc using 2 mm and 1 mm balls showed similar particle size distribution when compared with the case of using 0.1 mm balls. In both series, the intensity of the sub-microscale group was larger when using 2 mm and 1 mm balls than when using 0.1 mm balls, which indicates a larger particle size when using 0.1 mm balls.

The different particle size distributions with similar surface area could be observed if the aggregation of ultrafine particles occurred as a result of grinding. The larger particle size distribution indicates the more aggregated status of talc powder. In this study, the ground talc powder using 0.1 mm balls shows a larger particle size than other talc powders with similar surface areas, which indicates that it is easily aggregated due to grinding.

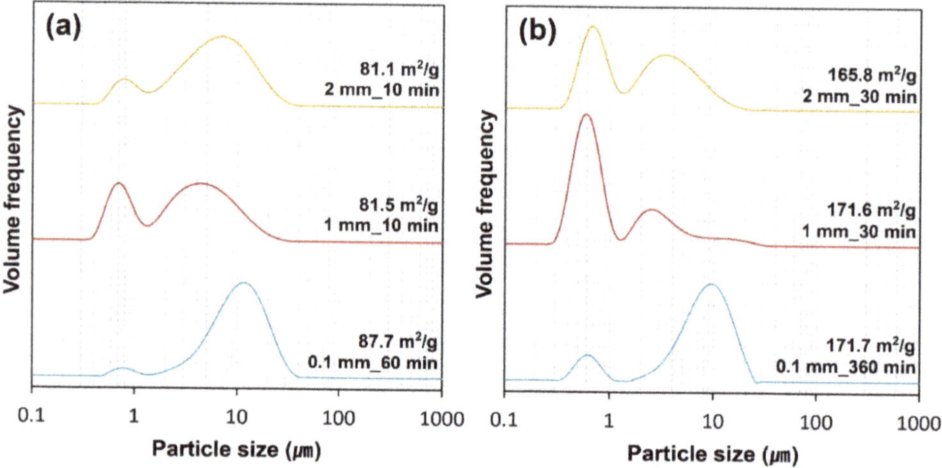

Figure 9. Comparison of particle size analysis results for ground talc powder having similar specific surface area. The BET specific surface areas are (**a**) 81–88 m^2/g and (**b**) 165–172 m^2/g.

The increased aggregation of ground talc with 0.1 mm balls than with 1 mm or 2 mm balls could be related to the degree of dispersion in the aqueous solution used in the particle size analysis. The aggregation of ultrafine particles generally occurs in order to reduce the high surface energy [34,35]. The talc powders with a similar surface area and different aggregation behavior indicate that these ground talc powders have a different surface energy and/or property. The results of XRD analysis showed that different crystallinity of ground talc powder depends on ball size. The crystallinity of ground talc is higher when using 0.1 mm balls than when using 1 mm and 2 mm balls. Crystalline talc is hydrophobic due to the siloxane sheet of the T-O-T layer. Hence, it aggregates well in aqueous solution [1,36]. On the other hand, disordered talc with lower crystallinity due to grinding is delaminated. Breakage of the layers exposes the hydroxyl groups in the octahedral layer. As a result, the talc is less hydrophobic due to the loss of crystallinity [8,37,38]. In an aqueous solution, the more hydrophilic talc is better dispersed.

As shown in Figure 3, the crystallinity of ground talc prepared using 0.1 mm balls during grinding is higher than that using 1 mm and 2 mm balls even though these talc powders have similar surface areas. The ground talc with relatively higher crystallinity (i.e., when using 0.1 mm balls for grinding) has higher surface energy than other ground talc powders (i.e., when using 1 mm and 2 mm balls for grinding). Thus, increased aggregation could occur in ground talc using 0.1 mm balls due to the loss of crystallinity and hydrophobicity.

4. Discussion

As described above, the grinding of talc using a high-energy ball mill results in differences in crystallinity, particle size, specific surface area, and aggregation behavior. Figure 10 shows the changes in the crystallinity, particle size, and specific surface area of talc, according to the ball size used in milling.

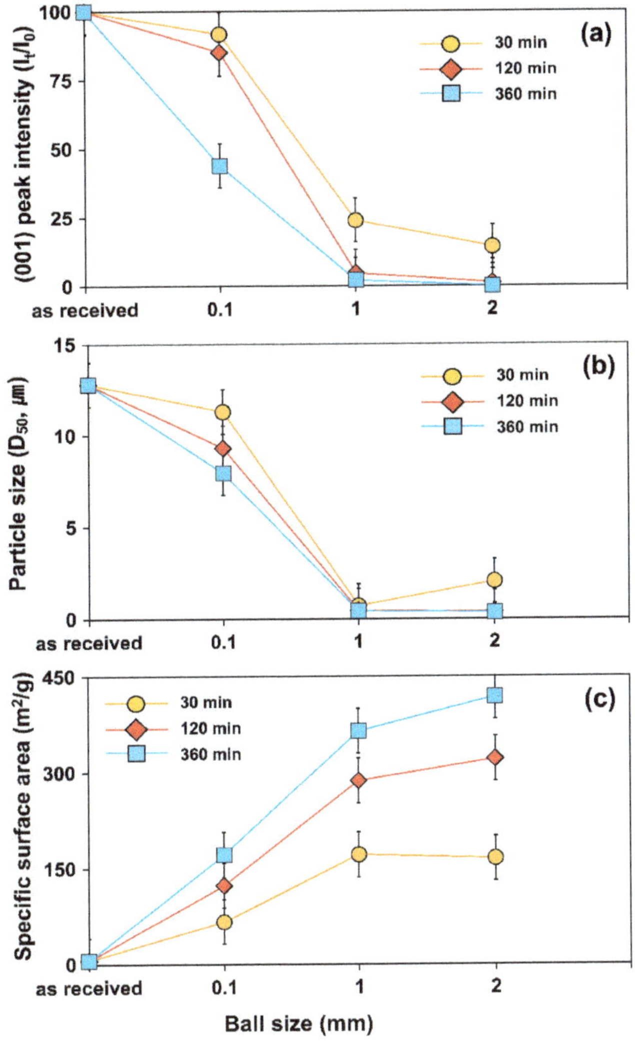

Figure 10. Effect of ball size for high-energy ball milling on (**a**) relative intensities of (001) XRD peak, (**b**) particle size (D_{50}), and (**c**) specific surface area.

4.1. Effects on Crystallinity of Ground Talc

The crystallinity of talc decreased upon grinding, and the degree of decrease varied depending on the ball size (Figure 10a). When 1 mm and 2 mm balls were used in grinding, the crystallinity decreased rapidly within 60 min of grinding, and delamination of (00*l*) planes mainly occurred. After 120 min of grinding, almost all XRD peaks disappeared. In contrast, when grinding was performed using 0.1 mm balls, not only was there a reduction in the crystallinity decrease, but also crystal surfaces other than the (00*l*) surface were fractured together. The crystallinity decreased continuously even after 120 min, but it was still much higher when compared with the case of using 1 mm and 2 mm balls. Compared with it prior to grinding, the (001) peak heights of XRD decreased to 91.7%, 23.9%, and 14.3% of the original size after 30 min of grinding when 0.1 mm, 1 mm, and 2 mm balls were used, respectively. After 120 min of grinding, it decreased to 85.1%, 4.7%, and 1.3%, respectively, and, after

360 min of grinding, it decreased to 43.9%, 2.2%, and 0.1%, respectively. When comparing the XRD patterns of ground talc with a similar specific surface area, the smaller the ball size used, the higher the final crystallinity of the talc is. These results indicate that the smaller ball size is appropriate to grind the talc while retaining the crystalline structure.

4.2. Effects on Comminution Rate and Limitation

The ball size in ball milling also affects the comminution rate and the limitation in particle size reduction. The increase in the comminution rate is clearly observed in the variation of particle size and BET-specific surface area, according to the ball size (Figure 10b,c). The BET-specific surface area of ground talc powders for 360 min shows approximately 419.1, 365.1, and 171.7 m^2/g using 2 mm, 1 mm, and 0.1 mm sized balls, respectively. The ESD for those talc powders correspond to 5.1, 5.9, and 12.5 nm, respectively. The particle size measured by the laser diffraction method also corresponds to the submicroscale when using 1 mm and 2 mm balls, but to the microscale when using 0.1 mm balls after grinding for 360 min. According to the variation of BET-specific surface area, the comminution rate increased with increasing ball size. At the early stage of ball milling, the comminution rate difference between using 1 mm and 2 mm balls is insignificant. The comminution rate when using 1 mm balls is slightly faster than when using 2 mm balls at the early stage of grinding, but both the comminution rate and the limitation in particle size reduction are more effective when using 2 mm balls in prolonged grinding.

4.3. Effects on Aggregation

The ball size used in ball milling also affects the aggregation properties of talc. The larger the ball size used for grinding, the less aggregation occurred for talc powders with similar specific surface areas. When 1 mm and 2 mm balls were used, aggregates of submicron particles were formed in most cases, as observed in laser diffraction particle size analysis. However, when 0.1 mm balls were used, aggregates of several micron particles were formed, as shown in Figure 9. The differences between the particle size analysis and the specific surface area analysis would be caused by the formation of aggregates of the sample due to ongoing grinding. This means that, as the ball size increases, the aggregates are formed more easily. When talc powders with similar specific surface areas were compared, the smaller the ball size is, the larger the aggregates formed are. The aggregation behavior of ground talc varies with the ball size used during milling, which is the result of different talc crystallinities. The loss of talc crystallinity results in a loss of hydrophobicity [8,37,38] and, consequently, better dispersion in an aqueous solution. The ground talc exhibits more hydrophilic properties when larger balls are used during milling.

5. Conclusions

In this study, talc was ground at a fast rotation speed of 2000 rpm using a high-energy ball mill, and ultrafine talc particles were obtained. Furthermore, the effects of ball size on the grinding efficiency and crystallinity of talc were investigated. When a larger ball size was used in ball milling, the crystallinity decreased faster, and the grinding efficiency and the specific surface area of the talc increased. In the case of crystallinity, the smaller the ball size was, the smaller the decrease in crystallinity was. Therefore, using a small ball size can produce talc nanoparticles with higher crystallinity. Upon the progress of milling, the aggregation of talc particles occurs and the degree of aggregation shows the relations with crystallinity of talc. Comparing the ground talc with similar specific surface areas, the more disordered the structure of the ground talc is, the fewer aggregation and more hydrophilic properties are observed.

Author Contributions: Conceptualization, H.N.K. Validation, J.W.K. and M.S.K. Investigation, H.N.K., J.W.K., M.S.K., B.H.L. and J.C.K. Writing—original draft preparation, H.N.K. and J.W.K. Writing—review and editing, H.N.K. Visualization, J.W.K. and M.S.K. Project administration, H.N.K. Funding acquisition, H.N.K. Please turn to the CRediT taxonomy for the term explanation.

Minerals **2019**, *9*, 668

Funding: The National Research Foundation of Korea (grant number NRF-2016R1C1B1010108) supported this work and the "Human Resources Program in Energy Technology" of the Korea Institute of Energy Technology Evaluation and Planning (KETEP) granted financial resources from the Ministry of Trade, Industry & Energy, Republic of Korea (grant number 20194010201730).

Conflicts of Interest: The authors declare no conflict of interest.

References

1. Brigatti, M.F.; Galan, E.; Theng, B.K.G. Structures and mineralogy of clay minerals. In *Developments in Clay Science*; Bergaya, F., Theng, B.K.G., Lagaly, G., Eds.; Elsevier: Amsterdam, The Netherlands, 2006; Volume 1, pp. 19–86.
2. Zazenski, R.; Ashton, W.H.; Briggs, D.; Chudkowski, M.; Kelse, J.W.; Maceachern, L.; McCarthy, E.F.; Nordhauser, M.A.; Roddy, M.T.; Teetsel, N.M.; et al. Talc: Occurrence, characterization, and consumer applications. *Regul. Toxicol. Pharmacol.* **1995**, *21*, 218–229. [CrossRef] [PubMed]
3. Castillo, L.; López, O.; López, C.; Zaritzky, N.; García, M.A.; Barbosa, S.; Villar, M. Thermoplastic starch films reinforced with talc nanoparticles. *Carbohydr. Polym.* **2013**, *95*, 664–674. [CrossRef]
4. Maiti, S. Mechanical, morphological, and thermal properties of nanotalc reinforced PA6/SEBS-g-MA composites. *J. Appl. Polym. Sci.* **2015**, *132*. [CrossRef]
5. Elfaki, H.; Hawari, A.; Mulligan, C. Enhancement of multi-media filter performance using talc as a new filter aid material: Mechanistic study. *J. Ind. Eng. Chem.* **2015**, *24*, 71–78. [CrossRef]
6. Galet, L.; Goalard, C.; Dodds, J.A. The importance of surface energy in the dispersion behaviour of talc particles in aqueous media. *Powder Technol.* **2009**, *190*, 242–246. [CrossRef]
7. Kim, J.W.; Lee, B.H.; Kim, J.C.; Kim, H.N. Particle size analysis of nano-sized talc prepared by mechanical milling using high-energy ball mill. *J. Mineral. Soc. Korea* **2018**, *31*, 47–55. [CrossRef]
8. Caban-Nevarez, R.; Perez, O.J.P. Size-dependence hydrophobicity in nanocrystalline talc produced by high-intensity planetary ball milling. *MRS Online Proc. Libr. Arch.* **2015**, 1805. [CrossRef]
9. Dumas, A.; Claverie, M.; Slostowski, C.; Aubert, G.; Careme, C.; Le Roux, C.; Micoud, P.; Martin, F.; Aymonier, C. Fast-geomimicking using chemistry in supercritical water. *Angew. Chem. Int. Ed.* **2016**, *55*, 9868–9871. [CrossRef]
10. Dellisanti, F.; Minguzzi, V.; Valdrè, G. Mechanical and thermal properties of a nanopowder talc compound produced by controlled ball milling. *J. Nanopart. Res.* **2011**, *13*, 5919–5926. [CrossRef]
11. Kano, J.; Saito, F. Correlation of powder characteristics of talc during planetary ball milling with the impact energy of the balls simulated by the particle element method. *Powder Technol.* **1998**, *98*, 166–170. [CrossRef]
12. Dellisanti, F.; Valdrè, G.; Mondonico, M. Changes of the main physical and technological properties of talc due to mechanical strain. *Appl. Clay Sci.* **2009**, *42*, 398–404. [CrossRef]
13. Filio, J.M.; Sugiyama, K.; Saito, F.; Waseda, Y. A study on talc ground by tumbling and planetary ball mills. *Powder Technol.* **1994**, *78*, 121–127. [CrossRef]
14. Sanchez-Soto, P.J.; Wiewióra, A.; Avilés, M.A.; Justo, A.; Pérez-Maqueda, L.A.; Pérez-Rodríguez, J.L.; Bylina, P. Talc from Puebla de Lillo, Spain. II. Effect of dry grinding on particle size and shape. *Appl. Clay Sci.* **1997**, *12*, 297–312. [CrossRef]
15. Borges, R.; Macedo Dutra, L.; Barison, A.; Wypych, F. MAS NMR and EPR study of structural changes in talc and montmorillonite induced by grinding. *Clay Miner.* **2016**, *51*, 69–80. [CrossRef]
16. Cayirli, S. Dry grinding of talc in a stirred ball mill. *EDP Sci.* **2016**, *8*, 01005–01013. [CrossRef]
17. Sakthivel, S.; Pitchumani, B. Production of nano talc material and its applicability as filler in polymeric nanocomposites. *Part. Sci. Technol.* **2011**, *29*, 441–449. [CrossRef]
18. Čavajda, V.; Uhlík, P.; Derkowski, A.; Čaplovičová, M.; Madejová, J.; Mikula, M.; Ifka, T. Influence of grinding and sonication on the crystal structure of talc. *Clays Clay Miner.* **2015**, *63*, 311–327. [CrossRef]
19. Christidis, G.; Makri, P.; Perdikatsis, V. Influence of grinding on the structure and colour properties of talc, bentonite and calcite white fillers. *Clay Miner.* **2004**, *39*, 163–175. [CrossRef]
20. Kano, J.; Miyazaki, M.; Saito, F. Ball mill simulation and powder characteristics of ground talc in various types of mill. Adv. *Powder Technol.* **2000**, *11*, 333–342. [CrossRef]
21. Ohenoja, K.; Illikainen, M. Effect of operational parameters and stress energies on stirred media milling of talc. *Powder Technol.* **2015**, *283*, 254–259. [CrossRef]

22. Cook, T.M.; Courtney, T.H. The effects of ball size distribution on attritor efficiency. *Metall. Mater. Trans. A* **1995**, *26*, 2389–2397. [CrossRef]
23. Choi, W.S.; Chung, H.Y.; Yoon, B.R.; Kim, S.S. Applications of grinding kinetics analysis to fine grinding characteristics of some inorganic materials using a composite grinding media by planetary ball mill. *Powder Technol.* **2001**, *115*, 209–214. [CrossRef]
24. Lameck, N.N.S. Effects of Grinding Media Shapes on Ball Mill Performance. Master's Dissertation, University of the Witwatersrand, Johannesburg, South Africa, 2006.
25. Katubilwa, F.M.; Moys, M.H. Effect of ball size distribution on milling rate. *Miner. Eng.* **2009**, *22*, 1283–1288. [CrossRef]
26. Mio, H.; Kano, J.; Saito, F.; Kaneko, K. Effects of rotational direction and rotation-to-revolution speed ratio in planetary ball milling. *Mater. Sci. Eng. A* **2002**, *332*, 75–80. [CrossRef]
27. Liao, J.; Senna, M. Thermal behavior of mechanically amorphized talc. *Thermochim. Acta* **1992**, *197*, 295–306. [CrossRef]
28. Filippov, L.O.; Joussemet, R.; Irannajad, M.; Houot, R.; Thomas, A. An approach of the whiteness quantification of crushed and floated talc concentrate. *Powder Technol.* **1999**, *105*, 106–112. [CrossRef]
29. Yekeler, M.; Ulusoy, U.; Hiçyılmaz, C. Effect of particle shape and roughness of talc mineral ground by different mills on the wettability and floatability. *Powder Technol.* **2004**, *140*, 68–78. [CrossRef]
30. Faraji, G.; Kim, H.S.; Kashi, H.T. *Severe Plastic Deformation: Methods, Processing and Properties*; Elsevier: Amsterdam, The Netherlands, 2018.
31. Akçay, K.; Sirkecioğlu, A.; Tatlıer, M.; Savaşçı, Ö.T.; Erdem-Şenatalar, A. Wet ball milling of zeolite HY. *Powder Technol.* **2004**, *142*, 121–128. [CrossRef]
32. Shin, H.; Lee, S.; Suk Jung, H.; Kim, J.-B. Effect of ball size and powder loading on the milling efficiency of a laboratory-scale wet ball mill. *Ceram. Int.* **2013**, *39*, 8963–8968. [CrossRef]
33. Boulvais, P.; de Parseval, P.; D'Hulst, A.; Paris, P. Carbonate alteration associated with talc-chlorite mineralization in the eastern Pyrenees, with emphasis on the St. Barthelemy Massif. *Mineral. Petrol.* **2006**, *88*, 499–526. [CrossRef]
34. Jung, S.J.; Lutz, T.; Boese, M.; Holmes, J.D.; Boland, J.J. Surface energy driven agglomeration and growth of single crystal metal wires. *Nano Lett.* **2011**, *11*, 1294–1299. [CrossRef] [PubMed]
35. Thielmann, F.; Naderi, M.; Ansari, M.A.; Stepanek, F. The effect of primary particle surface energy on agglomeration rate in fluidised bed wet granulation. *Powder Technol.* **2008**, *181*, 160–168. [CrossRef]
36. Yi, H.; Zhao, Y.; Rao, F.; Song, S. Hydrophobic agglomeration of talc fines in aqueous suspensions. *Colloids Surf. A Physicochem. Eng. Asp.* **2018**, *538*, 327–332. [CrossRef]
37. Zbik, M.; Smart, R.S.C. Influence of dry grinding on talc and kaolinite morphology: Inhibition of nano-bubble formation and improved dispersion. *Miner. Eng.* **2005**, *18*, 969–976. [CrossRef]
38. Kim, M.S.; Kim, J.W.; Kang, C.D.; So, B.D.; Kim, H.N. Variation of water content and thermal behavior of talc upon grinding: Effect of repeated slip on fault weakening. *J. Miner. Soc. Korea* **2019**, *32*, 201–211. [CrossRef]

Article

Grinding Behaviors of Components in Heterogeneous Breakage of Coals of Different Ash Contents in a Ball-and-Race Mill

Jin Duan [1], Qichang Lu [1], Zhenyang Zhao [1], Xin Wang [1], Yuxin Zhang [1], Jue Wang [1], Biao Li [2], Weining Xie [3,*], Xiaolu Sun [4] and Xiangnan Zhu [5]

[1]   Advanced Analysis & Computation Center, China University of Mining and Technology, Xuzhou 221116, China; 06172243@cumt.edu.cn (J.D.); lqctm@cumt.edu.cn (Q.L.); 06171870@cumt.edu.cn (Z.Z.); 06188807@cumt.edu.cn (X.W.); 06182151@cumt.edu.cn (Y.Z.); 06192029@cumt.edu.cn (J.W.)
[2]   Mining and Minerals Engineering Department, Virginia Tech, Blacksburg, VA 24060, USA; biaoli@vt.edu
[3]   Advanced Analysis & Computation Center, China University of Mining & Technology, Xuzhou 221116, China
[4]   School of Mining and Technology, Inner Mongolia University of Technology, Hohhot 010051, China; sunxl@imut.edu.cn
[5]   School of Chemical & Environmental Engineering, Shandong University of Science & Technology, Qingdao 266590, China; zhuxiangnan@sdust.edu.cn
*   Correspondence: aacc_xwn5718@cumt.edu.cn; Tel.: +86-0516-8359-2925

Received: 30 January 2020; Accepted: 2 March 2020; Published: 3 March 2020

Abstract: Coals used for power plants normally have different ash contents, and the breakage of coals by the ball-and-race mill or roller mill is an energy-intensive process. Grinding phenomena in mill of power plants is complex, and it is also not the same with ideal grinding tests in labs. The interaction among various coals would result in changes of grinding behaviors and energy consumption characterization if compared with those of single breakage. In this study, anthracite and bituminous coal of different ash contents were selected to be heterogeneously ground. Quantitation of components in products was realized using the relation between sulfur content of the mixture and mass yield of one component in the mixture. Product fineness t_{10} of the component was determined, and split energy was calculated on the premise of specific energy balance and energy-size reduction model by a genetic algorithm. Experimental results indicate that breakage rate and product fineness t_{10} of the mixture decrease with the increase of hard anthracite content in the mixture. Unlike the single breakage, t_{10} of anthracite in heterogeneous grinding is improved dramatically, and bituminous coal shows the opposite trend. The interaction between components results in the decrease of the specific energy of the mixture if compared with the mass average one of components in single breakage. Breakage resistance of hard anthracite decreases due to the addition of soft bituminous coal, and grinding energy efficiency of anthracite is also improved compared with that of single grinding.

Keywords: grinding behaviors; energy consumption characterization; sulfur content; heterogeneous breakage; split energy

---

1. Introduction

Particle size reduction is widely involved in various industries, especially in mineral processing. In a common comminution process for mineral liberation, size reduction of the raw ore sample is generally realized by the use of a crusher, ball mill and Isa mill in sequence, making it an energy-intensive process. Due to both the decline in ore grades and the increased complexity of the ore characteristic, it is expected that the energy demanded for achieving an ideal mineral liberation will be greatly increased [1–3]. Statistical data indicate that the grinding process consumes about 70% of the total energy in mineral preparation plants [4]. Hence, any improvement in the grinding process would lead to a significant reduction of energy required for mineral beneficiation. In China, nearly 60% of the raw coal is applied for electric power generation, and it should be ground to pulverized fuel (PF) for improving combustion efficiency [5]. Generally, ball-and-race mills or roller mills are used to produce the PF, of which >85% of fines are finer than 90 μm PF [6]. It is estimated that the spent energy in this process accounts for 0.5–2% of total electrical power of the coal power plant [7]. Though this proportion is not big, the total energy loss from coal power plants in China is tremendous given the total electric power generation capacity of China. Unfortunately, this issue has not been paid enough attention by electricity producers in China.

Normally, the heat efficiency of boilers used in power plants is closely related to the properties of coals that are burnt inside, therefore, the boiler in a power plant is designed according to the quality of coal. Note that in some parts of the world, supplied raw coals are from various mines, and they show obvious differences in quality. Under this situation, coal blending prior to combustion in a boiler becomes very necessary to ensure an optimum heat efficiency. This, in turn, brings up challenges on how to maintain a high efficiency for the grinding process in the power plants as coals of different sizes, densities, coalification degrees, and ash contents are ground in ball-and-race mills, resulting in particles fed onto the grinding table being ground heterogeneously. Interaction among different components would have an effect on the grinding behavior and energy consumption characterization of mixtures and components. Many investigations have been conducted to study the above-discussed issue faced by coal power plants; however, previous work was carried out on samples in narrow size or density and thereby the findings based on them are not applicable to real industrial process [8,9]. On the other hand, both the structure and grinding mechanism of a vertical roller mill employed in coal power plants are different from those of a conventional lab-scale mill. Regarding the first abovementioned issue facing lab-scale grinding research, Hardgrove mill or lab-scale roller mill were applied to simulate the grinding process of particles in industrial vertical spindle mill [10,11]. Related experiments were first conducted by Austin in 1981, in which a modified Hardgrove mill with a torque meter was used [12]. Based on the extensive grinding results, a model including particle breakage, internal and external classification of ground products was successfully applied to the lab-scale and industrial E mills, respectively [13,14]. Shi used a similar machine (JK Fine-particle Breakage Characteriser (JKFBC)) and applied the classical breakage model (developed from Drop-Weight Tests) to describe the energy-size reduction process [15]. Later on, particle properties were modelled into the classical breakage equation based on grinding tests of coal in JKFBC, which further extended the application scope of this model [16,17]. It appears that grinding in a lab-scale Hardgrove mill can simulate the grinding process of coal in a vertical spindle mill. It is noted that the materials used in the above studies were samples in narrow size or density. For the second issue, mixture breakage was initially conducted in the ball mill [18], and samples were pure minerals for the easy separation of progenies by float-sink test or chemical reaction [19,20]. All of these are conducive to analyze the breakage behavior of the component in the mixture; however, the breakage phenomenon was too ideal to draw some substantial conclusions. The key issue for the heterogeneous grinding of coals is the quantitation of components in the mixture. A great amount of research has been done in regard to overcoming this problem. Cho studied the grinding kinetics of the components in a binary mixture of 1.6 g·cm^{-3} sink anthracite and 1.4 g·cm^{-3} float bimanous coal in a ball-and-race mill, and two coals in ground products were separated by the float-sink tests [21]. Austin conducted the mixture breakage of anthracite with quartz, cement clinker and another two coals in a small laboratory ball mill under standard conditions, and found the acceleration of grinding rate [22]. Xie also ground anthracite with pure minerals in a ball-and-race mill and compared the changes of grinding behavior of components [23]. Float-sink test is a useful tool for the separation of coal from a mixture that has been subjected to grinding. While for different coals, size-reduction also leads to the liberation of associated minerals, and density distribution of products becomes wide, which may result in the density coincidence of mixture products. Unlike the pure minerals, coal is a complex material and contains both organic and inorganic substances. As such, which part of coal can be used for quantitation should be discussed. Since the species of minerals in coal are numerous, different associated conditions and selective liberation can result in the unpredictable distribution of minerals in products. Though almost all the inorganic elements in nature can be found in coal, they are usually concentrated in parts of products due to the selective liberation of minerals. Moreover, experimental errors caused by tedious sample preparation processes have negative effects on the accurate quantitation of some rare elements in coal. Hence, specific inorganic elements may also not be possible for quantitation. Based on the above discussion, properties of organic substance or organic elements are potential to distinguish different coals. Xie and his colleagues applied the characteristic ratio of XRD pattern to quantify components in progenies, and therefore, confirmed the grinding behavior of components. It is worth noting that this method can only be used for mixtures of coals with various coalification degrees and also requires the ash content of sample to be sufficiently low in order to avoid the negative effect on the analysis of 002 peak of XRD pattern [24]. For coals in the same coalification degree or higher ash content, another organic element should be selected.

In addition to the breakage behavior of the component in the mixture, energy consumption characterizations of the mixture and component are also important output for heterogeneous grinding. Energy consumed by the component can be calculated by energy split factor. Kapur and his colleagues provided energy split factor in terms of breakage rate functions and production rate of fines. This method was based on the assumption that breakage behavior of the component was environment-independent due to the similar grinding path on a triaxial composition diagram [25]. Xie and his colleagues calculated the energy split factor according to the mass and energy balance for the two-component breakage of coal with one mineral [23]. Combined with the product fineness of the component, energy-size reduction relation was established for the comparison of energy efficiency (product t_{10} for the same energy) between the mixture and single breakage [26].

Coals used in a previous study about mixture breakage by authors were of low ash content (2.62% and 3.17%), which was lower than that of the coals used in power plants. Hence, another two coals (2.96% and 35.27%) were chosen. A series of mixture grinding tests were conducted in a ball-and-race mill, and breakage rate, product fineness, and specific energy were determined. As the sulfur content of the mixture was linearly related to the mass yield of one component of the mixture, product fineness of each component after mixture breakage was quantified based on the above relation. On the premise of the classical breakage model and specific energy balance of the mixture, specific energy of the component was computed by genetic algorithm. Split energies and breakage parameters in the energy-size reduction model of components for various mixed conditions and grinding time were determined to indicate the interaction effect on the grinding energy efficiency of the component.

2. Materials, Equipment and Method

2.1. Materials, Equipment and Grinding Tests

Two coals including the anthracite and bituminous coal were used in this study. Anthracite was sampled from the clean coal stream of the Taixi coal preparation plant, while the bimanous coal was sampled from the middling coal stream of the Linhuan coal preparation plant. These samples were subjected to crushing and sieving tests, and particles of −2.8 + 2 mm were selected for grinding tests. Ash contents of anthracite and bituminous coal were 2.96% and 35.27%, respectively.

Energy-size reduction tests were conducted in a ball-and-race mill, namely Hardgrove machine, with the addition of a power meter, to investigate changes of grinding behavior and energy consumption characteristics of components during the heterogeneous breakage process. The structure parameters are as follows: diameter of grinding table 76.2 mm, diameter of grinding ball 25.4 mm, the number of grinding balls 8, table revolution rate 20 rpm, grinding force 284 N, and rated power 90 W. The power meter was connected in the electric circuit of Hardgrove mill. Resolution of the power meter was 0.01 W, and instantaneous power was recorded at the frequency of 1 s. Forty gram samples were prepared for each grinding test, and eight balls were put on the particle bed evenly. The non-load power was first recorded, and power consumption for the breakage of single coal and mixture was measured later. Power for grinding samples was determined by subtracting the non-load power from the gross one.

Grinding tests of single anthracite and bituminous coal were first conducted, and grinding time was designed as 10 s, 20 s, 30 s, 40 s, 50 s, 70 s, 90 s, 120 s, 150 s, 180 s, and 240 s. For each experiment, size distribution of ground products was confirmed by sieving tests, and yield of unbroken particles in the top size, and product fineness t_{10} (yield of progenies which were smaller than 1/10th of the mean size of feed) were determined. Mixture breakage tests were carried out on mixtures of anthracite (A) and bituminous coal (B) at mass ratios of 3:1, 1:1 and 1:3. Grinding time and treatment on ground products of heterogeneous breakage tests were identical to those of homologous breakage tests.

2.2. Quantification of Components in Heterogeneous Grinding Products

For the quantification of components in heterogeneous grinding products, the relation between some characteristic indexes with the mass yield of a certain component should be determined. Firstly, mixtures of anthracite and bituminous coal were prepared with mass ratios of 1:0, 4:1, 3:1, 2:1, 1:1, 1:2, 1:3, 1:4, and 0:1. Then, these materials were ground to −74 μm fines by a vibrating mill respectively, and element composition of ground products was analyzed by the X-Ray Fluorite Spectroscopy (XRF). In the XRF test, about 20 elements were detected, most of which were inorganic elements. Note that the distribution of inorganic elements in coal samples was not even, and minerals associated with coal were of different sizes. Hence, the selective liberation would happen along with grinding, which would result in the unpredictable distribution of elements in various narrowly-sized progenies. So, the X-Ray Diffraction of two coals were then conducted to assist the selection of a proper characteristic element. XRD patterns of two coals are shown in Figure 1. As shown, mineral information of anthracite in the XRD pattern was weak because the ash content was only 2.96%, and the main associated mineral was kaolinite. The main minerals associated in the bituminous coal were kaolinite, montmorillonite and illite. Evidently, element Si is the common inorganic element of these two coals, but it cannot be regarded as the indicator to distinguish the grinding behaviors of components in the mixture due to the selective liberation. Meanwhile, Figure 1 reveals that there is no pyrite detected by XRD. Therefore, it is easy to conclude that sulfur quantified by XRF exists in the organic macromolecular structure of coal in an organic form. Since the breakage process of organic components was non-selective, the uniform distribution of sulfur can be anticipated in each narrowly-sized particles. Hence, it was reasonable to choose the organic element of sulfur for quantification.

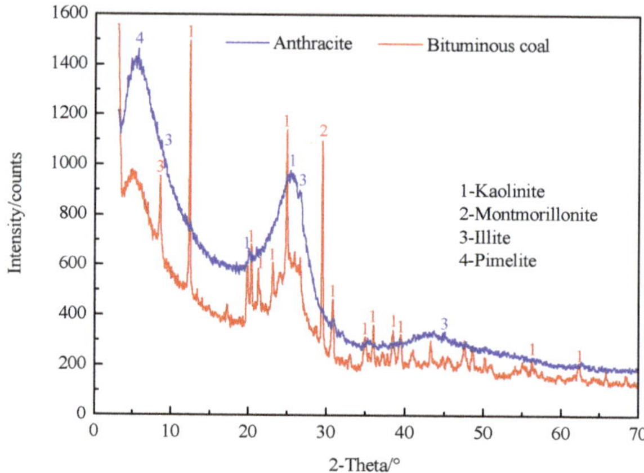

Figure 1. XRD pattern of anthracite and bituminous coal.

Figure 2 shows the relation between sulfur contents of mixtures and mass yields of bituminous coal. A linear equation was developed, with the R^2 of 0.98. Note that the characteristic size of t_{10} for $-2.8 + 2$ mm coals is 0.237 $(((2.8 \times 2)\char`^0.5)/10)$ mm. As it is difficult to accurately obtain these fines by sieving, progenies of -0.25 mm were used. For heterogeneous grinding tests of mixtures in various mass ratios of components and grinding time, -0.25 mm products were reground by a vibrating mill to -74 μm fines for the XRF measurement. With the help of the above empirical linear equation, product fineness t_{10} of each component in various heterogeneous grinding conditions was determined.

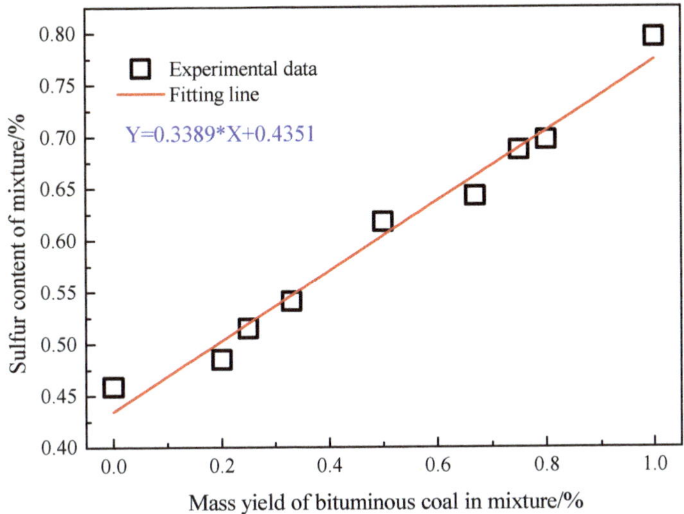

Figure 2. Relation between sulfur content of mixture and mass yield of bituminous coal.

3. Results and Discussion

3.1. Grinding Behavior of Components and Mixtures

Breakage of coarse particles to fines is a progressive process, which can be evaluated by two indicators, namely breakage rate of the top sized particles (BRTSP) and product fineness t_{10}. BRTSPs of mixtures at various mixed ratios (1:0, 3:1, 1:1, 3:1, and 0:1) are compared. Although the longest grinding time of this research is 240 s, it is shown from the size analyses of progenies that particles in the top size are already ground to fines before 90 s. Hence, yields of the top size particles at grinding time periods of 0 s, 10 s, 20 s, 30 s, 40 s, 50 s, and 70 s are plotted in Figure 3. As shown, for these semi-logarithmic curves, BRTSP shows the linear relation with grinding time, namely the first-order law. For the single breakage, yield of unbroken bituminous coal in the top size is a little lower if compared with that of anthracite for each grinding time. That is to say, bituminous coal has a fast breakage rate, and it is relatively easy to break. This is due to the fact that the ash content of the bituminous coal is much higher than that of the anthracite, so that liberated minerals accelerated the size-reduction of bituminous coal. Figure 3 also shows that with the increase in weight percentage of bituminous coal in the mixture, the yield of unbroken particles in the top size decreases, while the breakage rate of the mixture increases.

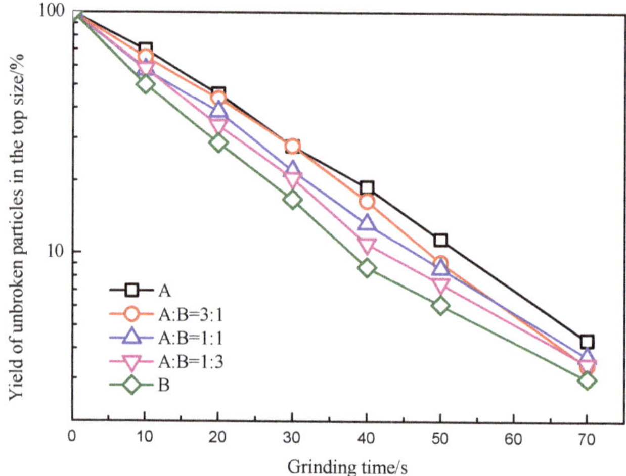

Figure 3. Yield of unbroken particles in the top size for various grinding times.

The results of product fineness t_{10} of mixtures with mixed ratios and grinding time are presented in Figure 4. Similar to the change trending of breakage rate, bituminous coal with a fast BRTSP shows a higher product fineness, and the difference in t_{10} between bituminous coal and anthracite is over 15%. Anthracite has a higher coalification degree, and it is difficult to be broken to fines, especially for those associated with small amounts of minerals. Predictably, t_{10} of mixtures increases with the small yield of anthracite in the mixture. In addition to the direct comparison of breakage rate and product fineness, t_{10} of components in heterogeneous grinding are also determined and compared. As mentioned above, −0.25 mm progenies yielded at various mixture breakage conditions are ground to fines for XRF measurement. Sulfur contents of these samples are shown in Table 1. Obviously, sulfur content increases with grinding time for these three mixtures, which indicates more anthracite fines in products. Based on data in Table 1 and the empirical equation in Figure 2, contents of two components in t_{10} of mixtures are determined. Aiming at the convenient comparison of t_{10} of components in heterogeneous and homogeneous grinding, values of product fineness t_{10} are calculated in terms of the percentage benchmark of the component itself according to data in Figure 4 and the mixed ratios. These data are shown in Figures 5 and 6 for anthracite to bituminous coal, respectively. Compared with the single breakage of anthracite, product fineness t_{10} is improved significantly, and a large value of t_{10} indicates the amount of anthracite in the mixture is small. If the content of anthracite decreases to 25%, t_{10} is more than twice that of homogeneous grinding. On the contrary, t_{10} of bituminous coal in mixture breakage is much lower than that in single breakage. For the hard anthracite, soft bituminous coal in the mixture not only improves grinding phenomena, but also benefits the transition of grinding energy to anthracite. That is why the product fineness of anthracite increases if compared with that of single breakage. In order to explain the interaction between two coals in mixture breakage, energy consumed characterization will be introduced in the next section.

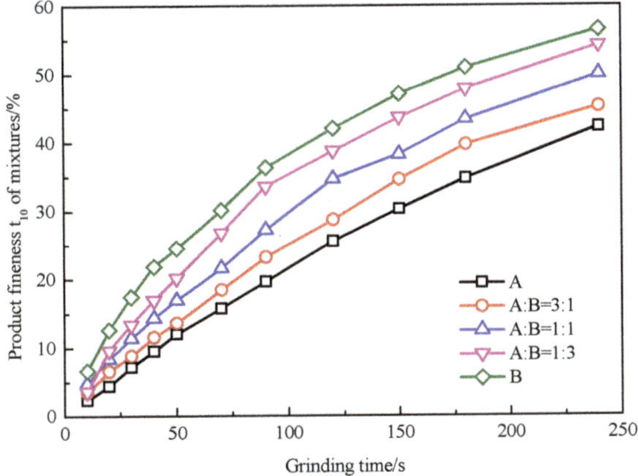

Figure 4. Product fineness t_{10} of mixtures for various grinding times.

Table 1. Sulfur content of −0.25 mm products yielded at various grinding times.

Grinding Time/s	Content of Sulfur/%		
	A:B = 3:1	A:B = 1:1	A:B = 1:3
10	0.507	0.648	0.687
20	0.502	0.618	0.649
30	0.503	0.597	0.650
40	0.502	0.573	0.623
50	0.503	0.565	0.632
70	0.504	0.561	0.639
90	0.501	0.564	0.637
120	0.502	0.561	0.638
150	0.502	0.563	0.638
180	0.501	0.563	0.635
240	0.501	0.559	0.635

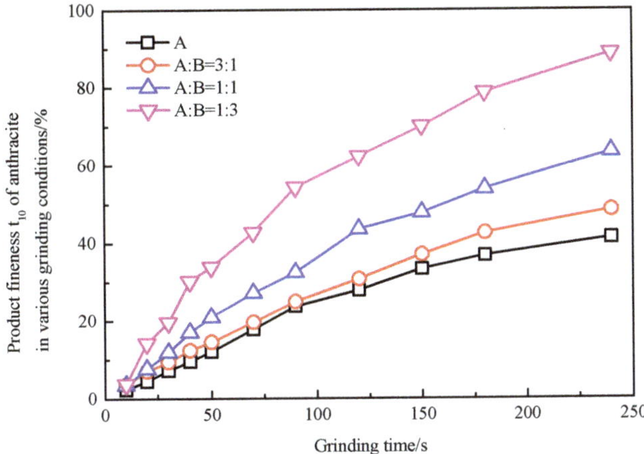

Figure 5. Product fineness t_{10} of anthracite for various grinding conditions and times.

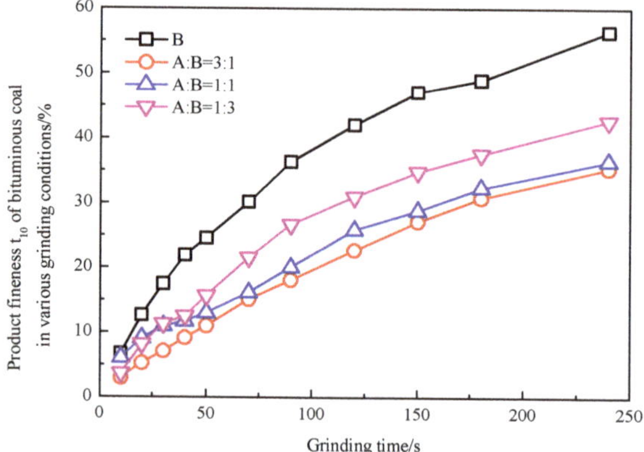

Figure 6. Product fineness t_{10} of bituminous coal for various grinding conditions and times.

3.2. Energy Consumed Characterizations of Mixtures and Components

The grinding process of mixtures should be evaluated not only by breakage rate and product fineness, but also by consumed specific energy. Here, recorded power (W) per second for the heterogeneous grinding of mixtures was converted to specific energy (kW.h.t^{-1}). Specific energy and product fineness t_{10} of each mixture breakage for various times are plotted in Figure 7. For the single breakage, bituminous coal shows a higher t_{10} with the same specific energy in comparison with that of anthracite. That is to say, the soft bituminous coal has a higher grinding energy efficiency. Predictably, fineness t_{10} increases with more bituminous coal in the mixture for the same energy input, just as shown in Figure 7. Previous breakage researches of narrowly-sized particles in the ball-and-race mill have illustrated the successful application of classical energy-size reduction model on experimental data [27], and indexes of particle properties are added to that model to improve the utilization [13]. This model is shown as follows:

$$t_{10} = A \times \left(1 - e^{-b \times E_{cs}}\right) \tag{1}$$

where t_{10} is the product fineness (%), E_{cs} is the specific energy (kW.h.t^{-1}), and A and b are breakage parameters.

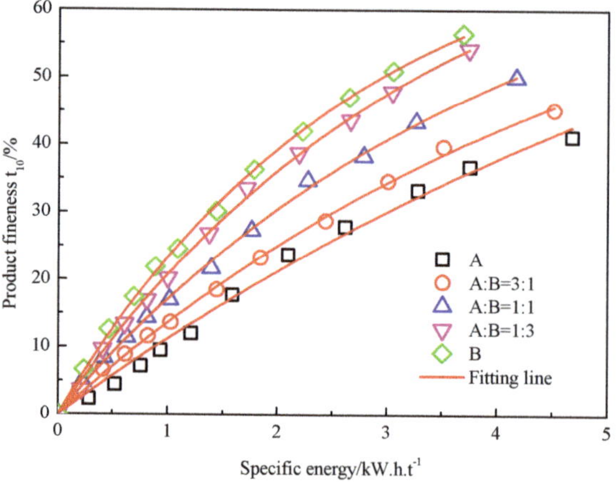

Figure 7. Specific energy vs product fineness t_{10} of mixtures.

In this paper, the classical model is also used to describe the heterogeneous grinding of two coals. Good fitted results, with R^2 above 0.98, are obtained for grinding tests of mixtures in three mixed ratios. Breakage parameters A and b are also determined for each mixture and shown in Table 2. The higher A*b value demonstrates the less resistance to being broken. Hence, bituminous coal, which has the highest A*b value, is easily broken. That is consistent with conclusions of breakage rate and product fineness. With the decrease of bituminous coal in the mixture, the indicator A*b decreases, and more energy would be consumed for yielding fines.

Table 2. Breakage parameters of mixtures.

Breakage Parameters	Mixture Conditions				
	A	A:B = 3:1	A:B = 1:1	A:B = 1:3	B
A	97.63	80.24	77.44	80.77	75.68
b	0.12	0.19	0.25	0.30	0.37
A*b	12.01	15.00	19.36	23.99	27.82

Product fineness t_{10} of component has been determined based on the characteristic index of sulfur content as mentioned above. However, the specific energy for yielding fines for each component at various grinding conditions and time is not clear. Recorded power and calculated specific energy are for mixture. Interactions between components can not only change the specific energy consumed by the component, but also have an effect on the specific energy of the mixture if compared with the mass average one. Table 3 lists both measured and calculated mass average specific energies of mixtures at various times. If component interaction does not happen in the mixture breakage, the specific energy should be close to the mass average one. However, data in Table 3 indicate that measured energy is a little lower than the mass average one. So, the mixture breakage is a potential method for energy saving. This difference illustrates that the component interaction may increase the grinding energy efficiency of one component and decrease that of another one. However, this speculation should be further verified by the specific energy split by each component.

Table 3. Comparison of measured and mass average specific energy.

Specific Energy/kW.h.t^{-1} Grinding Time/s	Measured Data					Mass Average Data		
	A	A:B = 3:1	A:B = 1:1	A:B = 1:3	B	A:B = 3:1	A:B = 1:1	A:B = 1:3
10	0.28	0.20	0.22	0.21	0.24	0.27	0.26	0.25
20	0.52	0.41	0.42	0.41	0.46	0.51	0.49	0.48
30	0.76	0.62	0.63	0.61	0.70	0.74	0.73	0.71
40	0.94	0.82	0.81	0.81	0.89	0.92	0.91	0.90
50	1.21	1.03	1.01	1.00	1.09	1.18	1.15	1.12
70	1.59	1.44	1.39	1.37	1.44	1.55	1.51	1.48
90	2.10	1.84	1.76	1.71	1.78	2.02	1.94	1.86
120	2.61	2.44	2.28	2.19	2.23	2.52	2.42	2.32
150	3.28	3.00	2.78	2.66	2.65	3.12	2.96	2.81
180	3.75	3.51	3.26	3.04	3.05	3.58	3.40	3.22
240	4.68	4.51	4.17	3.74	3.69	4.43	4.18	3.94

Authors have provided a method to determine the energy split factor of the component for the mixture breakage of coal with pure mineral in the Hardgrove mill [20,21]. In that study, interaction between components was reflected on the change of breakage parameters of mixtures with different mixed ratios. Though the energy balance was realized, it was not correct to calculate the specific energy of the component by the classical breakage model of the mixture. As a result, the energy-size reduction model with parameters for the mixture could also describe the breakage process of two components. In other words, coal, pure mineral and mixture showed the same resistance to being broken, which was not accurate enough. In this case, a new method was put forward to determine the specific energy of the component. First, the energy (kW.h) balance equation is divided by the mass of the mixture and converted to the specific energy (kW.h.t^{-1}) balance equation [26]. Second, it is assumed that the relation of the specific energy and product fineness of the component in the mixture breakage still follows the classical breakage model. This assumption is the connection of the known product t_{10} to the unknown specific energy of the component. Third, specific energies of components for various grinding conditions and time are calculated according to the specific energy balance of the mixture. Calculation of the third process is conducted by genetic algorithm (GA). Here, the GA toolbox in Matlab is used, and compiled programs are shown as the Appendix A in this paper. The initial population number is set as 50. Boundary conditions of parameters A and b in Equation (1) for components are [x_{max}, 100] and [0,1], respectively. In addition, the target error is set as 0.1%. If the population mean error is smaller than the target one, the optimum parameters A and b for each component are obtained. Then, specific energy of each grinding condition is calculated by Equation (1), with the results being listed in Table 4. The difference between the calculated and measured specific energy of the mixture is marginal except for the small specific energy of the short grinding time. These data would be used for the energy-size reduction model of the component. Note that the values of t_{10} for components in mixture breakage are in the percentage benchmark of the component itself, namely data in Figures 5 and 6. In addition, Figures 8 and 9 are the relation between t_{10} and specific energy of anthracite and bituminous coal, respectively. Energy efficiency of grinding anthracite increases with more bituminous coal being added in the mixture, and that of bituminous coal shows the opposite trend. These conclusions are similar with the changing law of product fineness t_{10}. Breakage parameters of components are also determined by Equation (1), as shown in Table 5. Breakage resistance of anthracite decreases, which indicates the soft bituminous coal improves the grinding phenomenon of anthracite in comparison with that in single grinding. While for bituminous coal, it acts as grinding media and therefore inhibits the breakage process.

Energy (kW.h) consumed by each component is calculated based on the specific energy (kW.h.t^{-1}) and mass of the component in the mixture. Figure 10 shows the content of energy consumed by anthracite as a function of grinding time. Compared with the mass content of anthracite in the mixture, a higher ratio of energy is split by it for mixtures of 1:1 and 1:3, which benefits the generation of fines. When the mass ratio in the mixture is 3:1, anthracite obtains less energy, however, the ratio is still above 70%. Particles used in this research are in the size of −2.8 +2 mm, which is relatively big. Hence, the population of bituminous coal particle in tests conducted at a large mass ratio of anthracite to bituminous coal is small. For the mixture of 3:1, not all the anthracite particles can be surrounded by the soft bituminous coal. This situation may affect the energy split of the component in the mixture.

Table 4. Specific energies of components at various grinding conditions and times.

Specific Energy/kW.h.t^{-1}	Mixture Conditions											
	A:B = 3:1				A:B = 1:1				A:B = 1:3			
Grinding Time/s	A	B	Calculated Mixture	Measured Mixture	A	B	Calculated Mixture	Measured Mixture	A	B	Calculated Mixture	Measured Mixture
10	0.22	0.31	0.24	0.20	0.20	0.29	0.25	0.22	0.10	0.20	0.17	0.21
20	0.43	0.55	0.46	0.41	0.44	0.44	0.44	0.42	0.40	0.46	0.45	0.41
30	0.58	0.75	0.62	0.62	0.69	0.63	0.66	0.63	0.57	0.65	0.63	0.61
40	0.77	0.99	0.82	0.82	0.94	0.71	0.82	0.81	0.94	0.73	0.78	0.81
50	0.92	1.20	0.99	1.03	1.16	0.86	1.01	1.01	1.08	0.93	0.97	1.00
70	1.29	1.71	1.39	1.44	1.56	1.14	1.35	1.39	1.46	1.34	1.37	1.37
90	1.71	2.09	1.81	1.84	1.96	1.41	1.68	1.76	2.05	1.74	1.81	1.71
120	2.22	2.74	2.35	2.44	2.85	2.00	2.42	2.28	2.55	2.09	2.20	2.19
150	2.86	3.42	3.00	3.00	3.16	2.37	2.76	2.78	3.15	2.44	2.62	2.66
180	3.50	4.03	3.63	3.51	3.95	2.73	3.34	3.26	4.05	2.70	3.04	3.04
240	4.30	4.87	4.44	4.51	5.25	3.11	4.18	4.17	5.73	3.25	3.87	3.74

Table 5. Breakage parameters of components.

Breakage Parameters	Anthracite				Bituminous Coal			
	Single	A:B = 3:1	A:B = 1:1	A:B = 1:3	Single	A:B = 3:1	A:B = 1:1	A:B = 1:3
A	97.6257	78.721	94.2295	99.958	75.684	73.043	56.1382	81.387
b	0.12305	0.223	0.21653	0.383	0.3676	0.136	0.3024	0.229
A*b	12.01	17.55	20.40	38.28	27.82	9.93	16.98	18.64

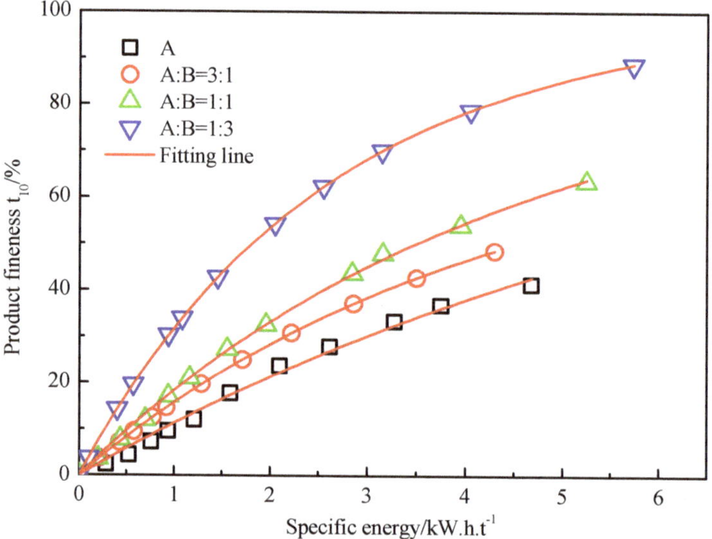

Figure 8. Specific energy vs product fineness t_{10} of anthracite in mixture breakage.

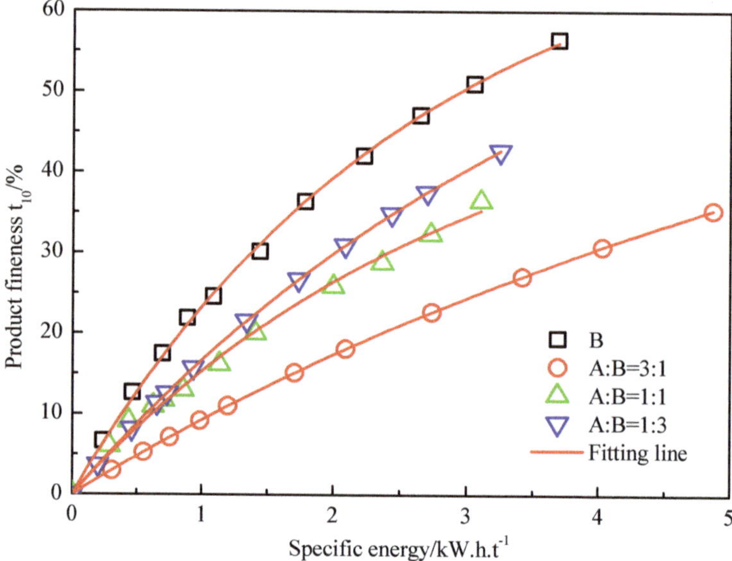

Figure 9. Specific energy vs product fineness t_{10} of bituminous coal in mixture breakage.

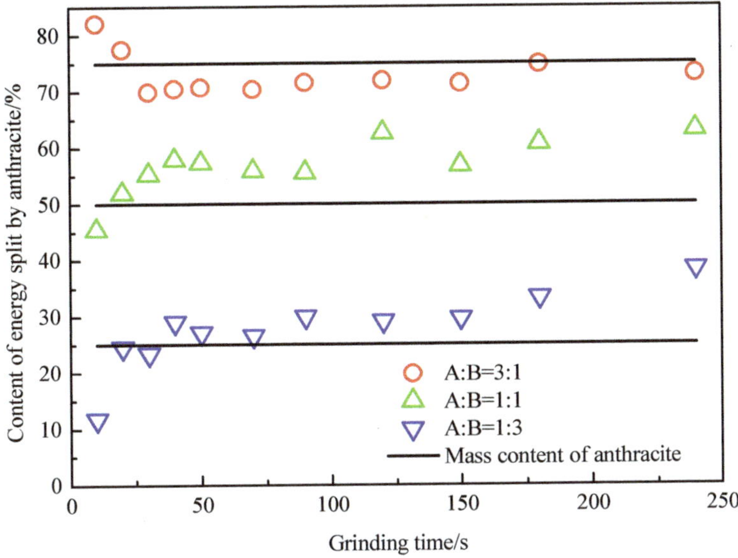

Figure 10. Content of energy split by anthracite in various grinding conditions and times.

4. Conclusions

This paper reports grinding behaviors of components in heterogeneous breakage of two coals of different ash contents in a ball-and-race mill. Quantitation of two coals is conducted by the relation of sulfur content of mixture and mass yield of bituminous coal. Values of product fineness t_{10} of components at various grinding conditions and time periods are determined, and specific energy and energy split by component are calculated by a genetic algorithm on the premise of energy balance of mixture, classical energy-size reduction model and t_{10} of components. The main conclusions of this paper are as follows:

(1) XRD results show that sulfur in two coals is in organic form, and its content is quantified by XRF. Sulfur contents of mixtures show a linear relation with the mass yield of bituminous coal, which contributes to determining the product fineness t_{10} of two coals at various grinding conditions and time periods.

(2) The breakage rate of the mixture obeys the first-order law. In the mixture breakage, the breakage rate increases with increasing the mass ratio of bituminous coal of the mixture. Compared with the single breakage of anthracite, t_{10} is improved significantly after mixing with the soft bituminous coal. If the content of anthracite decreases to 25%, t_{10} is more than twice that of homogeneous grinding, but t_{10} of bituminous coal is reduced.

(3) The classical energy-size reduction model can be applied for the mixture breakage of coals in the Hardgrove mill. Breakage indicator A*b of the mixture increases with adding more soft bituminous coal. Specific energy of the mixture is a little lower than the mass average one of components due to the component interaction in mixture breakage. The relation between t_{10} and the specific energy of the component indicates that energy efficiency of anthracite grinding increases during the heterogeneous grinding. Added bituminous coal surrounds anthracite particles and improves the grinding phenomenon if compared with the single breakage. Content of energy split by anthracite is bigger than the mass yield of it in the mixture, which indicates the easy transfer of energy to anthracite.

Author Contributions: Q.L., X.S. and W.X. designed experiments of this study, J.D., Z.Z., X.W., Y.Z. and J.W. conducted the heterogeneous and homogeneous grinding tests, and also analyzed size distribution of ground products, Q.L. and X.Z. conducted XRF measurement and analyzed data, W.X. and B.L. contributed to the writing the paper. All authors have read and agreed to the published version of the manuscript.

Funding: Work in this paper were supported by the National Natural Science Foundation of China (51904295), Science Foundation of Jiangsu province (BK20180647), Chinese Postdoctoral Science Foundation (2018M640538), and Research Program of Science and Technology at Universities of Inner Mongolia Autonomous Region (NJZY20083).

Acknowledgments: Authors are thankful for the technical support of the Advanced Analysis & Computation Center.

Conflicts of Interest: The authors declare no conflict of interest.

Appendix A

```
function z = hanshu(x)
syms A1 b1 A2 b2 y1 y2 x1 x2 r c;
syms x1_1 x1_2 x1_3 x1_4 x1_5 x1_6 x1_7 x1_8 x1_9 x1_10 x1_11;
syms x2_1 x2_2 x2_3 x2_4 x2_5 x2_6 x2_7 x2_8 x2_9 x2_10 x2_11;
A1 = x(1)
b1 = x(2)
A2 = x(3)
b2 = x(4)
y1 = [3.76 14.27 19.61 30.25 33.92 42.75 54.32 62.31 70.06 78.77 88.84]; 't10 of anthracite
y2 = [3.63 8.14 11.33 12.55 15.68 21.52 26.69 30.95 34.84 37.56 42.72]; 't10 of bituminous coal
x1 = -log(1 - y1./A1)/b1; 'specific energy of anthracite
x2 = -log(1 - y2./A2)/b2; 'specific energy of bituminous coal
r = 0.25. × x1 + 0.75. × x2;
c = [0.21 0.41 0.61 0.81 1.00 1.37 1.71 2.19 2.66 3.04 3.74];
z = abs(abs(c(1,1) - r(1,1))./max(c(1,1),r(1,1)) + abs(c(1,2) - (1,2))./max(c(1,2),r(1,2)) +
abs( (1,3) - r(1,3))./max(c(1,3),r(1,3)) + abs(c(1,4) - r(1,4))./max(c(1,4),r(1,4)) + abs(c(1,5) -
r(1,5))./max(c(1,5),r(1,5)) + abs(c(1,6) - r(1,6))./max(c(1,6),r(1,6)) + abs(c(1,7) - r(1,7))./max(c(1,7),r(1,7))
+ abs(c(1,8) - r(1,8))./max(c(1,8),r(1,8)) + abs(c(1,9) - r(1,9))./max(c(1,9),r(1,9)) + abs(c(1,10) -
r(1,10))./max(c(1,10),r(1,10)))
```

References

1. Kim, H.N.; Kim, J.W.; Kim, M.S.; Lee, B.H.; Kim, J.C. Effects of ball size on the grinding behavior of talc using a high-energy ball mill. *Minerals* **2019**, *9*, 668. [CrossRef]
2. Yang, J.; Shuai, Z.; Zhou, W.; Ma, S. Grinding optimization of cassiterite-polymetallic sulfide ore. *Minerals* **2019**, *9*, 134. [CrossRef]
3. Chelgani, S.C.; Parian, M.; Parapari, P.S.; Ghorbani, Y.; Rosenkranz, J. A comparative study on the effects of dry and wet grinding on mineral flotation separation—A review. *J. Mater. Res. Technol.* **2019**, *8*, 5004–5011. [CrossRef]
4. Dündar, H.; Benzer, H. Investigating multicomponent breakage in cement grinding. *Miner. Eng.* **2015**, *77*, 131–136. [CrossRef]
5. Wu, C.; Liao, N.; Shi, G.; Zhu, L. Breakage characterization of grinding media based on energy consumption and particle size distribution: Hexagons versus cylpebs. *Minerals* **2018**, *8*, 527. [CrossRef]
6. Xie, W.; He, Y.; Qu, L.; Sun, X.; Zhu, X. Effect of particle properties on the energy-size reduction of coal in the ball-and-race mill. *Powder Technol.* **2018**, *333*, 404–409. [CrossRef]
7. Shi, F.; Zuo, W. Coal breakage characterization—Part 1: Breakage testing with the JKFBC. *Fuel* **2014**, *117*, 1148–1155. [CrossRef]
8. Hu, Z.-F.; Sun, C.-Y. Effects and mechanism of different grinding media on the flotation behaviors of beryl and spodumene. *Minerals* **2019**, *9*, 666. [CrossRef]
9. Yilmaz, S. A new approach for the testing method of coal grindability. *Adv. Powder Technol.* **2019**, *30*, 1932–1940. [CrossRef]

10. Sato, K.; Meguri, N.; Shoji, K.; Kanemoto, H.; Hasegawa, T.; Maruyama, T. Breakage of coals in ring-roller mills part I. The breakage properties of various coals and simulation model to predict steady-state mill performance. *Powder Technol.* **1996**, *86*, 275–283. [CrossRef]

11. Reichert, M.; Gerold, C.; Fredriksson, A.; Adolfsson, G.; Lieberwirth, H. Research of iron ore grinding in a vertical-roller-mill. *Miner. Eng.* **2014**, *73*, 109–115. [CrossRef]

12. Austin, L.G.; Shah, J.; Wang, J.; Gallagher, E.; Luckie, P.T. An analysis of ball-and-race milling part 1. The hardgrove mill. *Powder Technol.* **1981**, *29*, 263–275. [CrossRef]

13. Austin, L.G.; Luckie, P.T.; Shoji, K. An analysis of ball-and-race milling part II. The bobcock E1.7 mill. *Powder Technol.* **1982**, *33*, 113–125. [CrossRef]

14. Austin, L.G.; Luckie, P.T.; Shoji, K. An analysis of ball-and-race milling part III. Scale-up to industrial. *Powder Technol.* **1982**, *25*, 127–134. [CrossRef]

15. Shi, F. Coal breakage characterization-Part 2: Multi-component breakage modelling. *Fuel* **2014**, *117*, 1156–1162. [CrossRef]

16. Shi, F. A review of the applications of the JK size-dependent breakage model Part 2: Assessment of material strength and energy requirement in size reduction. *Int. J. Miner. Process* **2016**, *157*, 36–45. [CrossRef]

17. Shi, F. A review of the applications of the JK size-dependent breakage model Part 1: Ore and coal breakage characterisation. *Int. J. Miner. Process* **2016**, *155*, 118–129. [CrossRef]

18. Tavares, L.M.; Kallemback, R.D.C. Grindability of binary ore blends in ball mills. *Miner. Eng.* **2013**, *41*, 115–120. [CrossRef]

19. Abouzeid, A.M.; Fuerstenau, D.W. Grinding of mineral mixtures in high-pressure grinding rolls. *Int. J. Miner. Process* **2009**, *93*, 59–65. [CrossRef]

20. Venkataraman, K.S.; Fuerstenau, D.W. Application of the population balance model to the grinding of mixtures of minerals. *Powder Technol.* **1984**, *39*, 133–142. [CrossRef]

21. Cho, H.; Peter, L. Investigation of the Breakage Properties of Components in Mixtures Ground in a Batch Ball-and-Race Mill. *Energy Fuels* **1995**, *9*, 53–58. [CrossRef]

22. Austin, L.G.; Bagga, P.; Celik, M. Breakage properties of some materials in a laboratory ball mill. *Powder Technol.* **1981**, *28*, 235–241. [CrossRef]

23. Xie, W.; He, Y.; Yang, Y.; Shi, F.; Huang, Y.; Li, H.; Wang, S.; Li, B. Experimental investigation of breakage and energy consumption characteristics of mixtures of different components in vertical spindle pulverizer. *Fuel* **2017**, *190*, 208–220. [CrossRef]

24. Lu, Q.; Xie, W.; Zhang, F.; He, Y.; Duan, C.; Wang, S.; Zhu, X. Energy-size reduction of mixtures of anthracite and coking coal in Hardgrove mill. *Fuel* **2020**, *264*, 116829. [CrossRef]

25. Kapur, P.C.; Fuerstenau, D.W. Energy split in multicomponent grinding. *Int. J. Miner. Process* **1988**, *24*, 125–142. [CrossRef]

26. Xie, W.; He, Y.; Ge, Z.; Shi, F.; Yang, Y.; Li, H.; Wang, S.; Li, K. An analysis of the energy split for grinding coal/calcite mixture in a ball-and-race mill. *Miner. Eng.* **2016**, *93*, 1–9. [CrossRef]

27. Shi, F. A review of the applications of the JK size-dependent breakage model part 3: Comminution equipment modelling. *Int. J. Miner. Process* **2016**, *157*, 60–72. [CrossRef]

MDPI
St. Alban-Anlage 66
4052 Basel
Switzerland
Tel. +41 61 683 77 34
Fax +41 61 302 89 18
www.mdpi.com

Journal Not Specified Editorial Office
E-mail: notspecified@mdpi.com
www.mdpi.com/journal/minerals